REVISING GREEN INFRASTRUCTURE

REVISING GREEN INFRASTRUCTURE

Concepts Between Nature and Design

Edited by
Daniel Czechowski
Thomas Hauck
Georg Hausladen

CRC Press is an imprint of the
Taylor & Francis Group, an **informa** business

CRC Press
Taylor & Francis Group
6000 Broken Sound Parkway NW, Suite 300
Boca Raton, FL 33487-2742

First issued in paperback 2017

© 2015 by Taylor & Francis Group, LLC
CRC Press is an imprint of Taylor & Francis Group, an Informa business

No claim to original U.S. Government works

ISBN-13: 978-1-4822-3220-2 (hbk)
ISBN-13: 978-1-138-89281-1 (pbk)

This book contains information obtained from authentic and highly regarded sources. Reasonable efforts have been made to publish reliable data and information, but the author and publisher cannot assume responsibility for the validity of all materials or the consequences of their use. The authors and publishers have attempted to trace the copyright holders of all material reproduced in this publication and apologize to copyright holders if permission to publish in this form has not been obtained. If any copyright material has not been acknowledged please write and let us know so we may rectify in any future reprint.

Except as permitted under U.S. Copyright Law, no part of this book may be reprinted, reproduced, transmitted, or utilized in any form by any electronic, mechanical, or other means, now known or hereafter invented, including photocopying, microfilming, and recording, or in any information storage or retrieval system, without written permission from the publishers.

For permission to photocopy or use material electronically from this work, please access www.copyright. com (http://www.copyright.com/) or contact the Copyright Clearance Center, Inc. (CCC), 222 Rosewood Drive, Danvers, MA 01923, 978-750-8400. CCC is a not-for-profit organization that provides licenses and registration for a variety of users. For organizations that have been granted a photocopy license by the CCC, a separate system of payment has been arranged.

Trademark Notice: Product or corporate names may be trademarks or registered trademarks, and are used only for identification and explanation without intent to infringe.

Library of Congress Cataloging-in-Publication Data

Revising green infrastructure : concepts between nature and design / edited by Daniel Czechowski, Thomas Hauck, and Georg Hausladen.
        pages cm
    Summary: "This book addresses the theme of "green infrastructure" from the perspective of landscape architecture. It brings together the work of selected ecologists, engineers, and landscape architects who discuss a variety of theoretical aspects, research projects, teaching methods, and best practice examples in green infrastructure"-- Provided by publisher.
    Includes bibliographical references and index.
    ISBN 978-1-4822-3220-2 (hardback)
    1. City planning--Environmental aspects. 2. Landscape architecture. 3. Urban ecology (Biology) 4. Sustainable urban development.

NA9053.E58R48 2014
712'.2--dc23   2014026531

**Visit the Taylor & Francis Web site at
http://www.taylorandfrancis.com**

**and the CRC Press Web site at
http://www.crcpress.com**

Contents

Preface...ix
Acknowledgments...xv
Editors and Contributors..xvii

SECTION I   Function and Process

Chapter 1   Green Functionalism: A Brief Sketch of Its History and Ideas in the United States and Germany...3

Thomas Hauck and Daniel Czechowski

Chapter 2   Carefully Radical or Radically Careful? Ecology as Design Motif....29

Greet De Block

Chapter 3   The City That Never Was: Engaging Speculative Urbanization through the Logics of Landscape.......................................47

Christopher Marcinkoski

Chapter 4   Landscape as *Energy* Infrastructure: Ecologic Approaches and Aesthetic Implications of Design.......................................71

Daniela Perrotti

Chapter 5   Landscape Machines: Designerly Concept and Framework for an Evolving Discourse on Living System Design..........................91

Paul A. Roncken, Sven Stremke, and Riccardo Maria Pulselli

Chapter 6   Problems of the Odumian Theory of Ecosystems.....................113

Georg Hausladen

SECTION II   Culture and Specificity

Chapter 7   The Garden and the Machine...137

Thomas Juel Clemmensen

v

vi Contents

Chapter 8 Infrastructure Design as a Catalyst for Landscape Transformation: Research by Design on the Structuring Potential of Regional Public Transport 153

Matthias Blondia and Erik De Deyn

Chapter 9 Beyond Infrastructure and Superstructure: Intermediating Landscapes 171

Sören Schöbel and Daniel Czechowski

Chapter 10 Landscapes of Variance: Working the Gap between Design and Nature 193

Ed Wall and Mike Dring

Chapter 11 Designing Integral Urban Landscapes: On the End of Nature and the Beginning of Cultures 209

Stefan Kurath

Chapter 12 Counterpoint: The Musical Analogy, Periodicity, and Rural Urban Dynamics 225

Matthew Skjonsberg

SECTION III Governance and Instruments

Chapter 13 A Transatlantic Lens on Green Infrastructure Planning and Ecosystem Services: Assessing Implementation in Berlin and Seattle 247

Rieke Hansen, Emily Lorance Rall, and Stephan Pauleit

Chapter 14 The Concept of "New Nature": A Paradigm Shift in How to Deal with Complex Spatial Questions 267

Susanne Kost

Chapter 15 Ecological Network Planning: Exemplary Habitat Connectivity Projects in Germany.. 285

Manuel Schweiger

Contents **vii**

Chapter 16 Planting the Desert: Cultivating Green Wall Infrastructure 301

Rosetta Sarah Elkin

Chapter 17 Designing for Uncertainty: The Case of Canaan, Haiti 323

Johann-Christian Hannemann, Christian Werthmann, and Thomas Hauck

SECTION IV   Applied Design

Chapter 18 Water-Sensitive Design of Open Space Systems: Ecological Infrastructure Strategy for Metropolitan Lima, Peru 355

Eva Nemcova, Bernd Eisenberg, Rossana Poblet, and Antje Stokman

Chapter 19 Green Infrastructure: Performance, Appearance, Economy, and Working Method 385

Paulo Pellegrino, Jack Ahern, and Newton Becker

Chapter 20 The Caribbean Landscape Cyborg: Designing Green Infrastructure for La Parguera, Puerto Rico 405

José Juan Terrasa-Soler, Mery Bingen, and Laura Lugo-Caro

Chapter 21 Forests and Trees in the City: Southwest Flanders and the Mekong Delta 427

Bruno De Meulder and Kelly Shannon

Index 451

Preface

> If we consider infrastructure as a constructed landscape of channels, pipes, grids, and networks that extend across vast territories and that precondition urban life, then we can borrow from several disciplines—urban geography, civil engineering, public administration, botany, horticulture—and combine that knowledge with biophysical resources to form the essential services of urban regions and construct new histories and new lineages.
>
> **Bélanger**
> *2012, 290*

Infrastructure increasingly plays a decisive role in urban design discourses. It no longer belongs exclusively to the realm of engineers (The Infrastructure Research Initiative at SWA 2011) but to "a crosscutting field that involves multiple sectors and where the role of designers is essential" (Shannon and Smets 2011, 70).

The term *infrastructure* in combination with prefixes like *green, landscape,* or *ecological* gains importance in landscape architecture and other disciplines related to the built environment, but there are multiple definitions and perspectives:

1. Handling the physical presence of infrastructure—its barrier effects, noise emission, and pollutions—by ways to conceive, to integrate, or to reconfigure infrastructures* or by redesigning infrastructural spaces.†
2. The use of natural assets to conserve ecosystem values and functions providing associated benefits to human society, expressed in "interconnected green space networks" that consist of "natural or restored ecosystems" and "landscape features" in a system of "hubs," "links," and "sites" (Benedict and McMahon 2006). Beyond providing ecosystem services (MEA 2005) and enhancing biodiversity and habitat connectivity, ecological infrastructure also can "act as a framework to define urban growth and urban forms across scales" (Yu 2010, 59), suggesting the following idea.
3. Landscape itself could "be considered infrastructure when acting as a kind of conveyance and distribution network, capable of moving people and supporting a variety of living systems" (Arquino 2011, 7). Even the whole "urban-regional landscape" can be conceived as infrastructure based on decentralized economic structures, which should take into account, for example, material cycles and watersheds (Bélanger 2009).

---

* Shannon and Smets (2010) differentiate between *hiding/camouflage, fusion* (reconfiguring the preexisting setting into a new composite landscape) and *detachment* (astonishing spatial expressions disregarding the natural context).
† E.g., Hauck et al. 2011.

In the current discourse on sustainability and how to deal with climate change and limited natural resources, a controllable performance is attributed to nature and landscape. The landscape is seen as a physical object, which is no longer "scenic" or "romantic" but "productive" and "well-designed." The capability to generate clean energy and healthy food, to clean water, to store rainwater, to protect against flooding, etc. is based on the idea of controlling processes in ecosystems so that the landscape works as a stable system serving human needs. Green infrastructure should be the technological means that not only provides ecosystem services but also ensures their production. "Ecology" as the guiding principle* should be the basis for a new technological approach that is more "careful," "softer," or "holistic" and produces "site-specific," "integrated," and "multipurpose" infrastructures.

Similar to the critique about the classic infrastructure of the industrial era, a discrepancy between technological progress and the quality of their spatial organization is stated within the development of that "alternative," "ecological," technological approach. Green infrastructure is also being developed in sectorial planning processes for optimization of effectiveness and efficiency. Design considerations play a secondary role. This lack of so-called architectural culture is not reducible to the creative indifference of engineers and ecologists who develop their design from the logics of the relevant technologies. They hardly possess the instruments to include greater spatial relationships in the design. On the other hand, architects, urban designers, and landscape architects are proposing "transdisciplinary" approaches to create appealing spaces within this new "eco-technological regime." They propose a wide range of conceptions as to how to collaborate and develop the interplay between design and ecology, but often they present nothing more than exuberant visions and utopias that are hardly functioning and incompatible with democratic decision making.

This gap between the technocratic and mostly uninspired approaches of ecologists and engineers and the pretentious aspirations of designers mirrors the discrepancy in the use of the term *ecology* in the professional culture.

But how can we close this gap? How do we handle the "convergence of landscape architecture, ecological planning and civil engineering" (Bélanger 2012, 314)? What are convenient terms and metaphors to communicate the interplay between design and ecology? What are suitable scientific theories and technological means? What innovations arise from multidisciplinary and cross-scalar approaches? What are appropriate aesthetic statements and spatial concepts? What instruments and tools should be applied?

This book asks these questions for discussion and presents innovative accesses in designing *green, landscape* or *nature* as infrastructure from different perspectives and attitudes instead of adding another definition or category of green infrastructure. The approaches range from retrofitting existing infrastructures through landscape-based integrations of new infrastructures and envisioning prospective landscapes as hybrids, machines, or cultural extensions. In four sections, this book presents the various relationships to "ecology" as well as the respective design or planning motifs and basic aesthetic attitudes necessary to differentiate and associate the contributions.

---

* Pierre Bélanger describes "ecology as the agent of urban renewal and expansion" (2012, 290) and as "urbanism's best insurance policy" (2012, 310).

Preface

Section I, Function and Process, discusses a "scientific" respective "functional" turn in landscape architecture. As an introductory piece, Hauck and Czechowski give a brief overview of the history of "green functionalism" in the United States and Germany. De Block reflects on the current relationship between infrastructure projects, design techniques, and imminent ecological apocalypse. The contemporary turn to parametric design is studied against the background of early 19th-century infrastructure interventions geared at curbing impending environmental crises in the emerging metropolis. The next three contributions are examples for how new "design logics" are deducted from "ecology" in order to meet both economic and environmental requirements and open new aesthetic relationships toward nature. Therefore, the authors use different theoretical frameworks. Marcinkoski formulates a "landscape-" and "ecology-driven" approach that uses "urban ecology" as an operative framework for developing urban settlements and infrastructures. Perrotti proposes a differentiation of the concept of energy to conceptualize landscapes as "energy infrastructure." Roncken et al. refer to "evolutionary thermodynamics" in order to design "landscape machines." Hausladen answers in a way critical to such attempts as he focuses on scientific as well as epistemic problems with the ecosystem theories of Howard and Eugene Odum, which are very popular within such approaches.

In Section II, Culture and Specificity, the contributors share a decidedly cultural perspective on nature as landscape. Their "ecological" view emphasizes the individual nature of specific local situations. Accordingly, their design ideas are based on a deep understanding of landscape structures, a historical analysis of landscape development, and a consideration of spatiotemporal changes in the landscape. From this cultural perspective, Clemmensen throws a critical light on the concept of landscapes as infrastructure and provides a cultural-historical exploration of the complex relationship between nature and infrastructure. Blondia and De Deyn work within this ambivalent relationship to investigate the possibilities of restructuring cultural landscapes with regional public transport infrastructures. Schöbel and Czechowski consider landscape as an intermediate level between infrastructure and superstructure. They argue that if landscape and open space are either only measured as functions or interpreted as abstract intellectual ideals, they lose their essential properties as social and spatial forms of nature. Compared to that, Wall and Dring use Fresh Kills on New York's Staten Island as an example in order to analyze two different approaches of creative intervention onto this landscape of waste: additive "elegant layered diagrams" as done by Field Operations with the *Lifescape* master plan and the concept of adaptive interventions as done by Matta Clark and Ukeles that interact with the disparate and unpredictable processes of the site. Kurath notes a dissolving of the cultural differentiation between nature and its counterpart society and so between landscape and urban realm. He provides ideas for designing a new "culture" beyond these categories: "integral urban landscapes." Skjonsberg analyzes a concept for designing the relationship of rural and urban as a dynamic balance using a musical analogy—contrapunctus—deducing different contrapuntal fields of application in landscape architecture and urban design.

Section III, Governance and Instruments, presents political ideas and programs defining social relations toward nature and their integration in different planning systems* as well as their impact on nature and society. The contributions deal with different ways of participation and cooperation within cities, regions, and nations.[†] Different spatial circumstances or contexts evoke complex spatial conflicts, revealing different understandings and priorities to protect or to develop nature or to establish so-called green infrastructures. Hansen et al. use a "transatlantic lens" on the implementation of goals that are formulated in the green infrastructure concept respective of the concept of ecosystem service in urban planning strategies, using examples of Berlin and Seattle. Kost is concerned with an approach called "new nature" that arose in the Netherlands as a strategy to handle complex questions in spatial planning. Schweiger focuses on ecological network planning by means of complex algorithms, using the example of habitat connectivity projects in Germany. Afterward, Elkin discusses how the global challenge of rapidly declining vegetative cover is being addressed by massive replanting projects that cross territorial, political, and cultural boundaries and presents two contemporary examples: "The Great Green Wall" in the Sub-Sahara and the "3 North Shelterbelt Program" in China. At the end of the section, Hannemann et al. introduce a neighborhood-led planning approach initiated in the emergent post-earthquake "city" of Canaan in Haiti, which strove not only for improvements on a household level but also aimed to tackle large-scale problems, such as the further occupation of flooding zones.

Section IV, Applied Design, describes projects implemented in local contexts to solve concrete problems or remediate malfunctions. Close to reality, these projects deal with the full scope presented and discussed throughout the book: the use of scientific knowledge, strategic thinking, communication with municipal authorities and local stakeholders, design implementation on site, and documentation and control of feedback and outcome with adequate indicators and metrics. Nemcova et al. respond to Lima's decreasing water resources with a conceptual strategy based on the country's flows of water and provision of ecosystem services. They demonstrate how the design of open space, while understanding the natural and man-made water infrastructure and the local ecology, can actively improve the urban water cycle, thus creating an essential socially valuable and economically feasible ecological infrastructure. Pellegrino et al. present an urban storm water plan in São Paulo and examine green infrastructure as a performance-based and performance-justified option to replace conventional gray infrastructure. Terrasa-Soler et al. propose a series of interventions to use the urban landscape of La Parguera, Puerto Rico, as a water treatment infrastructure that should improve urban quality and coastal ecosystem resilience and protect the shore against erosion and storm surges. Finally, De Meulder and Shannon discuss the interweaving of urbanism and forests by providing case studies in Flanders, Belgium, and the Mekong Delta, Vietnam. They state that, throughout history, forests almost always formed one of the major components of the planning of the territory.

The idea of this book arose during the international conference Designing Nature as Infrastructure at Technische Universität München that was organized and held by

---

* E.g., NYC High Performance Landscape Guidelines, http://www.nycgovparks.org/greening/sustainable -parks/landscape-guidelines.

† E.g., European Green Belt, http://www.europeangreenbelt.org/.

Preface

the editors in November 2012. The overwhelming interest in discussing ways nature is regarded in contrast to society, how human-natural systems could be organized, and how nature could be changed, optimized, or designed encouraged us to pursue this diverse and sometimes controversial discourse.

REFERENCES

Arquino, G. (2011). Preface to Landscape Infrastructure: Case Studies by SWA, The Infrastructure Research Initiative at SWA (Ed.), 6–7. Boston: Birkhäuser.

Bélanger, P. (2009). Landscape as infrastructure. *The Landscape Journal* 28(1): 79–95.

Bélanger, P. (2012). Landscape infrastructure: Urbanism beyond engineering. In *Infrastructure Sustainability & Design*, S. Pollalis, A. Georgoulias, S. Ramos, and D. Schodek (Eds.), 276–315, London: Routledge.

Benedict, M. and E. McMahon (2006). *Green Infrastructure. Linking Landscapes and Communities*. Washington, DC: Island Press.

Hauck, T., R. Keller, and V. Kleinekort (Eds.) (2011). *Infrastructural Urbanism: Addressing the In-Between*. Berlin: DOM Publishers.

The Infrastructure Research Initiative at SWA (Ed.) (2011). *Landscape Infrastructure: Case Studies by SWA*. Boston: Birkhäuser.

Millennium Ecosystem Assessment (MEA) (2005). *Ecosystems and Well-Being: Synthesis*. Washington, DC: Island Press.

Shannon, K. and M. Smets (2010). *The Landscape of Contemporary Infrastructure*. Rotterdam: NAi Publishers.

Shannon, K. and M. Smets (2011). Towards integrating infrastructure and landscape. *Topos* 74: 64–71.

Yu, K. (2010). Five traditions for landscape urbanism thinking. *Topos* 71: 58–63.

Acknowledgments

The editors thank the Technische Universität München, primarily the TUM Graduate School and the Graduate Centre for Architecture, the Faculty of Architecture, especially the Chair of Landscape Architecture and Public Space and the Department of Landscape Architecture and Regional Open Space, Stefanie Hennecke, Regine Keller, Martin Luce, Sören Schöbel, and Ingeborg Seimetz and Irma Britton, Kate Gallo, and Arlene Kopeloff at CRC Press.

Our special thanks go to all the contributors for their valuable writing and inspiring discussions.

Editors and Contributors

Daniel Czechowski (editor) is a registered landscape architect. He studied landscape architecture at Technische Universität (TU) Berlin. After several years of practical experience in landscape design in Berlin and Munich, he joined the Department of Landscape Architecture and Regional Open Space at Technische Universität München (TUM) in 2008 where he is a research and teaching associate. His research focuses on the (infra)structural qualities of landscape. He investigates formative landscape structures to represent, on the one hand, various types of urban landscapes and, on the other, to reveal a new design repertoire. He is co-editor of the reader *Euphorigenic Landscapes*, which is available online as an open-access publication in urban landscape studies.

Thomas Hauck (editor), Dr.-Ing., is a registered landscape architect and artist. He studied landscape architecture at the University of Hanover and the Edinburgh College of Art. He founded the interdisciplinary artists' group Club Real (2000). He was a researcher and an assistant professor at the chair of landscape architecture and public space at the Technische Universität München (TUM) from 2005 to 2014 and joined the Department for Open Space Planning at the University of Kassel in 2014. He is co-editor of the book *Infrastructural Urbanism: Addressing the In-Between*. Together with Cornelia Polinna, he has been the principal of Polinna Hauck Landscape + Urbanism in Berlin since 2007.

Georg Hausladen (editor) is a biologist with interests in the theory of ecology, vegetation ecology, and zoology at the Technische Universität München (TUM). Since 2010, he has been an assistant lecturer at the Department of Landscape Architecture and Regional Open Space at TUM. His PhD project focused on the theory of ecological engineering. He has worked freelance as a biologist in Munich since 2010.

Jack Ahern is a professor of landscape architecture and vice provost for international programs at the University of Massachusetts, Amherst. His research focuses on the integration and application of ecological knowledge in landscape planning and design with an emphasis on greenways, green infrastructure, and sustainable urbanism at multiple scales.

Newton Becker has a bachelor's degree in architecture and urban planning. He earned a PhD in landscape and environment at the School of Architecture and Urbanism, Universidade de São Paulo (FAUUSP), where he experimentally researched low impact development practices for urban stormwater management. He is currently an architect at the Universidade Federal do Ceará, Brazil, and a colaborative researcher at Laboratório Verde, a FAUUSP research group for environmental and ecological landscape design.

Mery Bingen lives and works in San Juan, Puerto Rico. She has a bachelor's degree in communications and visual arts from Sacred Heart University, San Juan, Puerto Rico. She studied architecture and holds a master's degree in landscape architecture from the Polytechnic University of Puerto Rico. Since completing her MLArch, she has collaborated on various competitions and projects, such as the Cultural Landscape Foundation's Landslide 2008 (Marvels of Modernism) and the La Parguera Green Infrastructure Plan with the Center for Watershed Protection and the Polytechnic University of Puerto Rico. She has been a member of the American Society of Landscape Architects since 2007. Currently, she works as a freelance landscape designer, graphic designer, and painter.

Matthias Blondia is an engineer-architect and spatial planner. He has been a researcher at the Department of Architecture, Urbanism and Planning at the University of Leuven since 2010 as coordinator of the Organizing Rhizomic Development along a Regional Pilot Network in Flanders (ORDERin'F) project. Previously, he was an infrastructure project advisor for the Flemish Government State Architect Office.

Thomas Juel Clemmensen earned a PhD in architecture and is a member of the Danish Landscape Architects Association. He is an associate professor at the Aarhus School of Architecture, where he teaches and researches in the fields of urban planning and landscape architecture. His main research interest is how landscape architecture can inform and improve current transformation processes in cultural landscapes. His publications include articles in the *Journal of Landscape Architecture* and the *Nordic Journal of Architectural Research*. Alongside his academic career, Dr. Clemmensen works as an independent consultant.

Greet De Block is a researcher and lecturer in the Department of Architecture at the University of Leuven. Her research focuses on the relationship between technology, space, and society, thus tying together the domains of STS, urbanism, architecture, and political geography. Her publications include "Planning Rural-Urban Landscapes: Rails and Countryside Urbanization in South-West Flanders, Belgium (1830–1930)," *Landscape Research* (published online March 18, 2013; doi:10.1080/01426397.2012.759917); "Designing the Nation: The Belgian Railway Project (1830–1837)," *Technology and Culture* (2011); with Janet Polasky, "Light Railways and the Rural-Urban Continuum: Technology, Space and Society in Late-Nineteenth-Century Belgium," *Journal of Historical Geography* (2011); and with Bruno De Meulder, "Iterative Modernism: The Design Mode of Interwar Engineering in Belgium," *Transfers Interdisciplinary Journal of Mobility Studies* (2011).

Erik De Deyn is an engineer-architect and landscape designer. He has been a researcher at the Department of Architecture, Urbanism and Planning at the University of Leuven since 2010 on the ORDERin'F project. He previously worked for several years as a designer in an architecture office.

Bruno De Meulder teaches urbanism at the University of Leuven. He is co-editor of the book series *UFO: Explorations in Urbanism* (Park Books, Zurich). His urban design research organization, Research Urbanism and Architecture (RUA), seeks innovative interplays of landscape, infrastructure, and urbanization that respond to contemporary challenges while critically valuing existing contexts.

Mike Dring is an architect, researcher, and master's program director in architecture at the Birmingham School of Architecture, Birmingham City University, where he leads Plastic, a design studio that explores the latent potentials of new towns in parallel with future models of architectural practice and production. He was a co-instigator of the Co.LAB Live Projects office at the School of Architecture and cofounder of "Modern Gazetteer" that he leads with artist Stuart Whipps as a cross-disciplinary exploration into the legacy of regional modernism. He is embarking on a PhD that will develop his ongoing research into infrastructural urbanism. He is scheduled to present a paper at the Association of American Geographers Annual Meeting 2014 entitled "Infrastructural Narratives: A History of a Node." He has been a longtime collaborator with Ed Wall.

Bernd Eisenberg, PhD, is a researcher at the Institute of Landscape Planning and Ecology at Stuttgart University. He studied landscape and open space planning in Hannover and earned his PhD on the topic of park metrics at the University of Stuttgart. He is involved in ongoing research activities related to the sustainability and resilience of cities (TURAS) and international educational cooperations. His research interests include quantification methods of open space qualities in general and, in particular, the relationship between spatial configuration and usability as outlined in the space syntax research field.

Rosetta Sarah Elkin is an assistant professor in landscape architecture at Harvard University's Graduate School of Design, where she teaches in the core studio sequence and leads seminars in representation and photography. She is a registered landscape architect and principal of RSE landscape, a design consultancy based in Cambridge and Amsterdam. Both her professional and academic work aim to reveal vegetative strategies from small-scale applications to large-scale innovation. Her work has been featured internationally, including Les Jardins de Metis, the Chelsea and Isabella Stewart Gardner Museum, and in publications such as *Topos*, *Lotus International*, and *Conditions* magazine. Her current research addresses geopolitical procedures that position vegetation as a resolution to global risk and climate change, supported by grants from Fonds BKVB, the Rockefeller Foundation, and the Graham Foundation for the Arts.

Johann-Christian Hannemann earned a bachelor's of landscape architecture and is currently finishing his master's in urbanism at Technische Universität München (TUM). In October 2012, he began researching for the transdisciplinary TUM research group Urban Strategies for Onaville, which is part of a vast urban agglomeration that developed after illegal land occupation following the devastating 2010 earthquake in Haiti. He focuses on participatory strategic spatial planning between community building, spatial urban interventions, and environmental risk mitigation. He conducted two extended research visits to Haiti, including participatory community-based work.

Rieke Hansen is a landscape architect with a focus on landscape planning. She graduated as an engineer for landscape and open space planning from the Leibniz University of Hannover and is an assessor of landscape planning. Since 2010, she has been a research and teaching associate at the chair for strategic landscape planning and management, Technische Universität München. As a doctoral candidate, she is part of the Urban Biodiversity and Ecosystem Services (URBES) project. Her research interests include governance of urban green space and reuse of brownfields as open space.

Susanne Kost studied at the Department of Architecture, Urban Planning, and Landscape Planning at the University of Kassel, Germany. In 2008, she finished her doctoral thesis, "The Making of Landscape: An Analysis of the Mentality of Making: The Netherlands as an Example." She works empirically in the field of perception, evaluation, and interpretation of landscape with a particular focus on planning, social, and aesthetic issues. Establishing connections between the genesis of (urban) landscapes and processes of modernization and their impact on land use and land management are important in her research.

Stefan Kurath, Prof. Dr.-Ing., studied architecture and landscape architecture in Switzerland and the Netherlands. He graduated with honors (summa cum laude) from the HafenCity University Hamburg in the Faculty of Urban Planning. He is a professor of architecture and urban design at the Institute of Urban Landscape at the Zürcher Fachhochschule (ZHAW). He is author of *Stadtlandschaften Entwerfen? Grenzen und Chancen der Planung im Spiegel der städtebaulichen Praxis* (2011) and editor of *Methodenhandbuch für Lehre, Forschung und Praxis in Architektur und Städtebau* (2013). He owns www.urbaNplus.ch and Iseppi-Kurath GmbH, Zurich and Grisons, with projects in the fields of residential building, public buildings, district and urban planning, and space research in the field of socio-technological processes and space configurations.

Laura Lugo-Caro is a licensed horticulturist and landscape designer with more than six years of professional experience. She has collaborated on various projects, such as the master plan for the Puerto Rico National Karst Park and the General Reconnaissance Survey of the Landscape Legacy of the Modern Movement in Puerto Rico. Her work has been published in the Sixth Landscape Biennial Barcelona and *Entorno* (the official journal of the Colegio de Arquitectos y Arquitectos Paisajistas de Puerto Rico). She has been invited as a lecturer and design juror to the University of Puerto Rico's School of Architecture and School of Interior Design and to the Polytechnic University of Puerto Rico's School of Architecture. In addition to landscape, she blurs the boundaries of her work by exploring other areas of design, including industrial design and jewelry design.

Christopher Marcinkoski is an assistant professor of landscape architecture and urban design at the University of Pennsylvania. He is a licensed architect and founding director of PORT Architecture + Urbanism, a leading-edge urban design consultancy. Prior to his appointment at Penn, he was a senior associate at James Corner Field Operations, where he led the office's large-scale urban design work,

Editors and Contributors xxi

including the QianHai Water City in Shenzhen and Shelby Farms Park in Memphis. He earned an MArch at Yale University and a BArch from Pennsylvania State University. He is the author of *The City That Never Was* (Princeton Architectural Press, 2015).

Eva Nemcova is a researcher and lecturer at the Institute of Landscape Planning and Ecology at the University of Stuttgart. She studied landscape architecture at the University of Brno, Czech Republic, and earned her degree in landscape and open space planning at the Leibniz University of Hanover, where she received the Alfred Töpfer Foundation Scholarship. She worked at the landscape architecture offices Station C23 architecture landscape urbanism (Leipzig), Stoss Landscape Urbanism (Boston), and osp urbanelandschaften (Hamburg) on projects dealing with integrated landscape planning and design. She is a member of Studio Urbane Landschaften. Since 2011, she has conducted research on water sensitive open space design in metropolitan Lima within the research project LiWa (Lima-Water).

Stephan Pauleit is a professor and chair for strategic landscape planning and management, Technische Universität München (TUM). He studied landscape architecture and landscape planning and is an expert in urban landscape planning and urban ecology. Prior to his appointment at TUM, he held positions as lecturer and professor, respectively, at the University of Manchester and the University of Copenhagen. He has broad experience in international research, and his interests include strategies for urban green infrastructure, urban adaptation strategies to climate change, urban forestry, and planting design.

Paulo Pellegrino, PhD, is a professor of landscape planning and design at the School of Architecture and Urbanism, São Paulo University, Brazil. He is an architect by training and a landscape architect by specialization. He is the coordinator of the Research Group Laboratório Verde, where he leads teams facing the development of cutting-edge research in the proposition, measuring, and assessment of landscape architecture projects, not only as extensive material artifacts, but complex systems analogous to living organisms, exhibiting many of the same characteristics. He participates in the emerging field of landscape design based on innovative infrastructure systems, adapting to flows and urban metabolisms, combining ecological and morphological systems of cities with social and cultural needs, and making them strategic for greater resilience and equity. His professional experience also includes the development of several pro bono projects and consultancies in the fields of landscape architecture, applied landscape ecology, urban waterways, parks and public squares.

Daniela Perrotti, PhD, is an architect and a postdoctoral fellow at the French National Institute for Agricultural Research (UMR SAD-APT INRA/AgroParisTech) and a lecturer at the School of Architecture and Society of Politecnico di Milano (Department of Architecture and Urban Studies). Her research focuses on the study of urban and landscape design strategies dealing with the notion of landscape as infrastructure from their historical premises to contemporary approaches. In her work, this notion is also intended to be a key factor for the acknowledgment of local and renewable energy potentials embedded in living systems and socio-ecological networks. Her postdoctoral project is supported by grants from the Ile-de-France Region (R2DS Network).

Rossana Poblet is a researcher and lecturer at the Institute of Landscape Planning and Ecology at Stuttgart University. She studied architecture and urban planning at Ricardo Palma University in Lima, Perú, and has a master's in urban renewal and an Erasmus Mundus double master's degree in international cooperation and urban development at TU Darmstadt and Sustainable Emergency Architecture at UIC Catalunya. She has been working with informal settlements, land administration, social housing, landscape, and urban planning in Peru, South Africa, and Kosovo, and she is part of the research project LiWa trying to develop ecological infrastructure strategies in desert cities like Lima, integrating planning and water management.

Riccardo Maria Pulselli, PhD, is an architect and a postdoctoral research fellow at the University of Siena, Italy. He is author of *City Out of Chaos: Urban Self-Organization and Sustainability* (2009) and *The Moving City: How to Explore Urban Kinetics* (2011). With colleagues at the Ecodynamics Group, he conducted studies on environmental monitoring and assessment methods applied to regional and urban systems, buildings and building technologies, agricultural chain processes, and manufacturing processes.

Emily Lorance Rall is a research associate and PhD candidate at the chair for Strategic Landscape Planning and Management, Technische Universität München (TUM). She holds undergraduate degrees in geography and psychology from the University of North Carolina at Chapel Hill and a master's in sustainable resource management from TUM, and she has worked for a variety of city and regional planning firms in the United States. Her research interests include urban green governance and salutogenic environments. Her doctoral research focuses on participatory approaches for cultural ecosystem service assessment.

Paul A. Roncken, Msc, is an assistant professor of landscape architecture at Wageningen University in the Netherlands. He focuses on design education and experiential aesthetics. In 2012, he founded the landscape machine design laboratory (landscapemachines.com) to continue developing new landscape products by living system design. He is the author of the monthly design blog "ideas on landscape and design" (paulroncken.com). As a practicing landscape architect, he coordinates professionals to act in multidisciplinary environmental projects. In 2015, his dissertation, titled "Shades of Sublime: About the Aesthetics of Landscapes and the Idea of the Sublime," will be published.

Sören Schöbel is a professor of landscape architecture and regional open space at Technische Universität München. After studying landscape architecture at the TU Berlin, he worked freelance beginning in 1995 and as a scientific assistant at the TU Berlin from 1998 to 2003, when he completed his doctoral work. His research focuses on the possibilities of and design methods for urban landscape architecture on a regional scale. He is author of *Windenergie und Landschaftsästhetik: Zur landschaftsgerechten Anordnung von Windenergieanlagen* (2012).

Manuel Schweiger graduated in landscape architecture and environmental planning from Technische Universität München. Since 2007, he has been a landscape architect for the ecological consultancy Planungsbüro für angewandten Naturschutz GmbH (PAN GmbH). A focus of his work includes species and habitat field surveys and management planning for the Natura 2000 network of protected sites. In addition, he has been involved in several Research and Development Projects (R+D projects) for the Federal Agency for Nature Conservation, for example, concerning policies related to indicator systems and climate-friendly peat land management. Since July 2014, he is a project leader at Frankurt Zoological Society for Wilderness Areas and National Parks in Germany and Europe.

Kelly Shannon teaches landscape architecture and urbanism at the Oslo School of Architecture and Design and at the University of Leuven. She is co-editor of the book series *UFO: Explorations in Urbanism* (Park Books, Zurich). Her urban design research organization, Research Urbanism and Architecture (RUA), seeks innovative interplays of landscape, infrastructure, and urbanization that respond to contemporary challenges while critically valuing existing contexts.

Matthew Skjonsberg is an architect and urban designer and a PhD researcher on the theme of periodicity and rural urban dynamics. He also is a founding member of the Laboratory of Urbanism at the Swiss Federal Institute of Technology–Lausanne (EPFL). He studied at Taliesin and Swiss Federal Institute of Technology (ETH-Zürich) and has taught at the University of Wisconsin, Rotterdamse Academie van Bouwkunst, and Laboratory Basel-EPFL. He founded collab architecture in 2001, and from 2007 to 2012, he was a project leader at West 8 in New York and Rotterdam. He has written a number of essays, including "Dancing with Entropy" (2012), "Magic Inc." (2011), "The Stolen Paradise" (2010), and "Second Nature" (with Adriaan Geuze, 2009).

Antje Stokman is a professor of landscape planning and ecology at Stuttgart University. She studied landscape architecture at Leibniz University Hanover and the Edinburgh College of Art. She has researched and lectured at Hanover University, TU Hamburg Harburg, Peking University, and Tongji University Shanghai, China, and, at the same time, practiced as a landscape architect in many national and international projects. She was awarded the Lower-Saxony Science Prize in 2009 and the Topos Award in 2011. She is currently a member of Studio Urbane Landschaften, the German National Advisory Board of Spatial Planning, and the Baden-Wurttemberg Sustainability Council.

Sven Stremke, PhD, is assistant professor of landscape architecture at Wageningen University in the Netherlands (tenured position). He and his team conduct research and design projects on sustainable energy landscapes. In 2012, he and Renée de Waal launched the NRGlab, a laboratory devoted to the relations between energy transition and landscapes. He recently co-edited a book dedicated to energy landscapes: *Sustainable Energy Landscapes: Designing, Planning and Management* (Taylor & Francis, 2012). In 2014, he has been appointed as principal investigator for energy at the newly established Amsterdam Institute for Advanced Metropolitan Solutions, a collaboration between MIT, TU Delft, and WUR.

José Juan Terrasa-Soler holds advanced degrees in ecology, environmental studies, and landscape architecture from the University of Michigan at Ann Arbor, Yale University, and Harvard University, respectively. He is a founding faculty member of the graduate program in landscape architecture at the Polytechnic University of Puerto Rico, San Juan, where he teaches design studios and courses on environmental systems and landscape architectural theory. His professional work as a consultant as well as his research and writing focuses on fluvial systems, landscape change, and the adaptation to the tropics of green infrastructure strategies. His design work has been published and exhibited in Spain, the United States, and Puerto Rico. He is a registered landscape architect and works as designer at the office of Marvel & Marchand Architects in San Juan, Puerto Rico.

Ed Wall is a landscape architect and urban designer. He is the academic leader for landscape at the University of Greenwich in London and is a visiting professor at the Politecnico di Milano. His research and practice focuses on the processes and forms of landscape, urbanism, and public space. In 2012, he co-organized the multi-institutional symposium Landscape and Critical Agency at University College London (UCL) the Bartlett, from which he is co-editing a collection of essays. He has written widely, including, with Tim Waterman, *Basic Landscape Architecture: Urban Design*. He is the founder of Project Studio, a platform for research and collaboration based in London. He has been a longtime collaborator with Mike Dring.

Christian Werthmann is a professor at the Institute of Landscape Architecture, Leibniz University, Hannover. He accumulated extensive professional and academic experience in Europe and the United States. In his academic work, his research encompasses the implementation of ecological infrastructure in heavily urbanized areas, especially in the nonformal cities of the Global South—a line of research he initiated as an associate professor at the Harvard Graduate School of Design. His most recent research is concerned with the development of landscape strategies for post-disaster management in Port-au-Prince in Haiti and preemptive urbanization strategies for the landslide prone hills of Medellin in Colombia.

Section I

Function and Process

1 Green Functionalism
A Brief Sketch of Its History and Ideas in the United States and Germany*

Thomas Hauck and Daniel Czechowski

CONTENTS

1.1 Differentiation of Green Functionalism in the United States 5
1.2 Differentiation of Green Functionalism in Germany 14
1.3 Green Infrastructure versus Landscape 19
1.4 Characteristics of Green Infrastructure 20
1.5 Characteristics of Landscape-Based Design Concepts 21
1.6 A Third Voice 23
1.7 Conclusion 25
Acknowledgments..26
References...26

Functionalistic aesthetics are holistic theories on beauty and therefore especially appreciated in those areas of art that aim at the so-called unity of art and life or of beauty and everyday life. These are applied arts, such as design and architecture, with their obligation toward the utility of their products, like the modern avant-gardes, due to their anti-aesthetic and anti-academic principles but also aesthetic traditionalists who believe that the beauty of an object is created through the process of optimizing its use over the course of generations. In the artistic expression of the protagonists of these "movements," the principles of functionalistic aesthetics are applied through *aesthetic ideas* and *patterns*, which create something like a genuine style: *aesthetic functionalism* (Eisel 2007). Based on the underlying aesthetic ideas and the political conception of the "greater good," this *style* is divided into at least two variations: *mechanistic* and *organicistic functionalism*. In both styles, beauty is understood as a quality of objects that stimulates the senses and produces pleasure. This is not merely a relation of cause and effect, but also a means to an end. Beauty's function is to point to the qualities of objects that are beneficial to the well-being of humans as well as to the whole world. The concept of the world as an

---

* For the historical sections of this article cf. Hauck 2014.

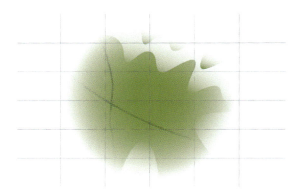

FIGURE 1.1 Subjective Landscape. Developed in painting, "Landscape" became a popular aesthetic practice for hikers, walkers, and other nature lovers but also an everyday viewing habit. The visual field defines the dimension of a landscape that is instantly composed by the viewer and gets transformed by changing viewing direction or position. This image visualizes the momentary landscape of a viewer not in motion but covering a visual angle of 360°. (Courtesy of T. Hauck.)

entity of interdependent parts—constructed as a *clockwork* or grown as *organism*—determines the specific aesthetic functionalism and what is understood as aesthetic qualities.*

One of the most important traditions of thought concerning aesthetic functionalism can be found, in contrast to what might be expected, in the relationship of society with nature as landscape. Here the conflict between the human craving for a holistic worldview and the dissection of the world, during the eras of the Enlightenment and Modernism, is expressed as a spatial conflict of urbanization and destruction of nature. Since the Enlightenment, the human craving for a holistic worldview can hardly be objectified in terms of generally accepted metaphysics anymore; it is only left to be aesthetically experienced, as, for instance, in the perception of a landscape (Ritter 1963) (Figure 1.1). Green Functionalism provides ideas about how to solve, or at least soften, the resulting spatial conflict through functionalistic landscape designs applying holistic and objectivistic principles of beauty. An artistic practice is the result that uses the *pattern* "Landscape"—originally developed in painting and having undergone further development arising from different aesthetic ideas—not only for the production of works of art, but also for the production of practical, useful things. In this new "format," what is central is not success with art collectors and the market; here the value of artistic representation is judged according to the possibilities of aesthetical experience within a practical frame of reference. In this sense, the beauty of a *landscape garden* and the management of rural property by its owner are closely related. Garden owners present themselves as the creators of a realized

---

* An influential aesthetic theory derived from mechanical principles is Edmund Burke's *A Philosophical Enquiry into the Origin of Our Ideas of the Sublime and Beautiful* from 1756; for an organicistic aesthetic theory, cf. Johann Gottfried Herder's *Kalligone* from 1800. On the significance of organismic theories in art around 1800 in Germany, cf. Ulrichs 2006.

utopia, which provides an indication of a desirable social relationship toward nature. This can only be conveyed if a garden (which is not an autonomous work of art anymore) is, at the same time, part of a well-managed rural property. Landscape ceases to be a work of art that can be aesthetically experienced and becomes an inhabited environment, a land that is worked. Landscape designers are, in this context, people who use their creativity to ensure that landscapes can develop their beneficial influence to their full extent so that the relationship between humanity and the landscape develops in a harmonic way.

These designers are artists as well as artist-engineers, who bring the demands of people and the beauty of the landscape into harmony. By *updating* the beauty of the landscape through their design, they are able to protect the landscape from destructive abuse. The artist-engineer believes that it is within her/his powers to produce objective beauty as an expression of a balanced relationship between nature and humanity or at least to be its custodian. This self-perception of the artist-engineer is connected with the mandatory commitment to help objective beauty emerge in every place.

Sections 1.1 and 1.2 are dedicated to presenting two concepts of landscape, which are based on functionalistic principles of beauty.

First is Frederick Law Olmsted's understanding of landscape as an apparatus to improve the urban condition, which we are here treating as a version of mechanistic functionalism, and second is the concept of landscape as a cultural entity, a concept that was developed in German landscape geography and is also based on objectivistic principles of beauty. In its realization as design, however, it follows the rules of an organicistic functionalism. Because of the differing ideas regarding the embodiment of the social relationship toward nature, the formal expression of the two variations took different paths. The mechanistic green functionalism provides the 19th-century metropolises with nature as *green infrastructure*, e.g., as park systems.

The aim of organicistic green functionalism is to reconcile *cultural landscapes* as expressions of harmonically evolved relationships between nature and humanity with the challenges of urbanization. Therefore, rather than creating a new form of nature, the new industrial artifacts need to be assimilated and integrated into cultural landscapes.

1.1 DIFFERENTIATION OF GREEN FUNCTIONALISM IN THE UNITED STATES

U.S.–American landscape architect Frederick Law Olmsted represents a conception of landscape that attributes functionality to *scenery*. It is perceived as a medium that exerts a positive effect on humans. Based on this concept, Olmsted thinks it advisable to construct a special infrastructure dedicated to supplying large towns with this medium. Olmsted transforms landscape—as *scenery*—from a *work of art* into a *technical object*. The *pattern* "Landscape" is given a new format, deducted from the conception of landscape as *garden* and as *scenery* in the tradition of the Picturesque. Olmsted uses the already established *pattern* of the English landscape garden and the *pattern* of the picturesque landscape and reforms them by means of the idea of *healthy open space*. Within this design concept, he is able to merge contemporary medicinal theories with liberal political ideas and thus to obtain a strategy of innovation in order to add current aesthetic relevance to his new conception of landscape. Due to Olmsted, the aesthetic object "Landscape" is transformed into a technical object that is no longer valued in an aesthetic and symbolic way but related solely to its utility. Consequently, Olmsted sees parks as technical facilities, as spatial systems, which provide the inhabitants of metropolises with the beneficial sensory stimuli of nature in the form of fresh air, light, vegetation, and beautiful views.

In this way, Olmsted follows the idea of 18th-century landscape gardening that a garden is supposed to exude an aesthetic impression, which contributes to the moral education and enlightenment of the recipient. With Olmsted, this pedagogical character of landscape is no longer based on a symbolic connection between beauty and moral ideas. It is derived from causal relationships on mental and somatic levels, which refer to "moral treatment" and "miasma" theories (Hewitt 2003, 4). During the American Civil War, in his position as general secretary of the sanitary commission, Olmsted was required to amend the poor state of health of the northern soldiers by means of controlling and improving the sanitary conditions of the troops. Through his cooperation with physicians on this matter, he gained knowledge of contemporary theories of his time on the possible causes of the spread of epidemic diseases, such as cholera. One of these theories stated that *miasma*, i.e., polluted, sickening air or gases, which originated in cramped, swampy places with bad air circulation, had advanced the outbreak of cholera. In his later calling as a landscape architect, he identified similar miasmatic places within the dense areas and cramped streets of industrial metropolises, which were also visited by plagues, and so he consistently recommended fresh air, sunlight, and vegetation* (Beveridge and Hoffman 1997, 127) to fight and prevent the emergence of miasma. "Again, the fact that with every respiration of every living being a quantity is formed of a certain gas, which, if not dissipated, renders the air of any locality at first debilitating, after a time sickening, and at last deadly; and the fact that this gas is rapidly absorbed, and the atmosphere relieved of it by the action of leaves of trees, grass and herbs, was quite unknown to those who established the models which have been more or less distinctly followed in the present street arrangements of our great towns" (Beveridge and Hoffman 1997, 127).

This critique of the living conditions in the metropolises does not lead Olmsted to an anti-urban approach, which might have resulted in proposing an alternative model of the "city without the city." On the contrary, he values the civilizing achievements[†] (Beveridge and Hoffman 1997, 171) of metropolises and accepts the liberal capitalist model of urban growth, only proposing measures to reform it: "He [Olmsted] suggested incremental growth while incorporating specific landscape typologies in accord with the prevailing medical etiologies as specific objectives meant to counter the evils. In particular three landscape typologies stand out in his writings: low density urban and suburban neighborhoods, large pleasure parks and smaller local parks, and tree-lined parkways with connecting promenades" (Hewitt 2003, 9). On the one hand, Olmsted sees the somatic impact of the environment as necessitating

---

* Frederick L. Olmsted: *Report of the Landscape Architects and Superintendents* (1868).
† Frederick L. Olmsted: *Public Parks and the Enlargement of Towns* (1870).

Green Functionalism

green areas in cities; on the other, he uses this fact to derive basic design principles necessary for the achievement of this beneficial sanitary impact: A park should be of sufficient size in order to be able to clean the air through sunlight and vegetation; wide open lawns to achieve air circulation and sunlight; and easy accessibility to areas from the various quarters of a city, especially for the poorer population in densely inhabited areas, in order to improve their health and even save their lives* (Beveridge and Hoffman 1997, 197). The third principle of design calls for the implementation of *parkways* as landscaped roads leading toward big parks and thus serving to improve their accessibility. Furthermore, these *parkways* can already enhance the therapeutic influence of the green areas on the way to the park.

The somatic impact as well as the psychological and psychiatric arguments are used by Olmsted to propagate his ideas. Here he refers to ideas of the *moral treatment* movement, which originated in England. This movement is dedicated to a more humane and reasonable treatment of "lunatics" and is based on the sensualist idea that the human mind is completely dependent on sensual impulses. Consequently, they try out therapies in which caring treatment, moral education, and good living circumstances, also expressed as a *healthy environment*, are the tools for healing their patients. This idea of the *healthy environment* for the *healthy mind* now is expressed for Olmsted in the traditional pattern "Landscape." Within the aesthetic idea of the *healthy open space*, Olmsted unites somatic and mental effects: "Evident in both his proposals and within the rationale of moral treatment theory is the separation and antithetical visual differentiation of therapeutic environments from pathogenic environments ..." (Hewitt 2003, 13). The mental effects of "Landscape," however, allow for the derivation of much more differentiated design principles than the somatic effects, and therefore, it is largely the mental effects upon which his new interpretation of "Landscape" is based.

Olmsted outlines the most important *design principles* for a big urban park as follows: "We want a ground to which people may easily go after their day's work is done, and where they may stroll for an hour, seeing, hearing, and feeling nothing of the bustle and jar of the streets, where they shall, in effect, find the city put far away from them. ... We want, especially, the greatest possible contrast with the restraining and confining conditions of the town, those conditions which compel us to walk circumspectly, watchfully, jealously, which compel us to look closely upon others without sympathy. Practically, what we most want is a simply, broad, open space of clean greensward, with sufficient play of surface and a sufficient number of trees about it to supply a variety of light and shade. This we want as a central feature. We want depth of wood enough about it not only for comfort in hot weather, but to completely shut out the city from our landscapes. These are the distinguishing elements of what is properly called a park"[†] (Beveridge and Hoffman 1997, 189). Within these spatial arrangements, we can see "Landscape" unfolding in its renewed form. It is the classical pastoral landscape—but inhabited by city dwellers of various social

---

* "The lives of women and children too poor to be sent to the country, can now be saved in thousands of instances, by making them go to the Park." Frederick L. Olmsted: *Public Parks and the Enlargement of Towns* (1870).

[†] Frederick L. Olmsted: *Public Parks and the Enlargement of Towns* (1870).

backgrounds within the liberal democratic and urban society of the United States—which is presented to the eye.

Olmsted deduces the *functionalistic principles* for his park design from the liberal ideas of natural equality, civil rights, and freedom and implements these employing the traditional design patterns of English landscape gardening to achieve a total work of art: the park. Olmsted's functionalistic design principles are the following:

First, there is *free movement*, uninhibited and agreeable, represented and facilitated in two ways: Through softly molded lawns and inviting movement and through paths, which make the whole park accessible and are separated according to the type of movement (on foot, by coach, on horseback) in order to optimize the path for the particular type of movement and to avoid mutual disruption. *Free movement* refers symbolically to the idea of freedom.

The second principle is called *easy access*, simple accessibility of the park for all classes of society. This is enabled by positioning parks within the urban context so that no great distances have to be traveled in order to reach them. Metropolises accordingly need a system of parks to supply this open space within their confines. Access is also facilitated through *parkways*, which serve as linear park connections between individual parks and create a tie to city areas not located in the vicinity of a park. *Easy access* for all citizens correlates with the liberal idea of equality in the form of equal civil rights for everybody.

The third principle is *free association*, the gathering of citizens for political, social, or religious reasons or for leisure activities on the big lawns and in park areas that are specifically dedicated to certain needs, such as meeting places, concert pavilions, etc. In this matter, the human being is addressed by Olmsted as a social creature, inherently free and equal, forming voluntary relationships according to its individual needs. This idea is answered by lawns spacious enough to accommodate different groups and congregations without disturbing each other and by a variety of possibilities offered to different user groups in other areas of the park (Roulier 2010).

These three ideas show clearly that Olmsted's design principles are symbolically related to the political ideas of liberalism but that, simultaneously, this analogy is being hypostatized using miasma and moral treatment theories. This process of hypostatization marks a change of design patterns from *landscape garden* to *park infrastructure* in the form of *urban park systems*. In the case of landscape gardening, the design principles of "Landscape" are symbolically related to the ideas of liberalism through the aim of design and imagery to incite ideas that lead to the education of the park visitor in Enlightenment terms. In Olmsted's park systems, this connection ceases to be only symbolic and associative; it becomes causalistic through the direct impact of park design on the mind and body of the visitor. Therefore, negative aspects of urban civilization can be compensated for and education in terms of Enlightenment and in building the American nation can take place simultaneously. The contradiction between the enlightening impact of metropolises on people, which is often emphasized by Olmsted, and metropolises' damaging effect on the individual can only be diminished through Olmsted's park systems, not resolved. Consequently, he continues developing a concept of urban living, which merges liberal city development and his park system: *suburban villages* or as named by Hayden: "picturesque enclaves" (Hayden 2004, 45).

Green Functionalism

Additionally, the German-speaking discourse about the significance and spatial arrangement of green areas in the growing metropolises around the turn of the 19th century was greatly influenced by the model of *urban* and *metropolitan park systems* developed by Olmsted and his "disciple," Charles Eliot. The significance of green areas was repeatedly emphasized in influential papers concerning urban design, e.g., by Reinhard Baumeister (1876), Joseph Stübben (1980, 439), and Camillo Sitte (1900), but no feasible form for the development and arrangement of these green areas was found in liberal urban development. The frequently discussed *park system* in the United States provided a model for solving this problem. The publications of Werner Hegemann, in particular, laid the ground for open space politics to adopt this model. During the competition for *Groß-Berlin* and the following *Allgemeine Städtebau-Ausstellung* (Hegemann 1911 and 1913) in the year 1910, Hegemann proclaimed the American park systems as exemplary models for German cities.* It is remarkable that Hegemann's descriptions of Olmsted's parks appreciate the landscape design, using phrases such as providing "delicious views" and "charming little enclaves" (Hegemann 1911, 7) for the urban visitor, making it sound as if he were addressing an antiquated style long since gone. Hegemann's main focus is more directed to the user value of the big lawn areas, sports fields, and playgrounds and the implementation of the open spaces (as parks dedicated to sports and play) within the urban context as a system, which is supposed to grant access from every given house within "stroller distance" (Hegemann 1911, 9). This functionalistic and infrastructural approach toward open space planning had a significant impact, especially during the interwar period in Berlin, with "Stadtbaurat" Martin Wagner as one of the protagonists. After the National Socialist takeover, it was abandoned because, in their ideology, the social relationship with nature was supposed to serve the goal of renewing and expanding German cultural landscapes and all "Volksgenossen" were supposed submit to this principle.

The theories and works of American landscape architecture in the 19th century, particularly those of Frederick Law Olmsted, extended their influence on methods and theories of landscape design far into the 20th century. It was a functionalistic-mechanistic position, which referred to *scenery* as providing free space for movement as well as a component of a sanitary infrastructure, serving the purpose of hygienically eliminating deficits (on a somatic and mental level). At the same time, scenery within the city was perceived as emancipatory open space facilitating political organization and private assemblies. In order to establish this green infrastructure, politicians and city planners provided suitable areas, which were supposed to be located in a sensible manner, ideally in the form of an open space system, in order to be accessible from all city areas. These areas were provided with landscape elements, which had, during the history of gardening (especially the English landscape garden), proved their value in achieving comfort and emancipation. The landscape architect had to organize the open space system and arrange suitable

---

* In the catalogue of the exhibition "Allgemeinen Städtebauausstellung" (Hegemann 1911 and 1913) and in the "Parkbuch" (Hegemann 1911), Hegemann introduced to the German city planning community parks and park systems planned by Olmsted Sr., the Olmsted brothers, and by Eliot (Metropolitan park system of Greater Boston) and the "Spielparksystem" (system of neighborhood parks) for Chicago planned by the Olmsted brothers and Daniel H. Burnham.

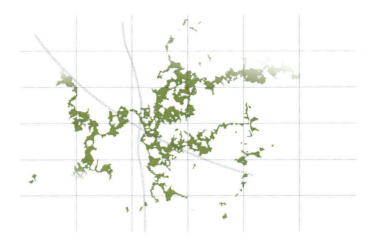

FIGURE 1.2 Green Infrastructure. "Landscape" is seen as a technical object that supplies (as a spatial system) the inhabitants of metropolises with the beneficial sensory stimuli of nature in the form of fresh air, light, vegetation, animals, and beautiful views. (Courtesy of T. Hauck. This drawing was inspired by Figure 15.2 in the article of Manuel Schweiger in this book.)

landscape elements in a sensible manner as scenery within these areas (Figure 1.2). This assignment adapted from urban engineering (including landscape architects) by urban planning and was further rationalized. The *picturesque* elements were schematized and "embedded" into rational planning procedures, which resulted in open space planning, emphasizing utility as realized, e.g., by Robert Moses—significantly inspired by German examples—in New York, which nonetheless did keep aspects of the pattern "Landscape" for recreational purposes.

Considering *scenery* as a natural resource, which should serve humanity, has characterized the landscape ideology of the U.S.–American regional planning since the 1920s, but especially during the period of the *New Deal*, a series of social and economical reforms inspired by Franklin D. Roosevelt as a consequence of the worldwide economic crisis of 1929. One package of measures was dedicated to developing technical and social infrastructures, such as motorways, hydroelectric power plants, schools, and public swimming pools under the direction of federal institutions, such as the Works Progress Administration (WPA), as well as state-run corporations, e.g., the Tennessee Valley Authority (TVA), or *state agencies*, such as the Texas Highway Department (THD). The construction of large infrastructure and its consequences for the landscape were considered a challenge but not an irresolvable conflict. Only correct management of resources was called for in order to keep the landscape as well as the ecosystem in balance or, even furthermore, to moderate the consequences of incorrect treatment due to inadequate natural scientific knowledge, such as *soil erosion* (Cushman 2000), and to eliminate natural disasters, such as flooding and malaria-infected water (cf. Black 2000). The designers of the TVA or THD could even integrate organicistic ideologies as proclaimed by the protagonists of the *conservation movement*, John Muir or Aldo Leopold (Black 2000, 72) without

Green Functionalism

having to leave the frame of rational planning due to the fact that they could, in the tradition of Olmsted, "embed" the beauty of "Landscape" in a pastoral schemata as *scenic resources* (Black 2000, 90) and, therefore, handle it in a rational context. Artistic intuition or innovation of any kind was not called for within the context of American holistic planning. Landscape architects, like ecologists, were part of the team of engineers, who planned their projects following functionalistic principles. "Organic technology" (Black 2000, 74) in the conception of *New Deal* landscape design represented a brilliant domination of nature.*

Architects such as Le Corbusier or Mies van der Rohe, who continuously increased their prominence within the urban planning discourse beginning in the 1920s, held a similar *functionalistic-picturesque* view of open space. The open space surrounding their *architectural machines* was conceived as free-flowing landscapes to be viewed from the building's windows in aesthetic detachment (Woudstra 2000). Landscape serves a similar purpose as conceived by Olmsted; only the urban concept is radically different: Olmsted attempted to compensate for the negative impact and hygienic deficiencies of liberal urban planning by means of creating open green space whereas Le Corbusier and Mies van der Rohe view the whole world as landscape, and therefore, it becomes the backdrop to functionalistic machines for living and working. This radicalization of the picturesque paradigm within urban planning can also be witnessed, e.g., in the works of Robert Moses. On the one hand, he remained true to Olmsted's principles through all the decades of his planning activities and created numerous *parkways* for New York, which were intended to bring nature to the densely built-up metropolis in the form of green corridors and, furthermore, to create paths toward *natural scenery*, like the beaches of Long Island, for the inhabitants of New York and Brooklyn. Concerning concrete urban planning projects, however, Moses reverted to the more radical urban development program of Le Corbusier: So-called "slum clearings" got rid of dense city quarters and replaced them with "towers in the park" conceived as machines for living set in scenic landscapes.

These functionalistic-picturesque positions of urban and regional planning were heavily criticized in the United States by a new generation of architects, who, since the 1930s, have accused them of being schematic rather than innovative, unsubstantial, and without any cultural or ecological references (Meyer 2000). In a negative adaptation of these positions, architect Frank Lloyd Wright devised an alternative, organic form of architecture, which involved the *genius loci* of the regional landscape in creating the design and thus integrated the architecture into the specific form of nature inherent to the local situation (Howett 1993). Landscape architects such as James Rose, Garett Eckbo, and Christopher Tunnard attempted to oppose Olmsted's and his followers' *picturesque schematism* with an innovative artistic position and tried at first to connect with the architectural functionalism taught by Gropius at Harvard. The resulting designs, however, could not stand their ground as independent, innovative contributions to landscape architecture because they suffered under the contradiction between the aesthetics of architectonical functionalism, which is based on design principles derived from building construction, and the tradition of

---

* On New Deal landscape ideology, see Chapter 16 by Rosetta Sarah Elkin.

landscape architecture, which uses nature (in the sense of natural things) as a source of design. But to erase the reference to nature would have meant abandoning an essential feature of the distinction toward architecture. Consequently, the landscape architects either based their thinking on rationalistic traditions within garden art, such as the French formal garden, or, if they wanted to stay on the progressive side, enhanced the emphasis on nature in their design concepts. These tendencies led to biomorphic stylistic elements in garden and park iconography but also, e.g., to the view that the usage of trees and bushes should allow for the development of their natural potential as individual plants (Treib 1993). Until the 1950s, this organicistic impulse in American landscape architecture stayed within the compounds of garden and park design.

Only the *ecologization* of landscape architecture was able to overcome the rigid design regime of functionalized scenery and the subordinate role of landscape architects as junior professional planners in *civil engineering* or urban planning projects. Ian McHarg has been a pioneer of landscape architecture ecologization since the 1950s. His book *Design with Nature*, published in 1969, had a tremendous impact on the profession (Spirn 2000). Influenced by the *organic planning* ideas of Lewis Mumford and Patrick Geddes, who considered planning a prerequisite for the harmonic and well-ordered development of urban and regional areas, McHarg was able to define the task of landscape architects in a more comprehensive way: "There clearly is a desperate need for professionals who are conservationists by instinct, but who care not only to preserve but to create and to manage. These persons cannot be impeccable scientists for such purity would immobilize them. They must be workmen who are instinctively interested in the physical and biological sciences, and who seek this information so that they may obtain the license to interpose their creative skills upon the land. The landscape architect meets these requirements" (McHarg 1969, 151). Instead of passively safeguarding *natural scenery*, it is the assignment of the landscape architect to actively manage *land use* in accordance with the natural circumstances. Spatial development (settling, transport, and economy) is not primarily based on economic interests, but on nature conceived as "interacting process, responsive to laws, constituting a value system, offering intrinsic opportunities and limitations to human uses" (McHarg 1969, 55). Usage possibilities and the limitations of nature (and therefore of "geographic landscape") are evaluated through analyzing the *physiographic region* by methods such as diagramed *overlays* of different factors of the region (physiography, hydrology, soils, plant associations, etc.). The assessment through these methods leads to the definition of *intrinsic suitabilities* for various land uses (agriculture, forestry, recreation, urbanization). Within the framework of a *matrix*, the different *land uses* are evaluated against the backdrop of *natural determinants* in order to identify an optimum of *multiple compatible land uses* (McHarg 1969, 127–144). Based on this, *ecological planning designs* for uses (e.g., urban development projects) should be created. These *designs* are supposed to continue the *adaptation** on other scale levels. Ecological planning and design is, for McHarg, the right aesthetic, technical, and moral measure to reconcile *man as planetary disease* (McHarg 1969, 43) with the laws of nature. "As a means of lending scientific integrity to his ecological approach,

---

* About McHarg's use of evolutionary analogies, cf. Spirn 2000, 109.

Green Functionalism

McHarg developed a scientific theory called *creative fitting* that both explained and validated designing with nature. McHarg's method was ecological not only because it used ecological data but also because the outcomes it produced matched the processes of adaptation and evolution. It helped determine where proposed human uses, such as buildings and roads, *intrinsically fit* on the land. Since this design method located the fittest environment for various land uses, it also fulfilled the basic principles of adaptation" (Herrington 2010, 6). According to McHarg, good design is a result of the successful adaption of human land uses to nature. The effect is beautiful nature as a stable ecosystem, and the perfect example for this design is the English landscape garden: "Moreover, to represent nature's design at a larger scale, McHarg consistently referred to 18th-century English landscape gardens, which he viewed as representing ecological concepts. Humans creating these gardens were *designing with nature*, while earlier Western gardens were *not designed with nature*" (Herrington 2010, 5). McHarg understands the human adaption to nature through design as the integration of humans into an *organism*, which incarnates itself in the form of the English landscape garden. Local characteristics and the singularity of *cultural landscapes* do not seem to attract his attention much, which might be due to the fact that the liberal political model of the United States, withstanding all criticism related to "American failure" (McHarg 1969, 77), remains the ideal one for McHarg and that this system does not express itself in unique and, consequently, site-specific cultural landscapes but in *picturesque scenery* as a ubiquitous concept of beauty, following the model of the English landscape garden.

McHarg's great influence on the self-perception and praxis of landscape architects as *ecological planners* was then countered by the critique of a new generation of landscape architects, who turned mainly against McHarg's ecological aesthetic and his rigid planning method.* Marc Treib comments, "McHarg's method insinuated that if the process were correct, the consequent form would be good, almost as if objective study automatically gave rise to an appropriate aesthetic. In response to his strong personality and ideas, landscape architects jumped aboard the ecological train, becoming analysts rather than creators, and the conscious making of form and space in the landscape subsequently came to a screeching halt" (Treib 1999, 31). As a protagonist of a new "artistic" self-conception of landscape architecture, James Corner challenges the idea of *scenery*: "... the landscape idea throughout much of this century has come mostly in the form of picturesque, rural scenery, whether for nostalgic, consumerist purposes or in the service of environmentalist agendas" (Corner 1999, 8). He turns against the rightly identified core of the hitherto existing tradition (and dialectics) of American landscape architecture in order to liberate himself from the weight of the *picturesque* design traditions as well as those of *ecological planning* and of *preservation*. Landscape architecture needs to dedicate itself again to visionary projects; it needs to experiment and invent, and most of all, the return to relevantly contributing to urban development is called for. The alternative

---

* Elizabeth Mossop (2006, 168) also states that "McHarg's methodology fails to account for the significance of design in the planning process, and his scientific rhetoric devalues the expression of arts and culture." Landscape architecture's focus on infrastructure as "the most important generative public landscape" (2006, 171) has the potential to bridge the gap between ecology and design, between the impact of McHarg's work and opposing art and design-focused approaches.

to *scenery*, which Corner proclaimed a progressive aesthetic idea for the profession, is now supposed to become something very "old": "Landscape" as "physical and social reality" (Trepl 2012, 162). "Inasmuch as landscape objectifies the world—in the form of 'scenery,' 'resource' or 'ecosystem,' for example—it sets up hierarchical orders among social groups, and among humans and nature more generally. One is always an 'outsider' as far as the beholding of manufactured landscapes goes, for to be 'inside' entails the evaporation of landscape into everyday place or milieu. It is in this deeper sense that landscape as place and milieu may provide a more substantial image than that of the distanced scenic veil, for the structures of place help a community to establish collective identity and meaning. This is the constructive aspect of landscape, its capacity to enrich the cultural imagination and provide a basis for rootedness and connection, for home and belonging" (Corner 1999, 12). Characterizing this process, it would be legitimate to refer to "organic turns" in American landscape architecture, which, in their contemporary form, are called *Landscape Urbanism*, described by Corner (2006, 29) as "a more organic, fluid urbanism," where "ecology itself becomes an extremely useful lens through which to analyze and project alternative urban futures."

Landscape Urbanism represents a counter position to picturesque (mechanistic) functionalism in the tradition of Olmsted*—currently expressed in the "greenway movement" (Little 1990, 26) and in communal "green" and "blue infrastructure" programs[†]—using the idea of a holistic development of cultural landscapes as milieu and environment for everyday life.

1.2 DIFFERENTIATION OF GREEN FUNCTIONALISM IN GERMANY

The German version of green functionalism is also based on objectifying the aesthetic field through functionalistic principles of beauty. These principles, however, are not deduced from a liberal worldview as in the United States, but are rooted in an organicistic conception of aesthetics—as, e.g., represented by the poet and philosopher Johann Gottfried Herder[‡]—which became the foundation for an idea of landscape as an object in the form of *cultural landscapes*. The actual *physical landscape* became the object of investigation for the newly emerging scientific discipline of geography[§] and, simultaneously, the arena for political disputation between preservers of tradition and their opponents, the advocates of technical progress. The custodians of cultural landscapes, being the inferior combatants in what they considered to be a "spatially" expressed political battle, attempted to mitigate the discrepancy between technical progress and the traditions of cultural landscapes on two levels: through integrating technical artifacts into cultural landscapes and through harmonizing spatial development by means of holistic or, rather, organic regional planning.

---

* Cf., e.g., Corner 2006, 24–26.
† Cf., e.g., Greater London Authority 2012 and the City of New York 2011.
‡ Cf., e.g., Kirchhoff 2005.
§ Cf. Eisel 1992.

Green Functionalism

In this new conception, landscape, constituted as an aesthetic object by the individual through perception, is transformed into a physical object, which can exist regardless of the observer. This geographical point of view is also useful outside of the discipline of geography because it is able to provide the traditional educated middle class, made uncomfortable by the impact of technical progress and the capitalist economy on everyday life, such as the rapid development of metropolises, the annihilation of social hierarchies, and the infrastructural reshaping of urban and rural areas, with a conservative utopia of a life in well-ordered structures and in accordance with the natural conditions—the idea of the *cultural landscape* (Figure 1.3).

The cultural landscape emerges as *unity of land and people* (Riehl 1851), as *heimat* evolved through the interactions of adapting to nature and cultivating it. A cultural landscape is not only beautiful; it is perfect. In Herder's aesthetics, it is the spatial equivalent of the perfect individual, who has cultivated himself within the framework of his possibilities and has, therefore, become who he is meant to be. A cultural landscape is made perfect "if a community fulfills their mandate to develop themselves and the nature of their *Lebensraum* to the extent of variety which is adequate to themselves and their natural environment, thereby creating an organic unity of land and people, a 'cultural landscape'. According to the genuine character of the people, the climate and specific history perfection will mean something different each time. … The difference between the agreeable, the (morally) good and the beautiful is made to disappear. To develop the cultural landscape is God's will, the result therefore good; it offers an agreeable environment and it is, because everything within is in perfect accordance, beautiful. *The perfection of cultural landscape is hence not (only) the one of an image, it is the perfection of reality*" (Trepl 2012, 158).

In the *landscape perception* of the German educated middle class (due to their conservative perspective), the individual appearance of the various cultural landscapes

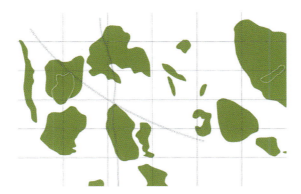

FIGURE 1.3 Cultural Landscape. "Landscapes" are seen as physical objects that exist regardless of the observer but emerge as *unities of land and people* through the interactions of adapting to nature and cultivating it. Cultural Landscapes are an inhabited environment, land that is worked, which expresses harmonic relationships between nature and humanity. (Courtesy of T. Hauck.)

consequently became the main criterion for the judgment of its perfection. A cultural landscape is deemed beautiful if the local people (through generations) have cultivated the site-specific nature according to its character. The modification of a cultural landscape through artifacts of technical progress and industrial production, which do not originate from a process of cultivating the characteristics of landscape but follow universal technical laws, which can be applied everywhere, is consequently judged as its *destruction*. As a reaction to this *destruction of Nature*, the Germany of the second half of the 19th century witnessed the rise of bourgeois "protest movements" with different agendas for the protection of *cultural landscape*, among others, *Heimatschutz* primarily represented by the musician Ernst Rudorff and *Naturdenkmalpflege* by the botanist Hugo Conwentz (Lekan 2004, 21). Striving to be politically relevant, the German nature protection movement used functionalistic and economic (recreation and encouragement of tourism) as well as symbolic and nationalistic (strengthening the patriotism of all classes of citizens) arguments. Only for the middle class did this strategy prove to be successful. For wide segments of the population, the demands of nature protection opposed their own interest in technical progress and economic growth, which were based on the hope of improving their living conditions (Lekan 2004, 36). The resulting ambivalence of the conservation movement toward technical progress, however, only led a few of its members, such as Rudorff, to a strict anti-urbanism. Other influential protagonists, such as Conwentz and Paul Schultze-Naumburg, accepted technical progress as a necessary evil, which had to be integrated into a reasonable development of cultural landscapes and proclaimed an alternative model of progress in *organischer Entwicklung* (Schultze-Naumburg 1908, 205; Eisel 2007, 69).

Based on the idea of finding a more reasonable form for modern progress and the perceived chaotic and messy urban growth, *Landesplanung* was established in the 1920s in the urban and industrial development areas of the Weimar Republic. The planning instruments were provided by the wartime economy of the First World War: "Although German military officials had failed miserably to provide adequately for the civilian population during the war, their efforts at corporatist economic planning and rationing offered new models of state intervention that fuelled the demand for better peacetime land use and natural resource administration" (Lekan 2004, 122). Planning associations, such as the *Siedlungsverband Ruhrkohlenbezirk* (SVR), in which the communities and districts of the Ruhr area came together to participate in greater planning activities, created economic and housing development plans, which were intended to regulate economic activities as well as settlement, transport, and open space development. The plans had numerous goals: economical facilitation (also to be able to make the reparation payments due to the First World War) and the reform of urban planning. Contemporary urban development was to be achieved—compatible with the demands of an industrial society—and that demand included the segregation of living and working areas, finding solutions for transport issues, and creating open spaces for the recreation of the working population. The request that cities and regions be adapted to the industrial development was connected to criticizing the liberal economic system, which did not allow well-regulated, harmonic growth.

The resulting deficit was to be removed by *Landesplanung*. "The economical development of the 19th century was taken as the cause for the present situation; the present was notably suffering from a loss of form, the spatial elements had started

Green Functionalism

to 'disturb' each other and there was a deficiency of 'order' and 'accordance.' Only anticipatory measures could help now: There was a need for a 'plan.'" (Leendertz 2009, 130). The lack of harmony also concerned the relationship between urban and rural areas. In the critics' eyes, this became obvious through the uncontrolled growth of cities into the surrounding landscape. *Landesplanung* should provide *reconciliation*: "*Landesplanung* aims at a harmonic unity of landscapes, forms of economy and housing. The forms of administration are lately being adapted to the prior mentioned in order to achieve perfect harmony between all expressions of human existence"* For the preservation movement, the evolving *Landesplanung* provided an opportunity to modernize itself as *Landespflege* and to leave the "romantic heritage" of the early days behind. *Landespflege* as a discipline was suitable for ensuring the *reconciliation* between technical artifacts and cultural landscape as, e.g., proposed by Lindner in *Ingenieurwerk und Naturschutz*: "Concerning all of this we cannot move forward without making a generous effort to balance on the one hand what needs to stay the most valuable treasure of our people's nature and on the other what must perforce be and become technology. ... We need complete insight and a total overview, in order to calibrate the planning and planning goals of such construction processes to national economics in the broadest sense and to the image of our homeland as the most significant expression of natural processes in the most precise way. Only this will be able to reconcile the relations of technology with the existing forms of landscape" (Lindner 1926, 7). *Landespflege*, in this context, sees itself as the discipline of reconciliation and compensation between the specific character of *cultural landscapes* and the industrially shaped appearance of technical artifacts. The objective of *Landesplanung* and *Landespflege* was the harmonization of spatial development in Germany using the idea of *organic planning*, which could be interpreted as the instrumental application of Herder's idea of *cultivation*. Technical progress should be integrated and therefore "tamed" as a new component of an organic spatial development with this kind of individualized planning and design.

This "taming" of progress through integration into individual cultures by means of *Landespflege* took a new direction under the rule of the National Socialists. The maintenance of *cultural landscapes*, which is based on the development of various *units of land and people*, was combined with industrialization to transform it into a project focusing on progress. The associated *landscape ideology* was thereby thrown into internal conflicts between the attempt to perfect the *cultural landscape*, which is the expression of "cultural work with and applied to nature" of the inhabiting (and therefore rooted) people, and the abstract domination of nature by technological progress. To solve these inner contradictions, the National Socialist' landscape designers introduced *racism* into their landscape ideology: If one specific *land and people unit*, namely the German, proves to be especially successful in the cultivation of natural circumstances, its duty is also to apply its inventive capacities to the development of cultural landscapes in other countries in the sense of a universal human mandate for *Kulturarbeit*.[†]

---

* M. Pfannschmidt (1929). Landeskunde und Landesplanung. In *Die Baupolitik 3* (Addendum to *Städtebau 24*), 53. Cited in Leendertz 2009, 134.
† Cf. Schultze-Naumburg 1908, 205.

The universal mandate thus created can be combined with technical progress and economic modernization, which now became the tool of a global *German cultivation mission*. The executor of *Kulturarbeit*, through application of rational means for the organic-intuitive development of landscape, was the *Landschaftsgestalter* as artist-engineer. This landscape ideology was the base of the demand of *Landespflege* to become a nationwide assignment because the whole country was supposed to be remodeled in line with this ideology. Through this ideology, it also became possible to view the conquering and exploitation of other countries as well as the annihilation and suppression of their population as a requirement for necessary *Kulturarbeit* and *Landschaftsgestaltung*. The abuse of the idea of cultural landscape as a justification for the displacement or annihilation of people was only possible if the landscape designers in question held strong racist views or were, at least, opportunistic.

In the work of the *Landschaftsgestalter*, the antinomy between organic-intuitive landscape development and economical and technical modernization could only be superficially resolved—again through integrating technical structures into the surrounding landscape, e.g., the *landschaftliche Eingliederung* of the *Reichsautobahn*.* The destruction of *cultural landscapes* through the construction of motorways, industrial plants, and military structures or through the impact of industrial forestry could only be mitigated through integration into the landscape and the individualization of technical artifacts by a design style called *Deutsche Technik*. During the implementation of *Landespflege* projects, the conflict, which, at the beginning of the 19th century, had already led to the evolution of the *Heimatschutz* movement, became apparent again. It was the conflict between the *perfection* of cultural landscapes and the invasion of technical infrastructures or economically rationalized cultivation. This conflict could be partially moderated by the *individualization* of technical artifacts and integrating design measures, such as landscape design, along motorways. But it did not succeed in "constructing" new cultural landscapes by means of *landschaftliche Eingliederung*. Here it becomes apparent that an innate inconsistency in an aesthetic of perfection, based on organicistic principles of beauty, is the reason that the growth of quasi-organisms like *cultural landscapes* can only be identified in retrospect when they have already evolved according to aesthetic ideas: An aesthetic object is hypostatized to an organic entity (landscape, city, face), and therefore, a unique development process is made apparent, which can also be underpinned with empirical data. Such a process is to be recreated by the artist-engineer through his intuition for the further development or new creation of *cultural landscapes*. He should quasi-empathize with an already existing organic entity and develop it further in its innate spirit or create an organic work of art, which, then again, becomes part of the whole organism. This organic creation principle conflicts with technical or, rather, industrial production. A technical application is universally suitable, not only, e.g., in the Black Forest area. The form of a technical product results, therefore, from the applied technology (everything else is called an ornament) and not from, e.g., site-specific natural conditions. Technical objects are indeed developed for specific contexts, but they are not tied to these contexts and are applied wherever they are useful. The conflict between cultural landscapes and ubiquitous technology, therefore,

---

* Scenic incorporation of the motorways of the Third Reich (translated by authors).

Green Functionalism

is the inescapable consequence of hypostatizing the aesthetic idea "Landscape" to organic entities by applying objectivistic principles of beauty.

After the Third Reich, *Landespflege* underwent modernization, mainly to overtly dissociate it from the National Socialist landscape ideology but also to resolve the contradictions between the authoritarian nature of *Landespflege* and democratic decision making. The modernization in a now-democratic society enhanced the scientific aspect of *Landespflege* and created a program of rational *ecological planning* (Körner 2001, 77 and 426), at first, under the influence of a *holistic ecological ideology*, proclaiming the protection of *national health* through a healthy natural environment. An attempt was made to analyze the beneficial impact of the aesthetic object "Landscape" to human physical and mental health with empirical methods and to deduce planning decisions from this data. This ecologization and scientification of *Landespflege* was opposed by landscape architects who continued to apply their artistic approach and who criticized the loss of individuality of Landscape and the disregard for its cultural significance. Due to the political necessity of rationally legitimized planning decisions within the framework of democratic decision making, the scientification of *Landspflege* received another boost in the 1960s, and it became a "consequent functional planning discipline" (Körner 2001, 428).

"This was to be facilitated through the alignment with precisely defined social, eventually economical user interests, the compilation of the functions of ecosystems and the exact quantification of landscape evaluation, in order to eliminate possible intuitive components of planning. Landscape was reduced to a resource for various uses …" (Körner 2001, 428). Above all, the recreational value of Landscape was to be argumentatively reinforced by empirical methods in order to gain political relevance. This can be seen most clearly in Hans Kiemstedt's procedure of evaluating the diversity of landscape structures within the framework of user value analysis. This was an attempt to turn the aesthetic value and symbolical significance of Landscape into a measurable quantity (Körner 2001, 428).

The development of *Landespflege* toward *instrumental landscape planning* was criticized by the representatives of *landscape architecture* because, while empirical methods were suitable for measuring ecological principles and human needs, concerning the quality of Landscape, these methods would only mirror already existing Landscape clichés of either the planners themselves or popular taste and, consequently, could not be the base for assessing the aesthetic quality of landscapes or the foundation for progress in this field. So it came as no surprise that, at the end of the 1990s, the "organic turn" of American *Landscape Urbanism* had a strong response in the field of German landscape architecture. "Designing landscape" emerged as a buzz phrase, and large-scale designs based on the *pattern* of Landscape became popular again. This "re-import" managed to reinforce an artistic counter position to landscape planning without adding the weight of the terrible ideological heritage of German Landscape Design.

1.3 GREEN INFRASTRUCTURE VERSUS LANDSCAPE

In the past few years, we have witnessed a renaissance of the concept of green infrastructures as a widely used planning instrument. At the same time, a recent discourse in landscape architecture has also produced some new functionalistic terms related to urban green and landscape.

The social background for this renaissance of infrastructural approaches in green planning is obvious, deriving from the perception of a so-called ecological crisis (Latour 2009) related to the detection of climate change and as a consequence of an emerging "culture of sustainability" (cf. Eisel and Körner 2006).

On an expert and planning level, there are three reasons:

1. The return of the "big plan" in urban planning in the form of strategic development plans after the planning crises (cf. SRL 2013). This also demands green planning on a metropolitan or regional scale, e.g., The All London Green Grid (Greater London Authority 2012) and PlaNYC (The City of New York 2011).
2. A more pragmatic approach in nature conservation. The preservation of nature on a holistic landscape level was not especially successful. Instead, a spatial formation, such as an ecological network, seems to be more appealing, especially to persuade the opponents of "preservationism"—the civil engineers. They are building so-called grey infrastructures. To complement these with green and blue infrastructure allows a win–win situation for planners and politicians.*
3. The concept of ecosystem services. In the past, green infrastructure was justified by miasma theory and the support for physical and mental health. These "soft" or even outdated justifications were replaced by ecosystem services providing "hard" quantifiable facts, such as carbon sequestration or water purification (MEA 2005).

However, the well-established and successful concept of *green infrastructure* is obviously not satisfying for all planners and architects. Otherwise, such a multitude of alternative concepts would not have emerged. The interesting point is that in these new concepts, e.g., *landscape as infrastructure* (Belanger 2009, 2012) and *landscape machines* (Roncken et al. 2011), the term "landscape" seems to play an important role while, at the same time, seeming to be irrelevant to green infrastructure concepts.

It is comparatively easy to recognize that the concept of green infrastructure as well as these new landscape concepts follow the tradition of the two previously described historical variations of green functionalism. What are the contemporary characteristics of these two positions as planning concepts, and what are their differences?

1.4 CHARACTERISTICS OF GREEN INFRASTRUCTURE

Green infrastructure planning is based on the identification of areas that supply relevant ecosystem services and then tying these valuable areas into a network to accumulate these services. This spatial organization makes it possible to spread these services over a maximum domain and to maximize the benefits for man and

---

* See, e.g., Chapter 15 by Manuel Schweiger.

environment. Areas are selected because of their utility and their position related to other areas. If some stakeholders disagree with a selection, the planner can propose some alternative areas, possibly with the same value, or an alternative area can be prepared to be as useful as the previous one. Thus, areas and their spatial formation are exchangeable and flexible because their value is abstract and not site-specific. These are excellent requirements for implementing green infrastructure in argumentative planning processes. The planning goals are the preservation, construction, and connection of habitats and/or urban green spaces. Because a network is a flexible form and the knots of a net do not have to be on specific places (but in specific relationships), it is possible to find alternative spatial compositions if, e.g., some selected areas are not available. The network is therefore a practical geometric model to create a spatial correlation. Before this background, the concept of green infrastructure and green infrastructure planning provides a pragmatic working method to preserve and develop green spaces well grounded on their functions for human well-being.

Aesthetics, however, is the weak point of the method. With a long tradition in 19th-century urban park planning and nature preservation, the concept has incorporated the traditional pattern of landscape aesthetics in its performance. As this aesthetic pattern is functionalized and naturalized in the method of green infrastructure planning, it is quite difficult for new aesthetic patterns to emerge. Aesthetic value is not necessary to support the establishment of green infrastructure; what counts are the ecosystem services they can provide.

1.5 CHARACTERISTICS OF LANDSCAPE-BASED DESIGN CONCEPTS

Landscape-based design concepts are, in many cases, connected with ideas of decentralization and site-specific infrastructure. The proponents understand their concepts as working methods to establish a new kind of infrastructure. They argue that there is a modernist kind of infrastructure that is centralized, mono-functional, separated from any context, and mainly based on nonrenewable energy sources. This old infrastructure must be retrofitted or substituted by infrastructure that is decentralized, multifunctional, multilayered, site-specific (interlinked with local ecosystems), and regionally renewable. The intention is to establish a technology regime that can be reconnected to the order of landscape or make it possible to create a new spatial pattern that will be perceived as landscape (Figures 1.4 and 1.5). The objective is to create landscapes with spatial patterns that are generated by infrastructures based on the site-specific natural resources, such as hydrological systems, wind conditions, etc.

In their concepts, the authors dissociate themselves from the romantic, bucolic, and picturesque pattern of Landscape (Belanger 2009, 92) and deliver manifestos for a new landscape aesthetics based upon efficiency, which constitutes an important difference from the concept of green infrastructure, in which beauty is only one of the many features that "green" has to offer anyway. This approach should mean more square meters of green, more beauty. The *new landscape functionalists* are following a different aesthetic concept, though. Here the aesthetic value of a landscape is grounded in its performance—it is beautiful because it is productive. But to unfold its beauty, the utility of a landscape has to be expressed in an optimal way, meaning

FIGURE 1.4 Landscape Infrastructure. "Landscapes" are seen as spatial patterns that are caused by decentralized infrastructures based on the site-specificity of natural resources like hydrological systems, wind conditions, etc. The aesthetic value of landscape infrastructure is grounded in the efficiency of its alternative infrastructure. (Courtesy of T. Hauck.)

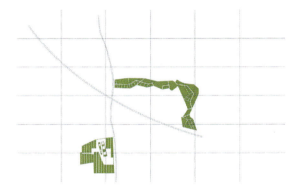

FIGURE 1.5 Landscape Machine. "Landscapes" are seen as well-designed biotechnological facilities, e.g., for water purification, fish, or energy production. The beauty of these landscapes arises from the productivity and efficiency of the production process on site. (Courtesy of T. Hauck.)

without frills and ornaments and redundant formality. This will only be possible if these new landscapes are publicly available and could be used for social activities and allowed to experience the production process that takes place on the site. Based on functionalistic aesthetic principles and new infrastructural approaches, landscape and infrastructure are combined into a new spatial entity.

The crucial question is whether or not this new infrastructural regime and its new landscape pattern is flexible and open enough to incorporate results of democratic decisions that contradict this new order, e.g., the decision to preserve traditional landscape patterns or to use "classical" centralized infrastructure like a pumped storage power plant. Even with design concepts that integrate infrastructure into

Green Functionalism

landscape patterns, it will still be necessary for a society to decide between practical, economic, and aesthetic values as we can witness in all landscape versus infrastructure discussions.

1.6 A THIRD VOICE

Aside from these two dialectic positions in green functionalism, there are other voices maintaining that the aesthetic, cultural, social, and, therefore, subjective components of Landscape and Nature, especially in questions related to planning and design cannot be neglected.* This critique is primarily based on the following two assumptions (1 and 2) and proposes a decidedly cultural and aesthetic position (3):

1. The application of the model "landscape" in its schematized form as "green" or "ecosystem" is an efficient method in open space and landscape planning to appease the social demands for nature that arise in political (democratic) processes. However, Green Infrastructure Planning provides only aspects of *square meters of green space per capita* and *ecosystem services* while concealing the aesthetical core of its program beneath rationalized methods and scientific methodology. Green Infrastructure Planning can therefore be considered successful in providing a planning process, which, in principle, allows for participation and traceability as well as in certain functional aspects, such as the technical element of ecosystem services. In particular, Green Infrastructure Planning does not seem to be able to reach satisfactory levels regarding aesthetics. Only in a few cases were the results addressing the needs for beautiful nature and landscape (as aesthetic qualities) convincing because aesthetic qualities had been part of the planning and had been discussed, but the outcome had already been considered sufficient or even high if the "green" somehow fit with the *pattern* "landscape." This approach is problematic, however, because most of the projects that are methodically oriented in this direction,[†] e.g., the majority of the greenway projects or urban open space systems (e.g., All London Green Grid), are not principally aimed at providing functional ecosystem service although this does play a role in the political argumentation. The main purpose is the "provision" of green for recreation and leisure in all its variations.
2. On the other hand, functionalistic concepts based on the pattern "landscape" are able to provide new aesthetic ideas[‡] for designing nature. However, they do not substantiate the quality of their ideas through subjective arguments (e.g., taste) but in functional arguments as a consequence of the efficiency or utility of objects and contexts, which constitute the specific landscape

---

* See Chapter 7 by Thomas Juel Clemmensen, Chapter 2 by Greet de Block, Chapter 9 by Sören Schöbel and Daniel Czechowski, and Chapter 10 by Ed Wall and Mike Dring.

[†] See Chapter 13 by Rieke Hansen, Emily Lorance Rall, and Stephan Pauleit.

[‡] See Chapter 3 by Christopher Marcinkoski, Chapter 4 by Daniela Perrotti, and Chapter 5 by Paul A. Roncken, Sven Stremke, and Riccardo Maria Pulselli.

in question. By means of a new aesthetic orientation in the profession and the improvement of popular taste, the landscape is supposed to be reunited once more with reality. In order to achieve this, first the reality of landscape development has to be acknowledged, and the own *affect structure* and the one of *popular taste* must be newly calibrated according to the current realities because beneficial entities (e.g., wind turbines) should also be beautiful. Second, this insight concerning reality is used to extract and refine new design principles in order to create an aesthetic principle of true beauty, which accords with current reality and which can be applied to recreate the *world as landscape.*

Skeptics consider it a weak point of these concepts that subjective (aesthetic) motives are being used as factual arguments and thus will become the base of a large-scale modification of the living environment of a multitude of people.

3. The most important criticism of functionalistic positions, however, concerns their willful ignorance of the aesthetic value of landscape structures gradually developed over the years. To tackle this problem, some designers integrate ideas like Corboz's *land as palimpsest* (cf. Corboz 1983) into their landscape concepts. The aim of what they believe to be a more sensible approach is that the outcome of landscape transformation would not be a new totality based on a new infrastructural regime but could be a process in which the layers and patterns of the existing landscape are respected and the change is perceived as enrichment and not destruction.

There is a broad spectrum of aesthetic ideas for the design of nature as can be seen with the functionalistic positions discussed, and many concepts contain overlays of the three ideal types of designing nature described. Some of the cultural/aesthetic positions lean toward organicistic functionalism, here represented by the historical German position of landscape design.

To conclude, we want to recapitulate the three positions presented in the discourse about planning and designing nature in their relation to each other:

We have presented the two versions of green functionalism approaches and how they have been carried out in an exemplary way in Germany and the United States. These two versions refer to each other in a critical way. The mechanistic position in the tradition of Olmsted can be conceived of as a rational methodology for the planning of nature. This position can be applied in democratic planning processes while the functionalistic landscape concepts, which are based on an artistic vision, are seen as totalitarian and undemocratic. Conversely, the organicistic position points out the aesthetical weakness and conventionality of rational planning processes of nature. A third voice now calls the entire functionalistic position into question by enhancing the spatial qualities in design and demanding also form-orientated approaches,*

---

* See, e.g., Frits Palmboom (2010, 42): "It is our strong conviction that within design form-orientated and process-orientated approaches cannot be driven apart from one another. Form is not purely expression of the process lying behind it. The direct identification of form with process and programme leads to a sort of neo-functionalism, (...)."

Green Functionalism   25

which is therefore not primarily determined by functional purposes but has its own end in itself in the sense of artistic practice.

1.7 CONCLUSION

Is it possible to continue developing spatial planning and design based on these three positions? We think it is but only if we use the aesthetic argument (the third voice) to resolve the antinomy between the two functionalistic positions. Both versions of green functionalism are right in their mutual critique of aesthetic weakness/conventionality versus the totalitarian approach/practical irrelevance. Consequently, neither of the opposing positions can be resolved through the arguments of the other side. But if the designing of nature and landscapes is seen as an aesthetic praxis, whose design principles are based on aesthetic ideas and not on (seemingly) objective criteria, there are possibilities for new aesthetic ideas concerning planning and the design of nature. Such an understanding of landscape architecture as aesthetic praxis provides both green functionalism positions with a range of options for action, which have, so far, been concealed behind the rhetorical veil of inherent necessities. Resolving the antinomy can therefore only occur if the aesthetic position is maintained and applied as follows:

1. For the mechanistic position of green infrastructure planning, it has become clear that the creation of green infrastructure is not primarily aimed at achieving functional ecosystem services but at the conservation and creation of beautiful nature. In order to achieve this within the framework of rational planning processes, the analogy of technical infrastructure to spatial systems is used. Consequently, the aesthetic value is accorded a different position within green infrastructure planning. So far, the aesthetic value of green infrastructure has been conceived of as an ecosystem service and has been more or less visibly integrated into the equation of overall performance of a green infrastructure. If this functionalistic concealment is abandoned and the aesthetic value of green infrastructure is positioned equally next to ecosystem services, the main focus of green infrastructure planning shifts away from the comprehensible calculation of rational alternatives to include aesthetic judgment of design ideas for nature. This means that participatory planning, e.g., in a planning dialogue, now includes, beyond the proposal and negotiation of practical interests, also the participation in evaluating aesthetic ideas (which also means everyone can contribute her or his own ideas). Design quality consequently plays a much greater role in green infrastructure planning, and the discourse on design variations and alternatives must take its rightful place because such a planning dialogue also includes debates on matters of taste.
2. With regard to the organicistic position, it becomes clear that the relationship between design and landscape cannot be merely functional, that this relationship is based on more or less innovative aesthetic ideas. These proposals, however, are not seen and presented as model innovations based on design ideas but rather as methods or techniques for achieving practical goals (e.g.,

optimization of ecosystem services or construction of decentralized, more sustainable infrastructures). The acknowledgment of the subjective decision of the designer in question as the base for these aesthetic innovations, instead of functional necessities, does not lead to artistic escapism because the matters in question are symbolic references, which contribute to relevant social issues of spatial development. To put the main focus on aesthetic praxis does not liberate the designer from having to make practical (moral) judgments (and having to defend these and act accordingly in his design practice) nor does it eliminate the possibility of formulating criticism. Therefore, it is time to make proposals concerning the rejuvenation of the historical pattern "Landscape" through innovative aesthetic ideas in order to satisfy the social need for beautiful landscape and nature. In the current process of spatial development, which is applying the old methods, this need can no longer be met satisfactorily anymore with regard to specific spatial situations.

Finally, considering the continuous transformation of landscapes and the challenges that await us, it comes down to the question of how options can be found for making the aesthetic praxis of designing landscapes a part of democratic planning processes without losing the persistence of the aesthetic in the course of the process. Through such a process, something would become possible that would be able to transcend aesthetic experience, going beyond the subjective exchange and reflection about it with others. This book is meant to contribute to this process.

ACKNOWLEDGMENTS

Thanks to Marianne Ramsay-Sonneck and Dr. Stephen Starck for supporting us in translating and proofreading.

REFERENCES

Baumeister, R. (1876). *Stadt-Erweiterungen in technischer, baupolizeilicher und wirthschaftlicher Beziehung.* Berlin: Ernst & Korn.
Belanger, P. (2009). Landscape as infrastructure. *Landscape Journal* 28: 79–95.
Belanger, P. (2012). Landscape infrastructure: Urbanism beyond engineering. In *Infrastructure Sustainability and Design*, ed. S. N. Pollalis et al., 276–315. London: Routledge.
Beveridge, C. E. and C. F. Hoffman (ed.) (1997). *The Papers of Frederick Law Olmsted: Writings on Public Parks, Parkways, and Park Systems.* The John Hopkins University Press.
Black, B. (2000). Organic planning: Ecology and design in the landscape of the Tennessee Valley Authority, 1933–1945. In *Environmentalism in Landscape Architecture*, ed. M. Conan, 71–95. Dumbarton Oaks.
The City of New York (2011). PlaNYC. A Greener, Greater New York. Update from 2011. http://www.nyc.gov/html/planyc2030/html/theplan/the-plan.shtml (accessed March 28, 2014).
Corboz, A. (1983). The land as palimpsest. *Diogenes* 31: 12–34.
Corner, J. (1999). Introduction. In *Recovering Landscape—Essays in Contemporary Landscape Architecture.* New York: Princeton Architectural Press.
Corner, J. (2006). Terra Fluxus. In *The Landscape Urbanism Reader*, ed., C. Waldheim 21–33. New York: Princeton Architectural Press.

Cushman, G. T. (2000). Environmental therapy for soil and social erosion: Landscape architecture and depression-era highway construction in Texas. In *Environmentalism in Landscape Architecture*, ed. M. Conan, 45–70. Dumbarton Oaks.

Eisel, U. (1992). Individualität als Einheit der konkreten Natur: Das Kulturkonzept der Geographie. In *Humanökologie und Kulturökologie. Grundlagen, Ansätze, Praxis*, ed. B. Glaeser and P. Teherani-Krönner, 107–151. Opladen: Westdeutscher Verlag.

Eisel, U. (2007). Emanzipation und Würde in einfachster Form—Die philosophische und politische Struktur funktionalistischer Ästhetik. In *Landschaft in einer Kultur der Nachhaltigkeit*, Band 2, ed. U. Eisel, and S. Körner, 56–75. Universität Kassel.

Eisel, U. and S. Körner (ed.) (2006). Landschaft in einer Kultur der Nachhaltigkeit, Band 1. Universität Kassel.

Greater London Authority (2012). Green Infrastructure and Open Environments: The All London Green Grid. http://www.london.gov.uk/sites/default/files/ALGG_SPG_Mar2012.pdf (accessed March 28, 2014).

Hauck, T. (2014). Landschaft und *Gestaltung*—Die Vergegenständlichung ästhetischer Ideen am Beispiel von "Landschaft." Bielefeld: Transcript Verlag.

Hayden, D. (2004). *Building Suburbia: Green Fields and Urban Growth*, 1820–2000. New York: First Vintage Books Edition.

Hegemann, W. (1911). Ein Parkbuch. Zur Wanderausstellung von Bildern und Plänen amerikanischer Parkanlagen, Berlin.

Hegemann, W. (1911 and 1913). Der Städtebau nach den Ergebnissen der Allgemeinen Städtebau-Ausstellung in Berlin nebst einem Anhang: Die internationale Städtebau-Ausstellung in Düsseldorf. 2 Vol., Berlin: Verlag Ernst Wasmuth A.-G.

Herrington, S. (2010). The nature of Ian McHarg's science. *Landscape Journal* 29: 1–10.

Hewitt, R. (2003). The Influence of Somatic and Psychiatric Medical Theory on the Design of Nineteenth Century American Cities. *History of Medicine Online*. http://priory.com/homol/19c.htm.

Howett, C. (1993). Modernism and American landscape architecture. In *Modern Landscape Architecture: A Critical Review*, ed. M. Treib, 18–35. Cambridge, MA: MIT Press.

Kirchhoff, T. (2005). Kultur als individuelles Mensch-Natur-Verhältnis. Herders Theorie kultureller Eigenart und Vielfalt. In *Strukturierung von Raum und Landschaft—Konzepte in Ökologie und der Theorie gesellschaftlicher Naturverhältnisse*, ed. M. Weingarten, 63–106. Münster: Verlag Westf. Dampfboot.

Körner, S. (2001). Theorie und Methodologie der Landschaftsplanung, Landschaftsarchitektur und Sozialwissenschaftlichen Freiraumplanung vom Nationalsozialismus bis zur Gegenwart. Berlin: Universitätsverlag der TU Berlin.

Latour, B. (2009). A Cautious Prometheus? A Few Steps Toward a Philosophy of Design (With Special Attention to Peter Sloterdijk). http://www.bruno-latour.fr/sites/default/files/112-DESIGN-CORNWALL-GB.pdf (accessed March 28, 2014).

Leendertz, A. (2009). Ordnung, Ausgleich, Harmonie-Koordinaten raumplanerischen Denkens in Deutschland, 1920 bis 1970. In *Die Ordnung der Moderne—Social Engineering im 20. Jahrhundert*, ed. T. Etzemüller, 129–150. Bielefeld: Transcript Verlag.

Lekan, T. M. (2004). *Imagining the Nation in Nature—Landscape Preservation and German Identity*, 1885–1945. Harvard University Press.

Lindner, W. (1926). *Ingenieurwerk und Naturschutz*. Berlin: Hugo Bermühler Verlag.

Little, Ch. E. (1990). *Greenways for America*. Baltimore, MD: John Hopkins University Press.

McHarg, I. L. (1969). *Design with Nature*. New York: The Natural History Press.

Meyer, E. K. (2000). The post–Earth Day conundrum: Translating environmental values into landscape design. In *Environmentalism in Landscape Architecture*, ed. M. Conan, 187–244. Dumbarton Oaks.

Millennium Ecosystem Assessment (MEA) (2005). *Ecosystems and Well-Being: Synthesis*. Washington, DC: Island Press.

Mossop, E. (2006). Landscapes of infrastructure. In *The Landscape Urbanism Reader*, ed. C. Waldheim 163–177. New York: Princeton Architectural Press.

Palmboom, F. (2010). *Drawing the Ground—Landscape Urbanism Today: The Work of Palmbout Urban Landscapes*. Basel: Birkhäuser.

Riehl, W. H. (1851). Die Naturgeschichte des Volkes als Grundlage einer deutschen Social-Politik. Erster Band. Land und Leute. Stuttgart: J. G. Cottascher Verlag.

Ritter, J. (1963). Landschaft—Zur Funktion des Ästhetischen in der modernen Gesellschaft. In *Subjektivität—Sechs Aufsätze*, J. Ritter, 141–163. Frankfurt/Main: Suhrkamp Verlag.

Roulier, S. (2010). Frederick Law Olmsted—Democracy by design. *New England Journal of Political Science* IV(2): 311–343.

Roncken, P. A., Stremke, S., and M. P. C. P. Paulissen (2011). Landscape machine: Productive nature and the future sublime. *Journal of Landscape Architecture* 6, Spring: 68–81.

Schultze-Naumburg, P. (1908). Kulturarbeiten, Band III: Dörfer und Kolonien (2. Auflage). München: Kunstwart Verlag.

Sitte, C. (1900). Großstadt-Grün. *der lotse. Hamburgische Wochenschrift für deutsche Kultur*, 1. Jg., Heft 5: 139–146 and Heft 7: 225–232.

Spirn, A. W. (2000). Ian McHarg, landscape architecture, and environmentalism: Ideas and methods in context. In *Environmentalism in Landscape Architecture*, ed. M. Conan, 97–114. Dumbarton Oaks.

SRL—Redaktionsgruppe des Arbeitskreises Städtebau (ed.) (2013). Der große Plan—Aktuelle Beiträge zum Städtebau. SRL Schriftenreihe 56. Berlin: SRL.

Stübben, J. (1980). Der Städtebau. Reprint from 1890, Braunschweig, Wiesbaden: Vieweg Verlagsgesellschaft.

Treib, M. (1993). Axioms for a modern landscape architecture. In *Modern Landscape Architecture: A Critical Review*, ed. M. Treib, 36–67. Cambridge, MA: MIT Press.

Treib, M. (1999). Nature recalled. In *Recovering Landscape—Essays in Contemporary Landscape Architecture*, ed. J. Corner, 29–43. New York: Princeton Architectural Press.

Trepl, L. (2012). Die Idee der Landschaft: Eine Kulturgeschichte von der Aufklärung bis zur Ökologiebewegung. Bielefeld: Transcript Verlag.

Ulrichs, L.-T. (2006). Das ewig sich selbst bildende Kunstwerk, Organismustheorien in Metaphysik und Kunstphilosophie um 1800. *Internationales Jahrbuch des deutschen Idealismus (Ästhetik und Philosophie der Kunst)* 4: 256–290.

Woudstra, J. (2000). The Corbusian landscape: Arcadia or no man's land? *Garden History* 28(1): 135–151.

2 Carefully Radical or Radically Careful?
Ecology as Design Motif

Greet De Block

CONTENTS

2.1 Introduction 29
2.2 From Engineering "Human" Territory to Encoding "Natural" Surfaces 37
2.3 Radically Careful or Carefully Radical? 40
2.4 Conclusion 41
References 45

2.1 INTRODUCTION

During the last decade or so, infrastructure has been successfully reclaimed by urban design as a structuring device capable of dealing with the current complex spatial condition, characterized by rural-urban hybridity, accelerating horizontal urbanization, neoliberal economic regimes, and rising environmental concerns. As an alternative to the inability of both modern(ist) design modes and traditional urban form "to produce meaningful, socially just, and environmentally healthful cities" (Waldheim 2013, 13), new *-isms*, such as *landscape urbanism* and *ecological urbanism*, are exploring new paths for the discipline by designing infrastructure systems as resilient frameworks guiding urban development within the context of ever-changing and fragile (ecological) processes. Indeed, as Pierre Bélanger reflects on postwar engineering, "(i)n the wake of over-planning, over-regulation and over-engineering of the past century," future design has to be oriented toward "the re-coupling, re-configuration, and re-calibration of these processes" through "the re-design of infrastructure" (Bélanger 2013, 24–25). As a common denominator of discourses breaking with functionalist, top-down planning and, more pragmatically, as one of the last resorts allowing public authorities to give structure to haphazard settlement (Shannon and Smets 2010, 9), infrastructure has emerged not only as the glue holding disperse urbanization together, but also as the object around which new visions about urbanization could be assembled in order to formulate novel grounds for the discipline.

More specifically, the tandem of (landscape) ecology and infrastructure—that is, *designing nature as infrastructure* as well as *designing city systems as biological metabolisms/metaphors*—is increasingly presented as system and strategy par excellence to address the complex urban condition, which is threatened by imminent global ecological crisis and enhanced by accelerating planetary urbanization and unpredictable capital flows. Indeed, green infrastructures or, more specifically, "underlying structures of topography and hydrology" are the major structuring elements of urban form, Elisabeth Mossop argues in her essay on landscapes of infrastructures for the *The Landscape Urbanism Reader* (Mossop 2006, 72). Robust, resilient landscape infrastructure systems encoded by the transformative power of self-organization and reciprocal interactions is becoming the universal guiding urban design. Or, in the words of Brian McGrath, "(b)ased on smart infrastructure, self-sufficiency and hybrid local models, highly adaptive design patterns take the form of responsive micropatches rather than overarching masterplans" (McGrath 2009, 48). Central within this disciplinary reorientation are digital design techniques, such as scenario-based modeling, pattern recognition, sensory mapping, indexing, and diagramming, which derive a sustainable local framework from the site—or, in the context of this book, even complex techniques of evolutionary thermodynamics generate "landscape machines" (see Chapter 5 by Roncken, Stremke, and Pulselli)—geared to supporting ecosystems as "batteries" for a variable (mostly market-driven) program. One of the archetypal examples of this parametric urbanism driven by ecology is the design of Groundlab for Shenzhen (China) focusing on the "ecological performance" of the river, thus proposing an urbanization model that is both "ecologically literate" and "economically viable" (Waldheim 2010) (Figure 2.1).

In an opinion piece for the special issue on city systems of *Architectural Design*, Colin Fournier explains this reliance on the parametric in infrastructure design as a reaction against top-down determinism of the modernist and engineer as well as a product of the commonly held view that city systems are becoming more complex over time and, hence, need more sophisticated conceptual models and analytical tools to understand and intervene in them (Fournier 2013). However, one could question this turn to sophisticated digital techniques and the grounds they rest upon as the discourse is reminiscent of earlier models of the systems engineer designing for cities, which were at least as complex as the current urban condition. He argues that the body of work of "new" landscape infrastructure is developing " 'all-knowing,' quasi-Orwellian operational models that will observe the city and the flows through its interconnected infrastructure systems in order to facilitate their control, rather than a more pluralistic discourse addressing topics other than the systemic and allowing different voices to be expressed" (Fournier 2013, 128). In this chapter, the critique hinting at similarities between the new *-isms* and the discourse of engineering will be further explored. The purpose of the chapter is to analyze the claim of new discourses to move infrastructure from the realm of engineering to urbanism and the concurrent methodological shift from determining top-down logics to bottom-up forces while these discourses might be drawing on similar—although with another degree of complexity—design techniques.

Rather than tying in with traditional tools of urbanism, producing fixed spatial layouts, it is my hypothesis that current design modes are more related to 19th- and 20th-century techniques of calculation and cartography—for instance, statistics, layering, and categorization—developed within the realms of engineering, medicine, sociology, and political economy to plan and govern transport infrastructure or

Carefully Radical or Radically Careful?

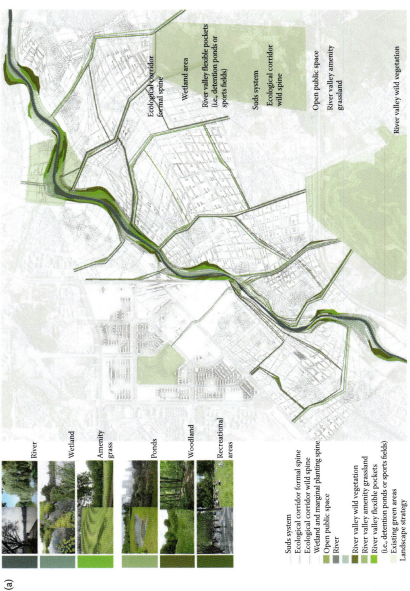

FIGURE 2.1 (a) Parametric infrastructure design driven by ecological performance: *Deep Ground* by Groundlab, Shenzhen (China). (Image courtesy of Groundlab.)

(b)

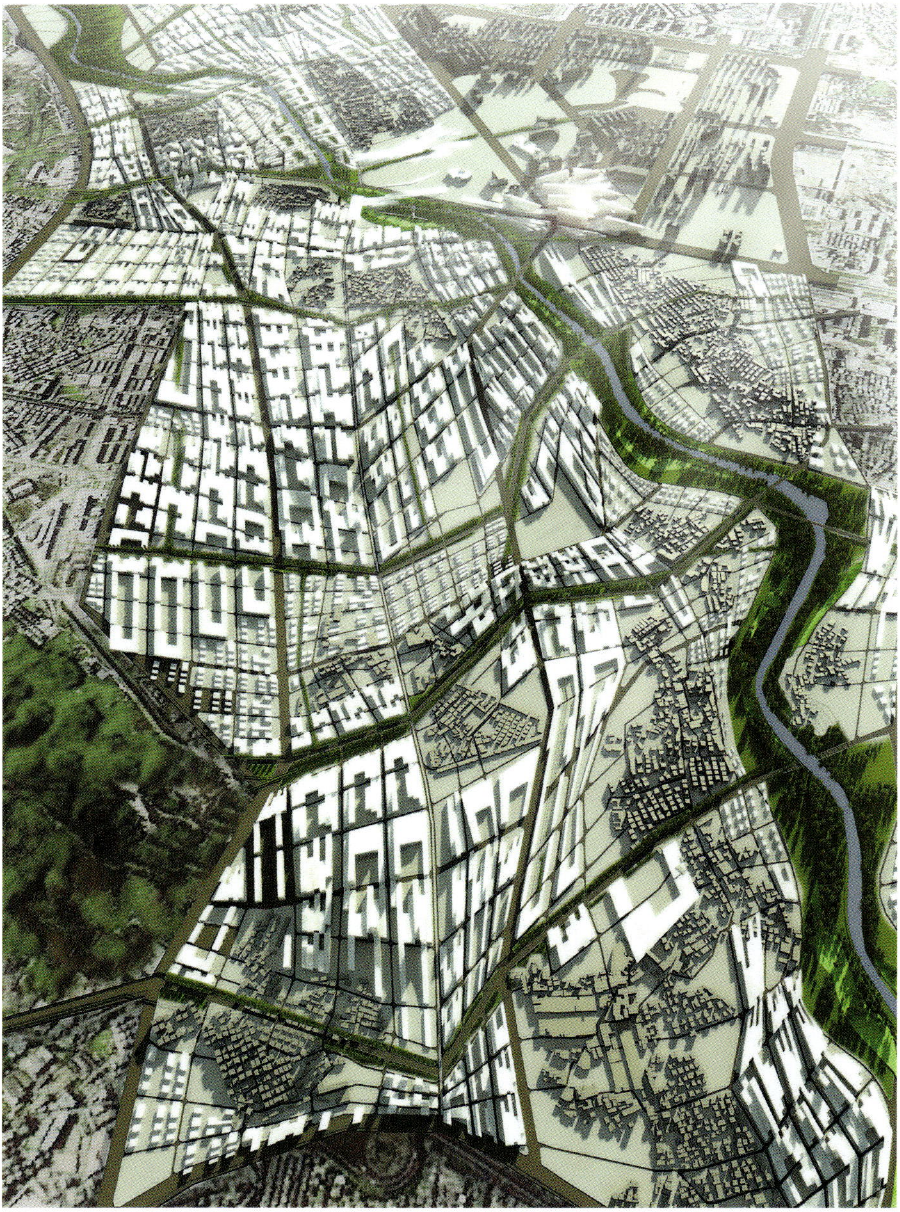

FIGURE 2.1 (Continued) (b) Parametric infrastructure design driven by ecological performance: *Deep Ground* by Groundlab, Shenzhen (China). (Image courtesy of Groundlab.)

Carefully Radical or Radically Careful?

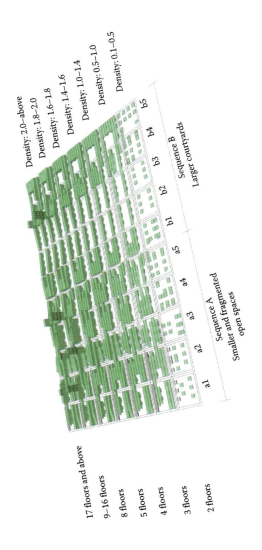

FIGURE 2.1 (Continued) (c) Parametric infrastructure design driven by ecological performance: *Deep Ground* by Groundlab, Shenzhen (China). (Image courtesy of Groundlab.)

service infrastructure, such as water supply systems, sewers, schools, and hospitals. Especially, the emergence of Quetelet's "social physics" or Villermé's "social ecology" could be an interesting point of comparison as it was established as a reaction to (the threats of) disease and revolt in the overcrowded and polluted industrial city during the early 19th century. Inspired by evolutions in biology, new methods of data collecting had to instigate interventions into the urban "organism," making it a more healthy and peaceful habitat for the human species (Rabinow 1989; Koch 2011). These earlier design modes generally made abstraction of local characteristics to formulate constructs of territorial organization and societal modernization mainly focusing on "human" ecology—e.g., social mechanisms of controlling and nurturing the masses (Foucault and Senellart 2007) (Figure 2.2a). Today, however, the engagement with sophisticated techniques deciphering sites in order to uncover and recover local, mostly "natural" ecological qualities is developing into a new paradigm (Fournier 2013; Picon 2010c) (Figure 2.2b).

As a historical positioning of these techniques has largely been left untouched,* today's technological (r)evolution determining infrastructure design is commonly approached from a deterministic point of view, thus resulting in a rather self-referential and often prolix discourse (see Kuhn 1962). One starts from a "new beginning" in urbanism, "against" traditional and modern urban form as well as "against" infrastructure engineering, instead of embedding these design ideas and techniques relating infrastructure, space, and society in a longer chronology and a wider range of disciplines. It appears as if designs/designers are submitted passively to technological change (Picon 2010a). The increasing role of anticipatory/projective knowledge in the production and governance of space and the persistence of an over-determined rhetoric of the new "-isms" make a historical-geographical reflection on design techniques of calculation and cartography timely, if not urgent. Tying in with Ian Thompson's analysis of *Landscape Urbanism* (2012) and the study of Thomas Kirchhoff et al. on *Landscape Ecology* (2013), both decoding the conceptual language of the "new" discourses, this chapter explores the recent reorientation of design to landscape ecological infrastructure against the background of historical ideas and techniques dealing with infrastructure, programmatic uncertainty, and imminent crisis to pierce today's predominantly self-referential discourse. For placing today's momentum (re)inventing design techniques, seemingly geared toward physical/natural qualities, in a tradition of technological innovation steering spatial transformation as an intermediary for social innovation could offer new perspectives to recent developments in urban design. It might even lead to (re)considering the way design techniques are embedded in and related to political and social contexts.

More specifically, the current relationship between design rationale and imminent ecological apocalypse is studied by turning to the past, looking at early 19th-century urban interventions geared at curbing an impending environmental crisis. Thus, the comparison is a theoretical experiment, let's say a first contribution to the "history of the present" (see Elden 2013 on Foucault) of landscape ecological infrastructure with the intention not to find "the origin" but to add some historical-theoretical sensitivity

---

* For a historical positioning on contemporary digital techniques in architecture, see Antoine Picon (2010a).

Carefully Radical or Radically Careful? 35

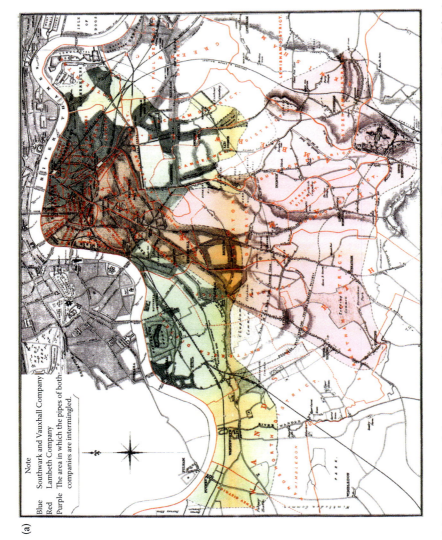

FIGURE 2.2 (a) John Snow's map relating water supply systems with human mortality. (From Koch, T., *Disease Maps. Epidemics on the Ground*, pp. 171 and 183. The University of Chicago Press, Chicago, 2011. With permission.)

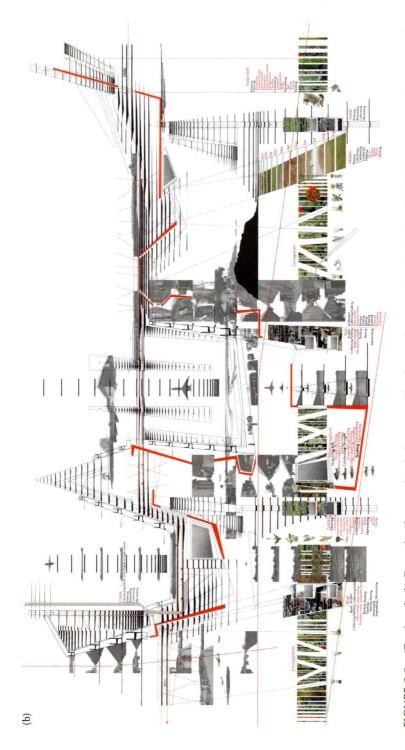

FIGURE 2.2 (Continued) (b) Example of a complex design proposition for an airport crossing by Mathur & da Cunha. (From Mathur, A. and da Cunha, D., *SOAK: Mumbay in an Estuary*, Rupa & Co., Calcutta, India, 2009. With permission.)

Carefully Radical or Radically Careful?

to the current discourse. Thus, the critical-historical analysis aims to contribute to the literature of *Landscape Urbanism* and *Ecological Urbanism*, to strengthen them as academic discourses, thus enhancing their outreach, as well as to highlight a number of questions that urgently ought to be addressed.

2.2 FROM ENGINEERING "HUMAN" TERRITORY TO ENCODING "NATURAL" SURFACES

Similar to today's turn toward infrastructure, previous regimes relating infrastructure, space, and society (e.g., railways, waterways, sewers) respond to a transformation of the socioeconomic and political context and associated ideas about modernization. Especially, the first half of the 19th century corresponds, *in abstracto*, to current design context as it was characterized by a combination of, on the one hand, a disciplinary crisis due to the mismatch between permanent architectural form and a rapidly changing reality and, on the other, the anticipation of an imminent environmental crisis. Authors such as Antoine Picon and Michel Foucault have explained that, traditionally, architecture and urbanism were the disciplines that offered spatial solutions for social problems, but from the 19th century onward, with the introduction of new technologies and the emergence of epidemics and revolutions spreading over large territories, the major problems of society were of a different type, and spatial intervention was entrusted to engineers by means of infrastructure (Foucault 1984; Picon 1992). Especially in France but also Belgium (e.g., De Block 2011, 2013) and many other nations, engineering had "an enduring connection" with social preoccupations (Picon 2007, 198). In contrast to the commonly held view that the engineering of infrastructure "was historically conceived in isolation, independent of the overall urban vision" (Hung 2013, 16), only determined by objective parameters such as safety, feasibility, and efficiency, Antoine Picon highlights that French engineers were even "somewhat messianic in their approach of technology as well as in their ambition to use it to service society" (Picon 2007, 202–203). The corps of the *Ponts et Chaussées* and the *Polytechniciens* were trained to develop overall planning policies that established a *medium* or *milieu*—that is, a multivalent framework in which a series of uncertain elements could unfold (Foucault and Senellart 2007, 20). Dynamic infrastructure frameworks had to instigate an implicit territorial transformation in order to generate a specific societal modernization.

Space was no longer conceived according to a static perception that would ensure the perfect solution *hic et nunc* or, in the words of Leonardo Benevolo, designing entailed no longer "to apply the plausible approximation of an absolutely invariable image to a very slow-moving reality" (Benevolo 1975, 12). The exponentially growing metropolis with problems of social segregation, high crime rates, and the constant threat of cholera outbreaks jeopardized "human ecology," an immediate concern that urgently demanded a completely new design method. In addition to these dire consequences of the Industrial Revolution, political revolutions had destroyed the urban design tradition, planning detailed monumental complexes and aesthetically refined landscape assemblages for the elites. Only the combination of scientific knowledge, methods of calculation and cartography, works for the public good, and the ambition of regenerating society by means of large-scale intervention embedded in engineering could tackle the problems of causality and circulation characterizing modern society. Engineers were assigned the task of equipping and managing or, rather, structuring of and mastery over the territory and *in extenso* society (Desportes and Picon 1997, 45). Consequently, the design of infrastructure networks and the embedded spatial development models for the territory became the object of persistent political attention and intervention in order to regulate (contested) modernization processes (Swyngedouw 1999; De Block and Polasky 2011).

When conceptually comparing early 19th century design to current historical momentum, many parallels can be traced in regard to the ideas and methods of "staging the ground for uncertainty" as well as the rising importance of the expert when faced with immediate threats—with the obvious differences that the role of the engineer is (re)claimed by the urban designer and more (or other) forms of life are incorporated into what Foucault would call mechanisms of security. However, when analyzing more fundamentally the socio-spatial construction of technology, a clear difference between earlier engineering and today's design is taking shape. Whereas the spatial project, or how one can shape society by the intermediary of a well-thought-out transformation of the territory through infrastructure, was implicit in the engineers' and politicians' practice and theory, the relationship between technology and space is again explicit. Current design is characterized by an explicit, inclusive urban strategy, formulating coherent spatial schemes of infrastructure and urbanization. The spatial and not only infrastructure policy *an sich* has become a full-fledged point of departure. However, in contrast to this prominent relationship between space and technology, "grand ideas" about society with related spatial ideologies are less explicit in contemporary practice.

The earlier strong ties between politicians and state engineers, both extensively formulating the constructs of technological innovation, territorial organization, and societal modernization, seem to be replaced by a focus on and accumulation of local interventions generated by what seem like more complex design techniques. Whereas in his essay "What ever happened to urbanism?" Rem Koolhaas' call for a "new urbanism" creating "enabling fields that accommodate processes" and manipulating "infrastructure for endless intensification and diversification" has been successfully answered, his concern about "taking positions" as the "most basic action in making the city" or "encoding civilizations on their territory" remains largely hidden under the zeal of "sophistication" (Koolhaas 1995). Consequently, the sociopolitical dimension of space or the notion of territory in general has become ambiguous in current design theory and practice. In his contribution to the conference on Ecological Urbanism at the GSD, Koolhaas is explicit about the direction in which the discipline needs to move: "We need to step out of this amalgamation of good intentions and branding and move in a political direction and a direction of engineering" (Koolhaas 2010, 70).

What could be considered as a complement or elaboration of Koolhaas' question about the disciplinary crisis of urbanism, Picon rephrased the question to "What has happened to territory?" (Picon 2010c). Although, in previous episodes, the territory was approached with a distant, "scheming gaze" as a project or resource mastered

Carefully Radical or Radically Careful?

by public authorities with a "panoptic-like overview," in current design theory and practice, space is no longer a predetermined project and "passive resource" (Picon 2010c, 94–99). The evolution in cartography and data-analyzing design techniques in general toward increasing complexity reflects this collapse between the rational and sensitive, between autonomy and environment within the spatial disciplines (Picon 2010c). James Corner elaborates on this evolution in his essay on mapping—the design mode that is currently becoming the leading "doctrine" in urbanism:

> [M]apping differs from "planning" in that it entails searching, finding and unfolding complex and latent forces in the existing milieu* rather than imposing a more-or-less idealized project from on high. (Corner 1999, 225–228)

Indeed, there is no longer an *exterior* from which space can be contemplated, as Peter Sloterdijk notes in his work on *Spheres* (Latour 2009b). "Dasein ist design," Henk Oosterling summarized it. This collapse between inside and outside is often dealt with in design by mapping "natural" inside forces and molding topographical inflections that challenge the distinction between landscape and architecture, between building and ground—verging on concepts such as *emergence* and *performance* formulated by authors such as Deleuze and Guattari (Picon 2010c, 98). Indeed, as Corner notes, mapping—the device *par excellence* of new *-isms*—is focusing on designing these envelopes or spheres by tracing "the potential for engaging and involving local intricacy," (Corner 1999, 224) mostly geared toward ecological performance. Instead of governing the territory, surfaces are parametrically encoded to allow "limited self-organization" as Martin Prominiski concludes his article "Designing landscapes as evolutionary systems" (Prominiski 2006). Fournier compares this parametrical, evolutionary approach with Jorge Luis Borges' parable *The map and the territory* in which the detail and the scale of the map merges with reality, thus warning of the fact that "we will have to accept that its internal logic will gradually escape our understanding … We will have to accept that we are out of control … Once the city-system and its model become fully sentient, we will enter … the Greek myth of Prometheus steeling fire from the Gods" (Fournier 2013, 129–131). Whereas the concept of territory used to be associated with administrative action aspiring a political goal or well-defined modernization project, the shift of infrastructure from engineering to urbanism, exploring and enabling the internal logics of the site, obscures the political forces and ideas at work in the complex "evolutionary" surfaces drawn out by designers.

Although there still are urbanists and architects who incline toward an "engineering" attitude and choose to design from the *outside* by superimposing structures in order to instigate clear territorial and societal transformation, a more careful attitude deriving (physical) landscape structures from the *inside* of the site is winning ground. Both approaches create a framework or *milieu*, facilitating uncertain programs, and *in se* rely on similar design techniques, but they are almost contradictory in their "design passion."

---

* James Corner does not refer to Foucault but to the general French meaning of medium, that is, "surrounding" and "middle," implying a field of connections.

2.3 RADICALLY CAREFUL OR CAREFULLY RADICAL?

> It is as though we had to combine the engineering tradition with the precautionary principle ... What is clear is that at this very historical juncture, two absolutely foreign sets of passions ... have to be recombined and reconciled ... We have to be radically careful, or carefully radical. (Latour 2009a)

Analyzing today's obsession with "design" in relation to the ecological crisis, Bruno Latour states that there were and still are two great narratives: one of emancipation, detachment, modernization, progress, founding, colonizing, and mastery, responding to an engineering attitude, and the other completely different, of attachment, precaution, entanglement, dependence, sustaining, nurturing, and caring, which could be categorized under what he defines as "design" (Latour 2009a, 2). According to Latour, design implies humility, an attentiveness to detail, a skill associated to arts and crafts (Latour 2009a, 3). It is defined by an absence of the heroic or, as Picon states, deciphering the context, making sense of information in order to formulate a contextual and ecological strategy is the new heroism (Picon 2010b).

Although this attentiveness to detail and the idea that "there is no way to escape our deeds" (Latour 2009a) is undoubtedly laudable, it is not unproblematic in a time when we are faced with such fundamental problems, and the scale of what has to be remade has become infinitely lager. Instead of the political project steering territorial intervention, carefulness is the *Leitmotiv*, "redesigning everything from chairs to climates" (Latour 2009a). In contrast to the infrastructure projects and discourse of Koolhaas advocating to take "insane risks" (Koolhaas 1995, 969) or the discourse of planning and political geography warning that we might be wasting a "good crisis" (Oosterlynck and González 2013), mapping is doing what no revolution has ever contemplated, namely the remaking of ecological systems, of life on Earth, by operating exactly opposite to modernizing/revolutionary attitudes. "It is as though we had to imagine Prometheus stealing fire from heaven in a cautious way!" (Latour 2009a, 3). In the light of what is currently at stake, namely our welfare, and, more fundamentally, the earth, Latour states:

> The new "revolutionary" energy should be taken from the set of attitudes that revolutionaries loathed most: modesty, care, precautions, skills, crafts, meanings, attention to details, careful conservations, redesign, artificiality, and ever shifting transitory fashions. We have to be radically careful, or carefully radical. (Latour 2009a, 5)

In terms of physical space, one could argue that we have already arrived upon the "carefully radical" with the launch of *Ecological urbanism* and worldwide landscape infrastructure projects in which the careful ecological is the leading principle motivating radical—i.e., large-scale—physical intervention as well as global implementation. Terms such as "ecosystem," "resilience," "CO_2," "bird habitat," and "bioremediation" no longer belong to the jargon of an obscure minority of tree huggers but are central in the "carefully radical" design of our future: Strikingly complex and precise digital mapping techniques, seemingly regardless of size and cultural context, derive sustainable frameworks from the site as a convincing rationale for the transformation of large areas. When considering "radical" in its political

Carefully Radical or Radically Careful? 41

meaning, questions could be raised. New discourses such as *Landscape Urbanism* and *Ecological Urbanism* are indeed radical regarding physical interventions, but one could state that its proponents are conservative when it comes to socioeconomic change. As Thompson notes, "One of the ironies of the 'battle of the urbanisms' is that New Urbanism and Landscape Urbanism are both uncritical of capitalist urbanisation and suspicious of governmental intervention" (Thompson 2012, 16). Moreover, the central position of the technique focusing on (pseudo)scientific rationales instead of socioeconomic problems/aspirations (might) instigate a transition to a populist stance within urbanism—a statement expressing one of the central concerns of this chapter related to arguments formulated within political ecology by academics such as Erik Swyngedouw:

> An extraordinary techno-managerial apparatus is under way ... with a view to producing a socio-ecological fix to make sure nothing really changes. Stabilizing the climate seems to be a condition for capitalist life as we know it to continue. (Swyngedouw 2010, 222)

2.4 CONCLUSION

In contrast to Latour's optimistic approach to the current "attentiveness to detail" of design techniques characterizing landscape infrastructure vis-à-vis an "engineering attitude," Koolhaas draws a line between two narratives in our (design) culture: "one of linear and reasonable progress" and "one of disasters and fundamental tensions between nature and mankind." The latter, he states, is fundamentally anti-modern as it relies on apocalyptic expectations depicting "nature as a kind of punishment of mankind" (Koolhaas 2010, 59). Similar to Swyngedouw's argument, Koolhaas brings the relationship between apocalyptic views underlying ecological design and conservative, even a-democratic, political views to the fore:

> Club of Rome is completely open about the fact that "global warming, water shortages, famine ... would fit the bill ... In searching for a new enemy to unite us." You could say that in the same year, they even suggested that "democracy in no longer well suited for the task ahead." (Koolhaas 2010, 65)

As an alternative, he proposes a design mode for landscape/ecological infrastructure that ties in with engineering and its fundamentally territorial outlook. Chapter 16 by Rosetta Elkin on cultivating green wall infrastructure offers an intriguing entry into the recoupling of engineering and landscape architecture/urbanism as it places contextual micro-scales and large cross-border sociopolitical structures within the same frame of reference for design. With less attention to the local, the North Sea Master Plan of OMA also serves as an example of this ecological urbanism driven by an engineering or, rather, socio-technical rationale as it envisions a cooperative international offshore development of renewable energy (OMA 2010) (Figure 2.3). As such, the design acts at the core of the global environmental issues, comparable to Buckminster Fuller's radical and simple schemes combining nature and network, beyond local "green-washing" of the urban tissue. In fact, these large-scale schemes

42 Revising Green Infrastructure

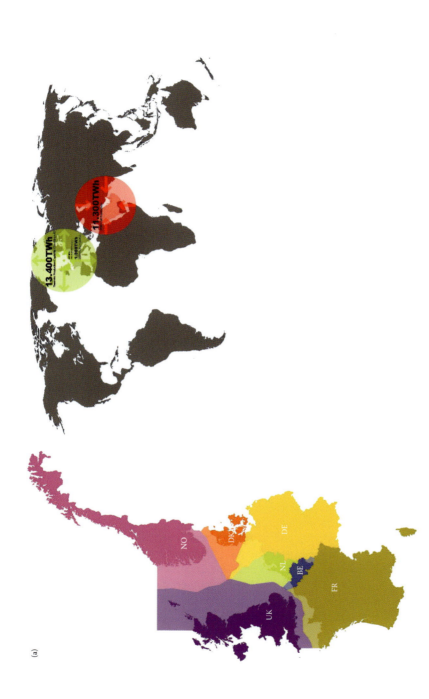

FIGURE 2.3 (a) Schemes for renewable energy production in the North Sea, by OMA. The wind farm development collectively creates a new and dynamic landscape at sea, beyond competing claims. The master plan attempts to reposition wind farm development in positive and productive relation to the cultural forces by which it is currently challenged. (Image courtesy of OMA.)

Carefully Radical or Radically Careful? 43

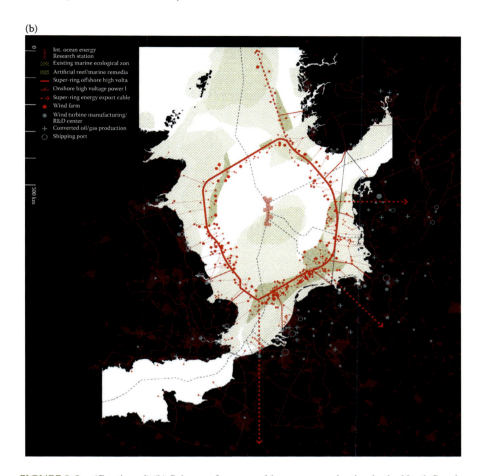

FIGURE 2.3 (Continued) (b) Schemes for renewable energy production in the North Sea, by OMA. The wind farm development collectively creates a new and dynamic landscape at sea, beyond competing claims. The master plan attempts to reposition wind farm development in positive and productive relation to the cultural forces by which it is currently challenged. (Image courtesy of OMA.)

of sharing resources tie in with a long tradition of transnational cooperation between engineers setting up international infrastructure networks. The recently published edited volume *Cosmopolitan Commons: Sharing Resources and Risks across Borders* (Disco and Kranakis 2013) studies some of these internationally engineered systems that are organized to protect both humans and nature. Although, in the book, it is not always clear who the "commoners" are and the spatial intentions or implications of the systems are not taken into account, the concept of infrastructure designed as (cosmopolitan) common could indeed open up new horizons for the landscape infrastructure today.

In the wake of Koolhaas' and Swyngedouw's analyses of the rhetoric on climate change strengthening post-political populism and the above historical reflection, this

FIGURE 2.3 (Continued) (c) Schemes for renewable energy production in the North Sea, by OMA. The wind farm development collectively creates a new and dynamic landscape at sea, beyond competing claims. The master plan attempts to reposition wind farm development in positive and productive relation to the cultural forces by which it is currently challenged. (Image courtesy of OMA.)

Carefully Radical or Radically Careful?

chapter concludes by formulating the questions: What about human in addition to natural ecology? Flows of power in addition to evolutionary flows of species? Ideas on radical advancement and societal progress in addition to a careful recovering of ecological performance to postpone apocalypse? These questions become particularly relevant when design rationales developed in the "Western world" are exported to developing countries in which social inequality in relation to (access to) landscape infrastructure is even more pertinent (e.g., Swyngedouw 2004; see Chapter 18 by Nemcova et al. and Chapter 20 by Terrasa-Soler et al.).

REFERENCES

Bélanger, P. (2013). Infrastructural ecologies: Fluid, biotic, contingent. In *Landscape Infrastructure. Case Studies by SWA. Second and Revised Edition*, 20–27. Basel: Birkhäuser Verlag.

Benevolo, L. (1975). *The Origins of Modern Town Planning*. Cambridge, MA: MIT Press.

Corner, J. (1999). The agency of mapping: Speculation, critique and invention. In *Mappings,* D. Cosgrove (Ed.), 213–252. London: Reaktion Books.

De Block, G. (2011). Designing the nation. The Belgian railway project, 1830–1837. *Technology and Culture* 52: 703–732.

De Block, G. (2013). Planning rural-urban landscapes. Rails and countryside urbanization in South-West Flanders, Belgium (1830–1930). *Landscape Research*. doi: 10.1080/01426397.2012.759917.

De Block, G. and J. Polasky (2011). Light railways and the rural-urban continuum: Technology, space and society in late nineteenth-century Belgium. *Journal of Historical Geography* 37: 312–332.

Desportes, M. and A. Picon (1997). *De l'espace au territoire. L'aménagement en France. XVIe-XXe siècles* Paris: Presses de l'école nationale des ponts et chaussées.

Disco, N. and E. Kranakis (2013). *Cosmopolitan Commons. Sharing Resources and Risks across Borders*. Cambridge, MA: MIT Press.

Elden, S. (2013). How should we do the history of territory? *Territory, Politics, Governance* 1: 5–20.

Foucault, M. (1984). Space, knowledge and power (1982), In *The Foucault Reader. An Introduction to Foucault's Thoughts*, P. Rabinow (Ed.), 239–256. London: Penguin Books.

Foucault, M. and M. Senellart (Eds.) (2007). *Security, Territory, Population. Lectures at the College de France. 1977–1978*. New York: Palgrave Macmillan.

Fournier, C. (2013). The city beyond analogy. *Architectural Design* 83: 48–53.

Hung, Y. (2013). Landscape infrastructure: Systems of contingency, flexibility, and adaptability. In *Landscape Infrastructure. Case Studies by SWA. Second and Revised Edition*, 14–20. Basel: Birkhäuser Verlag.

Kirchhoff, T., L. Trepl, and V. Vicenzotti (2013). What is landscape ecology? An analysis and evaluation of six different conceptions. *Landscape Research* 1: 33–51.

Koch, T. (2011). *Disease Maps. Epidemics on the Ground*. Chicago: The University of Chicago Press.

Koolhaas, R. (1995). What ever happened to urbanism (1994). In *S,M,L,XL,* OMA, R. Koolhaas, B. Mau, J. Sigler (Eds.), 958–971. New York: The Monacceli Press.

Koolhaas, R. (2010). Advancement versus apocalypse. In *Ecological Urbanism*, M. Mostafavi (Ed.), 56–72. Baden: Lars Müller Publishers.

Kuhn, T. (1962). *The Structure of Scientific Revolutions*. Chicago: University of Chicago Press.

Latour, B. (2009a). A cautious Prometheus? A few steps toward a philosophy of design (with special attention to Peter Sloterdijk). In *Networks of Design. Proceedings of the 2008 Annual International Conference of the History of Design,* J. Glynn, F. Hackney, V. Minton (Eds.). Boca Raton: Universal Publishers.

Latour, B. (2009b). Spheres and networks: Two ways to reinterpret globalization, *Harvard Design Magazine* Spring/Summer 2009.

Mathur, A. and D. da Cunha (2009). *SOAK: Mumbay in an Estuary.* Calcutta, India: Rupa & Co.

McGrath, B. (2009). New patterns in urban design. *Architectural Design* 79: 48–53.

Mossop, E. (2006). Landscapes of infrastructure. In *The Landscape Urbanism Reader,* C. Waldheim (Ed.), 163–179. New York: Princeton Architectural Press.

OMA (2010). Zeekracht. In *Ecological Urbanism,* M. Mostafavi (Ed.), 72–78. Baden: Lars Müller Publishers.

Oosterlynck, S. and S. González (2013). Don't waste a crisis: Opening up the city yet again for neoliberal experimentation. *International Journal of Urban and Regional Research* 37: 1075–1082.

Picon, A. (1992). *French Architects and Engineers in the Age of Enlightenment.* Cambridge: Cambridge University Press.

Picon, A. (2007). French engineers and social thought, 18–20th centuries: An archeology of technocratic ideals. *History and Technology: An International Journal* 23: 197–208.

Picon, A. (2010a). *Digital Culture in Architecture: An Introduction for the Design Professions.* Basel: Birkhäuser.

Picon, A. (2010b). Digital Culture in Architecture: An Introduction for the Design Professions. Lecture at Manchester University, November 2, 2010.

Picon, A. (2010c). What has happened to territory? *Architectural Design,* 94–99.

Prominiski, M. (2006). Designing landscapes as evolutionary systems. *The Design Journal* 8: 25–34.

Rabinow, P. (1989). *French Modern. Norms and Forms of the Social Environment.* Cambridge, MA: MIT Press.

Shannon, K. and M. Smets (2010). *The Landscape of Contemporary Infrastructure.* Rotterdam: Nai Publishers.

Swyngedouw, E. (1999). Modernity and hybridity: Nature, *regeneracionismo,* and the production of the Spanish waterscape, 1890–1930. *Annals of the Association of American Geographers* 89: 443–465.

Swyngedouw, E. (2004). *Social Power and the Urbanization of Water. Flows of Power.* Oxford: Oxford University Press.

Swyngedouw, E. (2010). Apocalypse forever? Post-political populism and the spectre of climate change. *Theory Culture, Society* 27: 213–232.

Thompson, I. H. (2012). Ten tenets and six questions for landscape urbanism. *Landscape Research* 37: 7–26.

Waldheim, C. (2010). On landscape, ecology and other modifiers to urbanism. *Topos* 71: 21–24.

Waldheim, C. (2013). Reading the recent work of SWA. In *Landscape Infrastructure. Case Studies by SWA. Second and Revised Edition,* 8–14. Basel: Birkhäuser Verlag.

3 The City That Never Was
Engaging Speculative Urbanization through the Logics of Landscape

Christopher Marcinkoski

CONTENTS

3.1 Situation at Hand 47
3.2 Disciplinary Contexts 52
3.3 Laboratory Madrid 55
    3.3.1 Entropy 59
    3.3.2 Fertility 59
    3.3.3 Agility 62
    3.3.4 Utility 65
3.4 Conclusion 65
References 67

3.1 SITUATION AT HAND

The last decade has witnessed the spectacular collapse of economies that became too reliant on urbanization as their primary means of economic production. From the ghost estates of Ireland to the artificial oases of Dubai to the sprawl of the U.S. Sunbelt to the unoccupied new towns scattered across western China to the abandoned infrastructures of Spain, speculative expansions of urbanized territories have proliferated in a wide range of economic, political, and geographic contexts (Buckley and Rabinovitch 2010; Campbell 2011; Glaeser 2010; Hardaway 2011; Lewis 2011; Medina 2012; Palmer 2011). I refer to this phenomenon as "The City That Never Was" and contend that it is a fundamental crisis facing contemporary city design and development praxis. As such, it is worth asking what the implications of this reality are for the planners and designers engaged in the implementation of these metropolitan initiatives and, in particular, how confronting this phenomenon might expand the potential efficacy of landscape-driven approaches to urbanization.

    Within this context, I would suggest that one of the most overlooked aspects of the recent global economic crisis is the significant role the urban design disciplines—architecture, landscape architecture, city planning, and civil engineering—have played in the production of this episode. Beyond discussions of banks, mortgage-backed securities, collateralized debt obligations, derivatives, credit default swaps, and "too big to fail," there is a clear body of evidence of design's complicity—whether intentional or not—in the production of this moment in history.

Certainly, liberalized banking policies enacted globally over the last 30 years have contributed to the increase in scale and frequency of these speculative initiatives (Krugman 2013). However, what is most striking about these proposals for metropolitan expansion is the increasing artificiality of their premise as well as the ubiquitous nature of their urbanistic proposition. That is, the complicity of the designer emerges from a pervasive reliance on increasingly standardized formats of settlement and easily replicable spatial products as the basis for conceiving of new urban expansions (see Figure 3.1) (Easterling 2012; Romer 2011).

What the urban design disciplines often ignore in their promotion of new metropolitan form is that these initiatives are rarely undertaken as a purely social or cultural enhancement for a population in need. Rather, they are more commonly initiated as a political or economic hedge by governments or private developers in search of destinations for surplus capital and accolades as a globally competitive world city (Hoffman and Wan 2013; Pettis 2013).* In the past, urbanization was viewed as a response to the social demands of economic growth. Today, however, this relationship has seemingly flipped: Urbanization is now seen as a means to growth rather than vice versa. Sometimes, these initiatives are wildly successful. Sometimes, they fail spectacularly. More often than not, they ultimately end up somewhere in between. It is this unpredictability that has the greatest potential for leveraging as we reconsider design and planning's relationship to these speculative initiatives.

The dominant media narrative today is that this phenomenon of speculative urbanization is unique to the first decade of the 21st century. However, analogous examples can be found throughout history, precipitated by a range of political actions and economic events (Sakolski 1932). For example, two of the three cases former Brown University economist Peter M. Garber cites in his 1990 essay "Famous First Bubbles" as the most frequently referenced speculative episodes in economic history are tied directly to real estate or land considerations: the 1720 *South Sea Bubble* in the UK and the 1719–1721 *Mississippi Bubble* in France. University of Southern California housing policy professor William C. Baer in his 2007 essay "Is speculative building underappreciated in urban history?" expands this collection of examples with his discussion of the speculative nature of housing construction in ancient Rome, 16th century Antwerp, and 17th and 18th century London. I would add to these examples the late 1880s *Los Angeles Land Boom and Bust* and the 1920s *Florida Land Rush* (see Figure 3.2), which, like the 18th-century examples cited above, offered investment opportunities tied directly to newly accessible exotic territories (Netz 1916; White 2009). However, perhaps the most relevant antecedent for our purposes is *The Panic of 1873*—one of history's most dramatic and long-lasting

---

* Although some have made the converse argument, i.e., that growth drives urbanization, it is clear that many policy makers are betting that causality runs the opposite way.

The City That Never Was

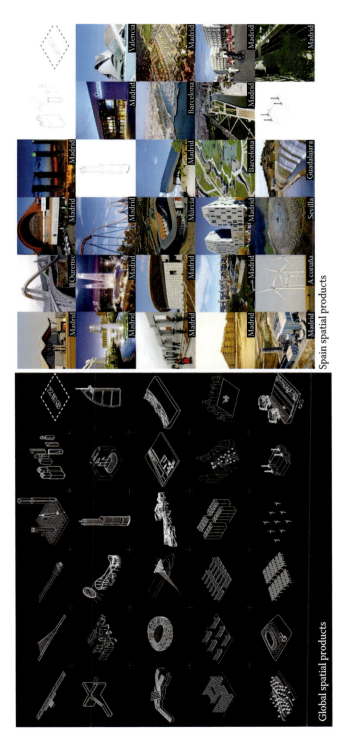

FIGURE 3.1 Global urban spatial products image matrix and corresponding Spanish projects. (Courtesy of the University of Pennsylvania *The City That Never Was* studio.)

FIGURE 3.2 Advertisement and photograph from the 1925 Florida Land Rush. Note the similarity in rhetoric in the advertisement and familiarity of the incomplete cityscapes as they relate to contemporary examples of urban speculation in previously exotic landscapes such as the UAE. (From *Florida News*, September 14, 1925, Baker Old Class Collection, Baker Library Historical Collections, Harvard University.)

examples of infrastructure-driven land speculation leading directly to a period of extended international financial and social crisis (Reynolds Nelson 2011).*

Despite these and other instances, the first decade of the 21st century has seen some of the most dramatic and consequential examples of speculative urbanization in history (Elsheshtawy 2013; Garcia 2010; Kitchin et al. 2010).† And yet notwithstanding the spectacular failures of these recent cases, other comparable initiatives continue to proliferate in places as economically and politically diverse as Turkey, Egypt, Panama, Brazil, Angola, Algeria, Saudi Arabia, Kazakhstan, India, Vietnam, and Malaysia (Archibold 2011; Fuller 2012; Medina 2012; Roubini 2013). And certainly no entity in history has shown a greater tendency toward speculative urbanization as present-day China (see Figure 3.3), a source of countless media reports of massive, vacant new towns and scale-replicas of cities such as London and Paris

---

* In a June 27, 2010 *New York Times* editorial titled "The Third Depression," the economist Paul Krugman notes the Panic of 1873 was one of only two eras in economic history so severe they were actually described as a "depression" during their occurrence.
† In the period following the 2007–08 downturn, Ireland, Spain, and Dubai became synonymous with overheated economies predicated on a continued expansion of their respective construction sectors and a continued influx of foreign populations to build and occupy these speculative property blooms.

The City That Never Was

FIGURE 3.3 Massive unnamed satellite development northeast of Xinyang in Henan Province, China, under construction since 2004. Image date: July 2012. (Courtesy of Google Earth/Christopher Marcinkoski.)

being rapidly reproduced throughout the country (Dyer 2010; Keeton 2011; Ming 2010). Clearly, speculative urbanization should be seen as an intensifying phenomenon demanding increased scrutiny irrespective of the superlative-laden prospectuses for projects or their seemingly sound prognoses of demand (Barboza 2011; Roubini 2011).*

In reflecting upon the arc of these examples, one might argue that, over the last 25 years, urbanization itself has become *the* preferred instrument of economic production and expression of political power for both established and emerging economies.† However, in contrast to my earlier assertion, many would continue to argue that the decision-making processes and policies that have led to these massive urbanization projects are beyond the traditional purview of design. Yet, when one surveys the examples cited, it is difficult to deny the fact that urbanization throughout history has proven to be a fundamentally speculative act—one characterized by risk, interruption, inflection, and failure (Shiller 2000). And, as such, we must question why the disciplines involved in the implementation of these projects continue to engage the task of planning and design for new settlement with little regard or acknowledgement of the capricious nature of this endeavor. This chapter therefore looks to confront the

---

* Determining the exact segment of China's GDP devoted to construction has proven difficult and varies by source. As such, 30% represents what is considered to be a conservative estimate. For example, a recent article in the *New York Times* (citing China's National Bureau of Statistics) estimates China's spending on urbanization in 2011 was near 70% of the country's total gross domestic product.
† For the purposes of this chapter, "speculative urbanization" is defined as *the undertaking of land acquisition and/or infrastructure and building construction in the pursuit of uncommon financial gains under the presumption of market demand despite the absence of a specific future tenant or consumer.*

urban design disciplines' tendency to treat territorial-scale expansions of settlement as instantaneous, totalizing ventures and proposes to use the increasingly speculative nature of contemporary urbanization as an opportunity to rethink future modes of urban design praxis.

3.2 DISCIPLINARY CONTEXTS

All ecological projects (and arguments) are simultaneously political-economic projects (and arguments) and vice versa. Ecological arguments are never socially neutral any more than socio-political arguments are ecologically neutral.

David Harvey
The Dialectics of Social and Environmental Change (1996, 182)

As global urbanization rates increase, how these areas of new settlements are organized, deployed, and managed is emerging as the critical question facing the city design and development disciplines (United Nations 2013). Given the volatile character of this urbanization, coupled with the disciplinary aspirations articulated within recent *Landscape Urbanism* and *Landscape Infrastructure* discourses, it would appear that the unpredictability seen in the deployment of these urban expansions is a milieu ripe for engagement through landscape and thereby ecological considerations (Belanger 2012; Waldheim 2006). However, I would contend that this engagement cannot be *landscape* and *ecology* defined solely through the lens of the biotic. Rather, these ecological considerations should be understood and employed as potentially operative political and economic devices for negotiating the speculative nature of contemporary settlement.

In the 2006 volume *The Landscape Urbanism Reader*, both Charles Waldheim's essay "Landscape as Urbanism" and James Corner's essay "Terra Fluxus" suggest that landscape's hypothetical capacity to navigate and subvert the hegemony of political, economic, and regulatory pressures on contemporary urbanization is a potentially potent disciplinary competence worth advancing. However, although both Corner and Waldheim evocatively describe this capacity, one can argue that a decade on from the initial publication of these essays, we have yet to see this facility tangibly manifest itself in practice. In fact, an increasingly common critique of the Landscape Urbanist discourse as simply a savvy rebranding of traditional landscape architectural capacities during a period of increased cultural awareness of environmental concerns could be understood as emerging directly from the perception of a rhetorical potential unfulfilled (Allen and McQuade 2011; Duany and Talen 2013). Certainly, it is this aspiration for a more politically operative role in the formation of the urban that I would argue has led to the emergence of the North American *Landscape Infrastructure* discourse out of Landscape Urbanism with its deliberate focus on relieving civil engineering and urban planning of their respective roles in the conceptualization and organization of the urban (Belanger 2012).

Ultimately though, if one considers the bulk of the projects being undertaken by the leading landscape architectural practices of today—Field Operations, West 8, MVVA, Agence Ter, Topotek, Gross Max, Turrenscape, SWA, etc.—they are, more often than not, engaged in elaborating the familiar public realm typologies of traditional landscape architectural practice (gardens, parks, plazas, promenades, etc.). These projects rightly tout their performative capacities from an environmental, social, and engineering perspective, but ultimately, this work has a relatively limited impact on the settlement patterns within which they are nested.* Spectacular waterfronts full of amenities for the bourgeois and large urban parks undertaken as political legacy projects are important contributors to a city's quality of life. However, they are of limited utility when it comes to the particular demands of managing and negotiating the speculative expansion of contemporary urban form. In fact, these kinds of civic amenity projects are often deployed as the alibi for or facilitator of these sorts of speculative development.†

Given that a significant portion of the landscape discipline's future work likely will be drawn from the rapidly urbanizing Global South, engaging questions of new settlement and urban expansion is an essential disciplinary project in need of advancement (Watson 2009). In fact, I would argue that the current landscape urbanist/landscape infrastructure discourse's preoccupation with the biotic is pursued at the expense of an occasion to truly assert disciplinary agency and expand professional scope as it relates to developing systems of operation for the soon-to-be urban. In essence, I am arguing for the elevation of landscape's disciplinary considerations of the economic and the political to the current level of the biotic and the social—embracing urban ecology in its broadest sense.

Specifically, direct engagement with the organization of new regimes of settlement and land occupation represents a unique opportunity to shift landscape architectural considerations from a primarily remediative mode of praxis toward a more projective, instrumental one tasked with the choreographing of future regimes of urbanization (Weller 2009). This is not just engagement with urbanization as a metaphorical process—a seed of Landscape Urbanism's emergence in the late 1990s (Shane 2006). But rather direct engagement with the political and economic acts (processes) related to the urbanization of a geographic territory.

Landscape Architecture as a discipline cannot continue to solely focus on reordering postindustrial sites or cleansing the environmental wounds of a modern capitalist society (retrofitting, reclaiming, remediating, recultivating, renovating, etc.) and expect to maintain its current cultural ascendancy. Instead, it should endeavor to open up new lines of inquiry that undertake the imagination of entirely new systems and strategies of land occupation that embrace the uncertainty of contemporary urbanization as their basis.

The more conservative disciplinary perspective might suggest that such an endeavor is too hubristic, that questions of settlement and metropolitan expansion are the domain of city planners, politicians, developers, and perhaps even architects. Yet I would argue landscape architecture is uniquely positioned for this work and

---

* One outlier to this trend is the role well-known landscape architectural practices are playing in the current work being done in the United States in response to 2012's "Super-Storm Sandy" in which some projects are deliberately exploring reorganizing or even entirely moving at-risk or already-impacted settlements. See, for example, the U.S. federal government's *Rebuild by Design* competition, http://www.rebuildbydesign.org.

† See, for example, the story of Irvine, California's Orange County Great Park or the widespread push by cities to replicate the real estate effects of New York City's High Line.

can, in fact, trace a strong line back to a historical role in the organization of settlement—particularly as it related to agriculture and land tenure. Although this former capacity certainly did not deal with the issues of hyper-dense urbanization with which we are confronted today, there is little validity in the argument that settlement organized and deployed primarily from a landscape perspective cannot accommodate, guide, or engender high-density configurations (Allen and McQuade 2011).* Thus, if McHargian planning (which continues to have far too much influence on contemporary landscape architectural practice) was concerned solely with the elaboration of *where or where not* to position settlement, the interest here is explicitly in *how* settlement itself is actually deployed.

It should be noted, however, that many involved in the conceptualization and planning of these contemporary expansions of settlement would argue that they are not failed or abandoned or incomplete but rather that the examples cited are simply in a state of delayed absorption (Cowen 2012).† That is, the market will eventually correct for whatever surplus is present as the cost of occupying the new settlement decreases. And eventually, they will be occupied as planned—i.e., the planning and the pro formas cannot possibly be wrong.

For example, Stephen S. Roach, former chairman of Morgan Stanley Asia and a senior fellow at Yale University, emphatically contends that concerns over contemporary Chinese ghost towns are overblown. Citing the relative vacancy of the Pudong New Area of Shanghai in the 1990s, now home to a population of roughly 5.5 million residents, Roach and others assert it is only a matter of time before these new vacant urban districts reach their projected populations. From this perspective, "China [simply] cannot afford to wait to build its new cities" with the country's urban population expected to expand by more than 300 million in the next 20 years (Roach 2012). Yet one cannot escape thinking of the repeated "adjustments" to the projected populations of Kangbashi, Ordos—the media's poster child for the exuberance of Chinese urbanization—seen over the last decade. From initial forecasts of more than one million residents to the most recent prognoses of the settlement's eventual population at around 400,000, current estimates suggest between 30,000 to 70,000 residents live in this new urban territory after more than a decade since its initiation (Bildner 2013). Returning to Corner and *Terra Fluxus*:

> ... it is increasingly the case that vast developer-engineering corporations are constructing today's world with such pace, efficiency, and profit that all of the traditional design disciplines (and not only landscape) are marginalized as mere decorative practices, literally disenfranchised from the work of [city] formation. (27)

---

* Despite this as a common critique of landscape's potential efficacy in organizing settlement and engaging the hyper-density of emerging cities in the Global South, the reality is quite the opposite. A particularly relevant example is James Corner Field Operations' ongoing work on the development of Qianhai Water City in Shenzhen, China, where a robust public realm and landscape infrastructure system (water fingers) organizes the entirety of the 4500-acre project.

† What is interesting about this point of view is its potential reframing through an ecological systems perspective in which "failure" in the pejorative sense is seen as invalid. For example, can an extended delay in the realization of an initiative for new settlement be embraced as a design opportunity or reframed as a productive event?

So the question remains: Are architects, landscape architects, urban designers, and city planners exempt from responsibility for the consequences of these speculative projects because the ventures originate from political and economic initiatives considered to be "above their pay grade?" Or does the escalation of this phenomenon actually provide a unique moment in which to step back and reconsider the disciplinary planning and design conventions upon which these new proposals for settlement rely?

I would suggest that this situation does imply an occasion to consider how one might plan and design for growth so that the urbanistic accoutrements necessary for a population are available only when they are *needed*—a kind of just-in-time urbanism that is nimbler and lighter yet retains the qualities of urban form and space demanded by a particular culture. And for the purposes of this volume, we might specifically consider how landscape systems and ecology-driven logics can, in fact, provide the infrastructures necessary for such an approach to urbanization.

3.3 LABORATORY MADRID

In order to explore this potential shift in disciplinary orientation, I directed a graduate design-research studio at the University of Pennsylvania (Penn) in the spring semester of 2012 that used the incomplete and unoccupied urban landscapes at the periphery of Madrid as a laboratory for considering the increasingly speculative nature of urbanization being seen globally. In this regard, Spain provided a particularly useful context given it represented an established western European democracy that experienced a rate of urbanization over a relatively compressed time period more commonly associated with emerging Asian economies than its own continental neighbors. Consider the following:

From 1998 to 2008, Spain experienced an unprecedented expansion of its urbanized territory (see Figure 3.4) (Garcia 2010). Between 2004 and 2008 alone, more than 2.4 million new residences were constructed across the country with current estimates of more than half of those continuing to remain unoccupied (Campbell 2011; McGovern 2008; Pollard 2009). Spain built more than 5000 km of new highway in the last two decades, leaving the country with the fourth-longest highway system in the world behind only the continental networks of the United States, China, and Canada (CIA World Factbook 2012; Eurostat 2011). Of Spain's 50 international airports, 15 qualify economically as ghost airports based on annual passenger numbers with the privately financed Aeropuerto de Ciudad Real perhaps the most conspicuous with zero commercial flights currently scheduled despite an investment of nearly 1.1 billion euros (Grupo AENA 2011; Moran 2012). Spain has constructed more than 2600 km of high-speed rail track, ranking behind only China in terms of total length (EuropaPress 2010; Sheehan 2012). And the Comunidad de Madrid nearly tripled the length of its metro system between 2000 and 2011, creating the eighth longest metro system in the world despite being only the 50th most populous metropolitan area (MetroMadrid 2012). This infrastructural investment was companioned by the construction of innumerable cultural and civic facilities throughout the country, including infamous white elephants such as Peter Eisenman's Galicia City of Culture and Santiago Calatrava's City of Arts and Sciences in Valencia (Moix 2010).

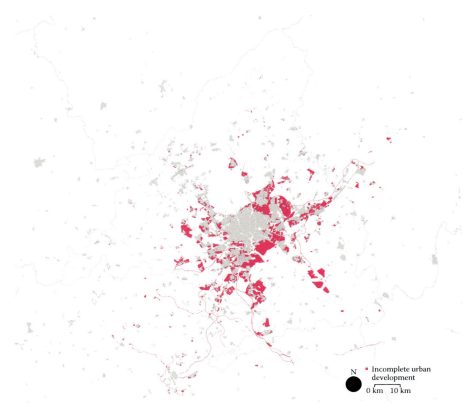

FIGURE 3.4 As of the end of 2012, more than 25% (in pink) of urbanized lands in and around Madrid are composed of partially vacant or incomplete urban development. (From SIOSE 2009/Madrid Regional Planning/Christopher Marcinkoski.)

Spain's investment in urbanization can be traced back to three political events: the Land Law of 1998, the introduction of the euro in 2002 (with its corresponding reduction in interest rates), and the liberalization of labor constraints that same year (Government of Spain, Head of State 1998; Ministry of Labor and Social Affairs 2002). The result was nearly 20% of Spain's economy being tied to construction-related industries—nearly four times the proportion seen in the United States and the UK, whose economies both experienced contemporaneous real estate bubbles (Government of Spain, Ministry of Development 2011). Although held up in the mid-2000s as a global exemplar of both economic and architectural innovation, Spain's urban growth policy led directly to the precipitous state of the country's current economy (Ouroussoff 2006; Riley 2006; Thomas 2012).

Given such a context, a particular interest of the work of the Penn studio was how landscape and/or ecological considerations could be leveraged to either retrofit this incomplete urban ground for future, unexpected uses (see Figure 3.5) or how designers could use the situation that emerged in Spain between 1998 and the present as

The City That Never Was 57

FIGURE 3.5 Traces of ubiquitous speculative development in the Madrid metropolitan region—in considering the speculative nature of contemporary urbanization, one might argue that, over the last 25 years, urbanization itself has become *the* preferred instrument of economic production. (From Plan Nacional de Ortofotografía Aérea, 2012/Christopher Marcinkoski.)

the basis for developing fresh, new strategies for settlement and land occupation. The premise of the studio was therefore quite simple: *Conventional approaches to urbanization are increasingly producing incomplete or abandoned settlements that have enormous social, environmental, and economic consequences. How might we use what occurred in Spain as a platform for developing new priorities and logics of urbanization that challenge these singular, monofunctional, totalizing conventions?*

For the purposes of the studio, the ecology-driven design logics we were most interested in, vis-à-vis thinking about new forms of urban settlement, included the following: (1) *The value of bigness,* that is, the implication of use diversity and multiple micro-sets of activity within a territory and, in particular, the interface or overlap of seemingly incompatible programs and land uses; (2) *Instability/Unpredictability,* not as liability or something to be assuaged but rather as a condition ripe for instrumentalization as a design opportunity; (3) *Limits of Control,* understanding design as a participatory agent within an urban environment rather than the overseer of said environment; (4) *Resilience/Adaptability,* building capacity to accommodate and endure disturbances to a system; (5) *Self-organization,* states of urbanization characterized by often unexpected change, moving systems from simple to more complex states; (6) *Interdependence* or embracing the comingling of the biotic and the urban, rather than treating them as oppositional binaries; (7) *Patchworks or Nonlinear organization,* the simultaneous occurrence of multiple states within a whole or the rejection of the monofunctional single-state; (8) *Scenarios rather than sequences,* elaborating multiple potential outcomes rather than designing a linearly phased ideal state; (9) *Sudden state-change,* the inevitability of urban form to shift in terms of its orientation or health; and (10) *Design as facilitator or catalyst,* influencing or inflecting an open urban system through design intervention rather than trying to elaborate and implement a closed system.*

It is this last "logic," the catalytic potential of landscape-driven design interventions, which is of greatest interest within this work. Rather than urbanization continuing to rely on heavy, permanent, singular infrastructural devices that control growth, the potential here is for hybrid landscape systems to be deployed as more nimble, catalytic scaffolds that influence and inflect the processes of urbanization, embracing the volatile, speculative nature of these endeavors. While in no way fully formed, it is this potential to have a greater role in the organization and deployment of new settlement that makes the question of the role of landscape-driven urbanization so ripe for consideration.

Participants in the studio included master's students in landscape architecture, architecture, and city planning, typically in the final year of their studies. This mix of disciplinary capacities was exceptionally useful in developing more complex and sophisticated responses to the studio prompt. However, it also meant that very few of the projects could be characterized as purely a "landscape" project or an "infra-

---

* These "logics" are drawn from a number of sources, including the writings of James Corner, C. S. Holling, William B. Honachefsky, Nina Marie Lister, Mohsen Mostafavi, Frederick Steiner, and Robert Ulanowicz, among many others, and represent more of an interpretation than a precise transfer of definition.

The City That Never Was

structure" project. Rather, the studio produced "urban" projects useful for both provocation and critical reflection.

In reviewing the work produced, certain themes or considerations repeatedly emerge. These themes or lenses, as I prefer to consider them, include notions of *Entropy, Fertility, Agility, and Utility.* And while somewhat imprecise and broad in definition, they are quite useful as considerations for evaluating proposals for new urban settlement vis-à-vis negotiating the speculative nature of these initiatives. For the purposes of this chapter, I will briefly sketch out each of these four lenses as they relate to the question at hand.

3.3.1 ENTROPY

While there has been a broad cultural focus on the "waste" that emerges from anthropogenic settlement, the majority of this discourse has been oriented toward the desire to limit the production of these byproducts or design away waste altogether. This aspiration has typically manifested itself through proposals for the optimization of existing urban systems, increases in efficiency through the introduction of new systems, or the removal of existing systems altogether (Braungart and McDonough 2002; Penn Institute for Urban Research 2008). However, as Alan Berger (2006), among others, has noted, such a desire is perhaps a fool's errand. Urbanization, by its very nature, is inefficient, messy, irrational, and rhizomatic. Therefore, regardless of aspirations otherwise, attempts at engineering "optimized" formats of settlement or devising new physical or political devices of control, urbanization inevitably produces waste, entropy, and disorder.

Yet one can argue that "waste" is a subjective evaluation or, at the very least, one too often used in the pejorative. If considered in a projective, anticipatory way, the collateral byproducts of urbanization actually have the capacity to become new resources or commodities if we reorient our approach to the planning and design of new settlement. This notion of waste as a commodity or resource is not about recycling per se or a naively earnest sense of environmental stewardship. Rather, it is about an acknowledgement that urbanization is a fundamentally disruptive act. The ability to anticipate the multiplicity of potential outcomes of this disruption, as well as project potential strategies for leveraging this interference should become a core competency of contemporary urban design. In particular, there is great potential in imagining an urban landscape conceived of as emergent from waste rather than perpetually in conflict with it (see Figures 3.6 and 3.7).

3.3.2 FERTILITY

Given the prominent role discussions of environmental and ecological concerns presently hold in contemporary discourse surrounding urban planning and design, it is disappointing to see the majority of work emerging from these discussions is, more often than not, performing roles of mitigation or remediation rather than more primary operations of organization and catalyst. Mobility and housing demands, in particular, continue to dominate the discussion of new urban form, essentially foreclosing on the possibility of systems of agriculture, food, water, or natural systems becoming the basis of new hybrid formats of settlement. This is a missed opportunity.

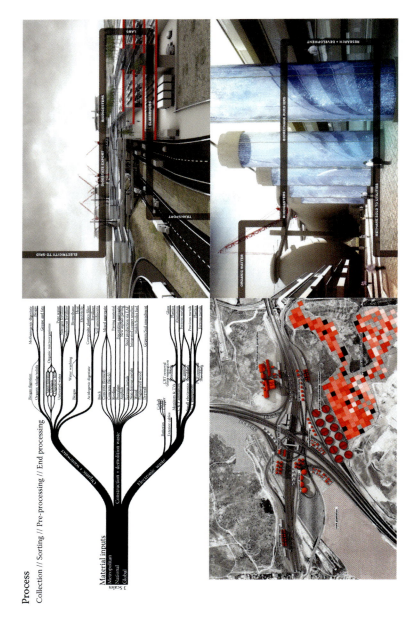

FIGURE 3.6 *On-Ramp Off-Shoots* uses the underutilized toll roads built around Madrid as sites of a new waste-based economy. Utilizing the connectivity of these corridors, the project proposes a new network of waste material processing, storage, and research. What makes the project most compelling is that the waste-based economic strategy is bundled with a scalable platform for education, including vocational training, postgraduate research, and start-up incubators. (Courtesy of the University of Pennsylvania *The City That Never Was* studio.)

The City That Never Was   61

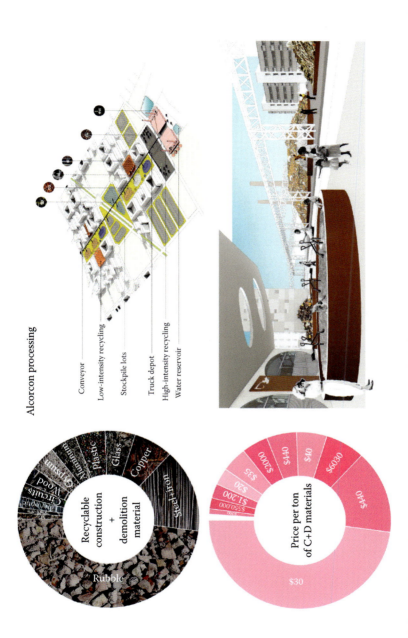

FIGURE 3.7 *Ouroboric Urbanism* uses the current situation in Madrid as an occasion to test urbanistic processes of downgrading and upgrading. The project is predicated on an ongoing material transaction that takes incomplete areas of the metropolitan region offline (those measured to have the highest ecological value) in order to concentrate investment in those areas of the metropolitan region most likely to find near-term urbanistic success. (Courtesy of the University of Pennsylvania *The City That Never Was* studio.)

62 Revising Green Infrastructure

Understanding landscape considerations as only a response to urbanization is a reductionist approach to land occupation. On the contrary, ecological principals have the capacity to become the fertile substratum of new urban form. For example, the strategic deployment of constructed ecologies of vegetal, aquatic, or climatic production can generate catalytic urban effects in which issues of desertification, water scarcity, air quality, heat island, or habitat erosion serve as vehicles for reorganizing anthropogenic occupancy. These same considerations also offer the means by which to reorient existing territories toward a similar logic of production (see Figure 3.8). Rather than continuing to consider landscape regimes as responses to or balances against urbanization, I would argue for imagining entirely new formats of settlement initiated and driven by productive regimes of ecology and landscape rather than continuing to rely on 19th-century formats of urbanization, smartly dressed in festive wreaths of green (Branzi 2006; Waldheim 2010).

3.3.3 AGILITY

Agility implies both speed and nimbleness, two characteristics rarely thought of in relation to accepted cultural ideas about urban form. Yet speed and nimbleness are exceptionally useful traits to pursue in engaging the speculative nature of contemporary urbanization.

On one hand, one of the primary challenges of planning and implementing new settlement is the mandate to deliver imagable, territorial-scale urbanization in the expedited fashion demanded by today's highly mediated, transactional economy (McKinsey Global Institute 2010; Oosterman 2012). Pursuit of the *Global City* designation, in particular, as a model of both economic and urban growth has led to a widespread oversubscription to generic development formulas presumed to be capable of replicating established urbanistic successes through the redeployment of the spatial products upon which those successes rely (Goldman 2011). This accelerated form of urbanization is, more often than not, driven by the demands of financial investors or political statecraft rather than actual population or market demand (see Figure 3.9). As such, these proposals for new settlements are rarely calibrated to the particulars of a place but, more commonly, are driven by a kind of unimaginative checklist model of planning.

On the other hand, considering the organization and shaping of settlement from alternative points of view related to characteristics such as impermanence, nomadism, periodicity, or disposability fundamentally change the criteria we use to evaluate an urban expansion. These are seemingly radical propositions for sure, but attributes such as these are worth consideration as ways of engaging the demands for rapid urbanization.

In both regards, the nature and scope of the initial infrastructures deployed is of the utmost interest in guiding future settlement. Consider, for example, the outcomes and implications of initiating a project through site-clearing for hard, permanent monofunctional infrastructures, such as roads and sewers, versus surgically shaping of the ground to accommodate soft, supple urban infrastructures related to water management, carbon sequestration, and biodiversity. The urban form that emerges from these two approaches to initiation is fundamentally different even if all work

The City That Never Was 63

FIGURE 3.8 *A Modest Proposal* proposes a muscular metropolitan reforestation strategy, which functions as both an interim landscape regime as well as a longer-term urban structure. The project leverages the surplus of cleared and irrigated land present in the incomplete settlements surrounding Madrid in order to implement a grassroots initiative that would reintroduce a tree canopy and understory to the Madrid region that has not been seen for centuries. (Courtesy of the University of Pennsylvania *The City That Never Was* studio.)

FIGURE 3.9 *Reoccupying Airports* tests the capacity of actually proposed "quick fix" programs by combining them into seemingly hyperbolic configurations and deploying them into the site of Aeropuerto de Ciudad Real. The project is obviously not interested in providing a solution. Rather, what it does do is critique the idea that 21st-century urbanization can somehow succeed as a transportable, reproducible, standardized product. (Courtesy of the University of Pennsylvania *The City That Never Was* studio.)

The City That Never Was   65

that follows is exactly the same. As such, the capacity to organize and realize new settlement without relying on the artificiality of standardized compositions as well as more incremental or nimble strategies for build-out should be primary aspirations of any new approach to planning and implementing settlement.

3.3.4 Utility

The final lens brings us back to the stated topic of this volume: infrastructure or, as I have chosen to recast it here, utility. The failure of the Spanish urban growth model and, thus, other examples of incomplete speculative development globally can be characterized as a fundamental failure of infrastructure—if not the systems themselves, then, at the very least, the way these systems are conceived and deployed. Economic growth—and thereby urban growth—is predicated on the deployment of urban infrastructures as the chief catalysts of future expansion. For example, new development in Spain prior to the collapse ostensibly could not be marketed or sold until essential services (water, telecommunication, sewer, gas, electricity, and roads) were completely installed (López and Rodríguez 2011). Yet despite this critical role, the basic physical form and subsequent settlement organized by these utilities has remained ostensibly unchanged for the better part of a century.

Politically charged and socio-impactful, these services—with their cumbersome hardware and rigid networks—are too often considered simply a given of urbanization. Moreover, the enormous physical weight and economic cost of urban infrastructure has cast the potential to modify or rethink these systems as too great of a risk. Rather, alterations to these systems tend to take on one of two forms: aesthetic embellishment or complete perceptual suppression. In fact, the redundant, overscaled, underused infrastructures that criss-cross Spain suggest that continued investment in these familiar metropolitan utilities in their current form provides no guarantee of urbanistic, economic, or social success.

Given the inevitability of speculative urbanization and the ever-changing demands placed upon urban form, the work of the studio reconsidered these essential urban utilities in one of two ways: (1) in a "light" or "disposable" form in which failure or obsolescence is anticipated (see Figure 3.10) or (2) as something capable of being synthesized with other urban systems in order to produce a greater repertoire of affect. In both cases, the familiar formats of settlement that emerge from the morphology and configuration of these familiar infrastructures are discarded in favor of new opportunities for organizing settlement.

3.4 CONCLUSION

Moving forward, it is clear that the increasing demands for urbanization implied by global population projections cannot adequately be dealt with through a continuation of business-as-usual development practices dominated by the superlative-driven megaproject, ubiquitous housing array, or iconic cultural institution. Yet neither can they be negotiated solely from the point of view of the biotic or environmental. Rather, urbanization at this scale demands a new set of design priorities—priorities that acknowledge the political and economic demands of these initiatives but also their near-inevitable likelihood of

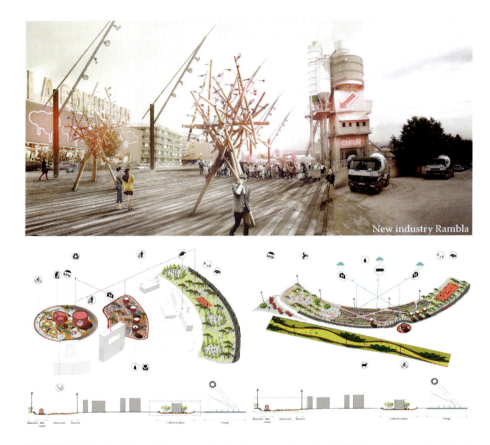

FIGURE 3.10 *Peripheral Opportunism* looks to reorient Madrid's incomplete urban landscapes through the deployment of "light, catalytic infrastructures." These interventions generate new collective spaces and events that serve as provisional civic experiments. When successful, these experiments can guide longer-term, more capital-intensive investment once the Spanish economy begins to recover. (Courtesy of the University of Pennsylvania *The City That Never Was* Studio.)

interruption, inflection, and, in many cases, failure. Certainly, the four themes outlined in Sections 3.3.1 to 3.3.4 are quite broad and, in many ways, reside in a state of ongoing development. Yet perhaps they will never be fully formed because their usefulness is not as a transportable guidebook or ideology but rather as lenses through which new proposals for settlement can emerge and be evaluated.

At a point in time when there is widespread consensus on the ascendency of the city as, to borrow Ed Glaeser's (2011) description, "humankind's greatest invention," it is worth pausing to consider whether the formats of urbanization upon which the last century has relied represent the appropriate way forward for conceiving of and deploying 21st-century urban settlement. Or are designers at a moment in history that demands they stop focusing solely on the desired city of the future and instead turn their attention toward anticipating the city that never was?

The City That Never Was

REFERENCES

Allen, S. and M. McQuade, eds. (2011). *Landform Building: Architecture's New Terrain.* Zurich: Lars Mueller.

Archibold, R. C. (2011). Bursts of economic growth in Panama have yet to banish old ghosts. The *New York Times*, December 13.

Baer, W. C. (2007). Is speculative building underappreciated in urban history? *Urban History* 34(2), August.

Barboza, D. (2011). Building boom in China stirs fears of debt overload. The *New York Times*, July 6.

Belanger, P. (2012). Landscape infrastructure: Urbanism beyond engineering. Conference primer, *Landscape Infrastructure, Harvard Graduate School of Design*, March 23–24.

Berger, A. (2006). *Drosscape: Wasting Land in Urban America.* New York: Princeton Architectural Press.

Bildner, E. (2013). Ordos: A ghost town that isn't. *The Atlantic*, April 8. http://www .theatlantic.com/china/archive/2013/04/ordos-a-ghost-town-that-isnt/274776/ (accessed August 24, 2013).

Branzi, A. (2006). *Weak and Diffuse Modernity: The World of Projects at the Beginning of the 21st Century.* Milan: Skira.

Braungart, M. and W. McDonough. (2002). *Cradle to Cradle: Remaking the Way We Make Things.* New York: North Point Press.

Buckley, C. and S. Rabinovitch. (2010). Special report: China bets future on inland cities. *Thomson Reuters*, August 3. http://www.reuters.com/article/2010/08/03/us-china -urbanisation-idUSTRE6721D320100803 (accessed August 24, 2012).

Campbell, D. (2011). Economic crisis: The pain in Spain. *The Guardian*, December 3. http:// www.guardian.co.uk/world/2011/dec/04/economic-crisis-the-pain-in-spain (accessed January 28, 2012).

CIA World Factbook. (2012). Roadways. https://www.cia.gov/library/publications/the-world -factbook/fields/2085.html#sp (accessed January 31, 2012).

Corner, J. (2006). Terra Fluxus. In *The Landscape Urbanism Reader.* C. Waldheim, ed., 21–33. New York: Princeton Architectural Press.

Cowen, T. (2012). Two prisms for looking at China's problems. The *New York Times*, August 11. http://www.nytimes.com/2012/08/12/business/two-ways-to-see-chinas-problems -economic-view.html?_r=0 (accessed September 14, 2013).

Duany, A. and E. Talen. (2013). *Landscape Urbanism and its Discontents: Dissimulating the Sustainable City.* Gabriola Island: New Society Publishers.

Dyer, G. (2010). China: No one home. *The Financial Times*, February 21. http://www.ft.com /cms/s/0/47cfb09c-1f0f-11df-9584-00144feab49a.html (accessed August 24, 2012).

Easterling, K. (2012). Zone: The spatial softwares of extrastatecraft. *Design Observer*, June 11. http://places.designobserver.com/feature/zone-the-spatial-softwares-of-extrastatecraft /34528/ (accessed July 15, 2012).

Elsheshtawy, Y. (2013). Urban dualities in the Arab world: From a narrative of loss to neo-liberal urbanism. *The Urban Design Reader—Second Edition.* New York: Routledge.

EuropaPress. (2010). Los Reyes inaugurarán el AVE a Valencia, y los Príncipes la conex-ión a Albacete. October 12. http://www.europapress.es/economia/transportes-00343 /noticia-economia-ave-reyes-inauguraran-ave-valencia-principes-conexion -albacete-20101210123611.html (accessed February 2, 2012).

Eurostat. (2011). Transport networks concentrated around economic hubs: Inland transport regional infrastructure. http://www.eds-destatis.de/de/downloads/sif/KS-SF-11-005-EN-N.pdf.

Fuller, T. (2012). In Vietnam, growing fears of an economic meltdown. The *New York Times*, August 22. http://www.nytimes.com/2012/08/23/business/global/23iht-vietnam23.html? pagewanted=all (accessed December 12, 2012).

Garber, P. (1990). Famous first bubbles. *The Journal of Economic Perspectives* 4(2).

Garcia, M. (2010). The breakdown of the Spanish urban growth model: Social and territorial effects of the global crisis. *International Journal of Urban and Regional Research*, December 34.4.

Glaeser, E. (2010). Housing hangover in the Sun Belt. The *New York Times*, January 12.

Glaeser, E. (2011). *Triumph of the City*. New York: The Penguin Press.

Goldman, M. (2011). Speculative urbanization and the making of the next world city. *International Journal of Urban and Regional Research* 35.5, May.

Government of Spain, Head of State. (1998). BOE—Official State Gazette, Number 89. April 14. http://www.boe.es/boe/dias/1998/4/14/pdfs/A12296-12304.pdf.

Government of Spain, Ministry of Development. (2011). Statistical yearbook. http://www.fomento.gob.es/MFOM/LANG_CASTELLANO/ESTADISTICAS_Y_PUBLICACIONES/INFORMACION_ESTADISTICA/EstadisticaSintesis/Anuario/default.htm (accessed August 18, 2012).

Government of Spain, Ministry of Labor and Social Affairs. (2002). BOE—Official State Gazette, Number 284. November 27. http://www.boe.es/boe/dias/2002/11/27/pdfs/A41643-41650.pdf.

Grupo AENA. (2011). Annual Report. http://www.aena.es/csee/ccurl/695/965/2011-annual-report.pdf.

Hardaway, R. M. (2011). *The Great American Housing Bubble*. Santa Barbara, CA: Praeger.

Harvey, D. (1996). The dialectics of social and environmental change. *Justice, Nature and the Geography of Difference*. Cambridge: Blackwell.

Hoffman, A. and G. Wan. (2013). Determinants of urbanization. *Asian Development Bank Economics Working Paper*, Series No. 355, July.

Keeton, R. (2011). *Rising in the East—Contemporary New Towns in Asia*. Amsterdam: Sun Architecture publishers.

Kitchin, R. et al. (2010). A haunted landscape: Housing and ghost estates in post-Celtic Tiger Ireland. *National Institute for Regional and Spatial Analysis Working Paper*, Series No. 59, July.

Krugman, P. (2013). This age of bubbles. The *New York Times*, posted August 22. http://www.nytimes.com/2013/08/23/opinion/krugman-this-age-of-bubbles.html?_r=0 (accessed August 23, 2013).

Lewis, M. (2011). When Irish eyes are crying. *Vanity Fair*, March. http://www.vanityfair.com/business/features/2011/03/michael-lewis-ireland-201103 (accessed July 26, 2012).

López, I. and E. Rodríguez. (2011). The Spanish model. *New Left Review 69*—May–June.

McGovern, S. (2008). From boom town to ghost town. *BBC News*, posted August 27. http://news.bbc.co.uk/2/hi/7584097.stm (accessed January 28, 2012).

McKinsey Global Institute. (2010). *Farewell to Cheap Capital: Implications of Long-Term Shifts in Global Investment and Saving*. New York: McKinsey Global Institute.

Medina, S. (2012). A ghost city in Angola, built by the Chinese. *The Atlantic—Cities*, July 17. http://www.theatlanticcities.com/design/2012/07/ghost-city-angola-built-chinese/2608/ (accessed August 10, 2012).

MetroMadrid. (2012). Rail Network. http://www.metromadrid.es/en/conocenos/infraestructuras/red/index.html (accessed February 6, 2012).

Ming, X. (2010). Beijing district releases official housing vacancy rates, *Market Watch: The Wall Street Journal*, August 26. http://articles.marketwatch.com/2010-08-26/markets/30799388_1_vacancy-rate-chaoyang-district-andy-xie (accessed August 24, 2012).

Moix, L. (2010). *Arquitectura Milagrosa*. Barcelona: Anagrama.

Moran, L. (2012). Spain's ghost airport. *The Daily Mail*, July 9. http://www.dailymail.co.uk/news/article-2170886/Spains-ghost-airport-The-1BILLION-transport-hub-closed-just-years-thats-falling-rack-ruin.html (accessed July 23, 2012).

Netz, J. (1916). The great Los Angeles real estate boom of 1887. *Annual Publication of the Historical Society of Southern California* 10(1/2).

Oosterman, A. ed. (2012). City in a Box. *Volume 34*, December. Amsterdam: Archis.

Ouroussoff, N. (2006). Exhibition review: On Site—A Survey of Spain, Architects' Playground. The *New York Times*, February 10.

Palmer, A. (2011). Bricks and slaughter. *The Economist*, May 3. http://www.economist.com/node/18250385 (accessed August 16, 2012).

Penn Institute for Urban Research. (2008). *The Penn Resolution: Educating Urban Designers for Post-Carbon Cities*. Philadelphia: University of Pennsylvania School of Design.

Pettis, M. (2013). The urbanization fallacy. *China Financial Markets Blog*, August 16. http://blog.mpettis.com/2013/08/the-urbanization-fallacy/ (accessed August 17, 2013).

Pollard, J. (2009). Political framing in national housing systems: Lessons from real estate developers in France and Spain. *The Politics of Housing Booms and Busts*. New York: Palgrave Macmillian.

Reynolds Nelson, S. (2011). A storm of cheap goods: New American commodities and the panic of 1873. *The Journal of the Gilded Age and Progressive Era* 10(4), October.

Riley, T. (2006). Contemporary Architecture in Spain: Shaking Off the Dust. *On-Site (exhibition catalogue)*. New York: Museum of Modern Art.

Roach, S. S. (2012). China is okay, *Project Syndicate*, August 29. http://www.project-syndicate.org/commentary/china-is-okay-by-stephen-s--roach (accessed September 3, 2012).

Romer, P. (2011). The World's First Charter City. *TED Talk*, June. http://www.ted.com/talks/paul_romer_the_world_s_first_charter_city.html (accessed August 3, 2012).

Roubini, N. (2011). Beijing's empty bullet trains: Is China investing way too much in its infrastructure? *Slate*, posted April 14. http://slate.com/id/2291271/ (accessed June 6, 2011).

Roubini, N. (2013). Back to housing bubbles. *Project Syndicate*, posted November 29. http://www.project-syndicate.org/commentary/nouriel-roubini-warns-that-policymakers-are-powerless-to-rein-in-frothy-housing-markets-around-the-world (accessed December 12, 2013).

Sakolski, A. M. (1932). *The Great American Land Bubble: The Amazing Story of Land-Grabbing, Speculations and Booms from Colonial Days to Present Time*. New York: Harper & Brothers.

Shane, G. (2006). The emergence of landscape urbanism. *The Landscape Urbanism Reader*. New York: Princeton Architectural Press.

Sheehan, T. (2012). Spain's high-speed rail system offers lessons for California. *Fresno Bee,* January 15. http://www.fresnobee.com/2012/01/15/2683896/spains-high-speed-rail-system.html (accessed February 3, 2012).

Shiller, R. J. (2000). *Irrational Exuberance*. Princeton: Princeton University Press.

Spain National Institute of Statistics via Eurostat. 2012. Construction new orders indices by country, period and type of building. http://www.ine.es/jaxi/tabla.do?path=/t07/a081/e01/l1/&file=03002.px&type=pcaxis&L=1 (accessed August 24, 2012).

Thomas Jr., L. (2012). Fears rising, Spaniards pull out their cash and get out of Spain. The *New York Times*, September 3.

United Nations Department of Economic and Social Affairs World Population Prospectus. (2013). The 2012 Revision: Highlights and Advance Tables. http://esa.un.org/wpp/Documentation/pdf/WPP2012_HIGHLIGHTS.pdf (accessed July 28, 2013).

Waldheim, C. (2006). Landscape as urbanism. In *The Landscape Urbanism Reader*, ed. C. Waldheim, 35–53. New York: Princeton Architectural Press.

Waldheim, C. (2010). Notes towards a history of agrarian urbanism. *Design Observer*, November 4. http://places.designobserver.com/feature/notes-toward-a-history-of-agrarian-urbanism/15518/.

Watson, V. (2009). Seeing from the South: Refocusing urban planning on the globe's central urban issues. *Urban Studies* 46(11), October.

Weller, R. (2009). *Boomtown 2050: Scenarios for a Rapidly Growing City.* Perth: University of Western Australia Publishing.

White, E. N. (2009). Lessons from the Great American Real Estate Boom and Bust of the 1920s—NBER Working Paper No. 15573. *National Bureau of Economic Research,* December.

4 Landscape as *Energy Infrastructure*

Ecologic Approaches and Aesthetic Implications of Design

Daniela Perrotti

CONTENTS

4.1 Introduction 71
4.2 Why Landscape as an Energy Infrastructure? 72
    4.2.1 Taking Energy Flows into Account within the Field of Landscape Infrastructure 72
    4.2.2 Energy and Potential Energies through the Lens of Ecology 73
4.3 How Can Design Capitalize on Energy in the Landscape Infrastructure? 76
    4.3.1 Landscape Infrastructure Design as a Project of Energy Exchanges and Connections 76
    4.3.2 Gardening with and Not against the Biological Engine of Living Systems 78
4.4 What Aesthetic Qualities of Landscape Are Understood as Energy Infrastructure? 80
    4.4.1 Experiencing Energy Exchanges and Connections within Landscape Infrastructure 80
    4.4.2 Closing the Energy-Nature Loop 81
4.5 Conclusion 86
References 89

4.1 INTRODUCTION

With the goal of bridging the gap between ecology and economy vis-à-vis the contemporary crisis, over the last decades a few emergent viewpoints in the debate on urban infrastructure have tried to contextualize infrastructural objects within a wider epistemological framework and a more comprehensive approach to urban and landscape practice (Perrotti 2011). They emphasize the perspective of re-bundling and redesigning essential urban services (e.g., water resources, waste cycling, energy generation, food cultivation, mass mobility, network communication) as *living landscapes*. These viewpoints focus on synergies and geographical, economic, and ecological interconnections between green, gray, and blue networks within metropolitan regions. Indeed, these synergies seem to better support fluid, dynamic patterns of urban growth (i.e., the flow of waters, waste, energy, and food, which mostly transcend geopolitical borders) instead of reproducing or consolidating the vertical, centralized, and inflexible structure of modern *"industrial"* cities.

In the 1990s, in his renowned article "Infrastructure as landscape, landscape as infrastructure," landscape architect Gary Strang (1992) described urban infrastructure as a landscape of systems, services, scales, resources, flows, processes, and dynamics, which support and cultivate urban economies. Since then, other practitioners in the field of urban and landscape design have tried to lay the foundation for establishing a complementary vision that takes landscape itself into account as a kind of "conveyance or distribution network," the function of which is to move people and to support heterogeneous living systems (Aquino 2011, 7).

Other designers' visions, such as that of Pierre Bélanger, have focused on the "horizontal nature" of the field of landscape infrastructure and the synthetic capacity to combine infrastructure and ecological processes. This field is considered both as system and scale. As an open system of endogenous and exogenous flows, it moves from the biomolecular to the global geographic by way of urban and ecological regions; its scale is operationalized through "ecological intelligence" (Bélanger 2012, 280).

Besides, in his 1999 article "Infrastructural urbanism," Stan Allen defines infrastructures as "artificial ecologies": As they organize the flow of energy and resources on the site, they "create the conditions necessary to respond to incremental adjustments in resource availability and modify the status of inhabitation in response to changing environmental conditions" (Allen 1999, 56).

By using the adjective "artificial," Allen shows that the notion of "ecology" may be used as a metaphor to explain the way of functioning of those infrastructures that constitute the *hard* systems of urban support (e.g., roads, sewers, bridges, waste treatment, and water delivery systems). Nevertheless, the landscape infrastructure approach moves beyond the mere metaphorical level because it takes also account of "natural-based" ecologies, i.e., the *softer*, leaner infrastructures that compose the living networks in the biosphere. These infrastructures constitute a support system for the process of "modernization," which incorporates nonformal logics associated with an operative view of ecology, involving softer morphologies, such as flow patterns, organizations, and synergies (Bélanger 2012).

4.2 WHY LANDSCAPE AS AN ENERGY INFRASTRUCTURE?

4.2.1 TAKING ENERGY FLOWS INTO ACCOUNT WITHIN THE FIELD OF LANDSCAPE INFRASTRUCTURE

The acknowledgement of the hard and soft systems of the grey and green/blue infrastructural networks in the urban landscape in light of their energy potentials requires an integrated vision of both "artificial" and "natural-based" ecologies. The constructed landscape of channels, pipes, grids, and networks that support the supply and

Landscape as *Energy* Infrastructure

distribution of energy across vast metropolitan regions (the "hardware") represents only one (dark) side of the energy supply system of urban societies. Indeed, the landscape infrastructure field is permeated by a soft capillary tissue composed of other kinds of energy and material flows. When we examine them through the lens of ecology, these flows represent a fundamental and structural component of the urban ecosystem.

The "ecologic approach" sheds light on the *energy potentials** inherent to landscape infrastructures; moreover, it might open up new perspectives when we take into account the potential for landscape to work as an energy infrastructure. If landscape itself, whether urban, agricultural, or "natural," is seen as an integrated infrastructural system (Perrotti 2012), its "*infra*-structural nature" refers to its ability to act as a supporting structure (in the etymological sense of the word infrastructure), which defines a "set of systems, works, and networks underpinning modern societies and economies"[†] (Bélanger 2009, 92). This set of systems not only functions as support for the development of human activities, but also for the life of all the other species that coexist within the same environment. Hence, landscape infrastructure is the *medium* through which the flows of energy may function as *drivers* for the distribution of the vital resources for human, animal, and vegetal species within the environment.

4.2.2 ENERGY AND POTENTIAL ENERGIES THROUGH THE LENS OF ECOLOGY

In his renowned text of 1971, *Environment, Power and Society*, Howard T. Odum provides an exhaustive description of the complex web of flows of energy and materials in environmental systems and of the way in which living systems work in their constant interactions with solar radiation through the cycles of primary production and respiration. In sunlight, the living photochemical surfaces of plants become charged with food storage from the onrush of solar photons, and oxygen is released. "During the day, while oxygen is generated, a great sheet of new chemical potential energy in the form of new organic matter lies newborn about the earth" (Odum 1971, 13). This *potential energy* is not dispersed in the atmosphere but rather reused and capitalized in numerous useful tasks. Odum (1971, 13) illustrates the activity of living systems as a great "breath" taken during the daytime, a great "exhalation" released when the sun passes into shadows before the night (Figure 4.1), and finally a "heat engine" that works in the darkness. Operating between high- and low-temperature regions, the "biospheric" machine creates the earth's great wind and water current system.

---

* For a complementary vision, see the position of Nadaï (2012) on the "technological potential" of wind energy (in the sense of "success" in installing wind power capacity) as a territorial feature, i.e., embedded in the territory. This potential exists neither before nor outside the planning processes but rather emerges from these latter (or from "social processes" in general). Otherwise, it could be assimilated to an "authoritative notion" of technological potential, which would be associated with a "generic" technology (see *, p. 81).

† As Bélanger underlines, the term "infrastructure" comes from the Latin "infra" ("below, underneath, beneath") and "structure" ("action or process of building or construction") (Merriam Webster). See also the definition of "infrastructure" in the Italian *Lexicon of Architecture and Urban Study*, edited by Paolo Portoghesi (1969, 189, translation by the author): "A term used in politics to indicate the projects which humans carry out to support their economic and political structures. ... More strictly speaking, and particularly in urban planning, the term is used to indicate the works that are necessary for the relational life of said political structures, i.e., the entirety of the lines and nodes which constitute the reticular system of connections and exchanges, the various types of systems for water and energy supply or waste disposal ..."

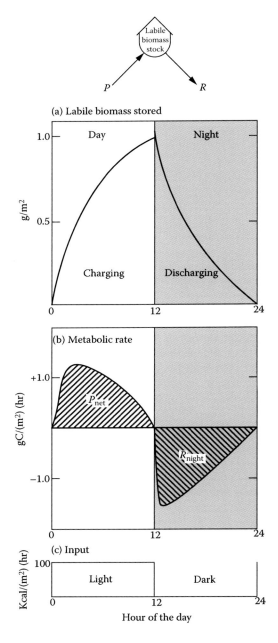

FIGURE 4.1 Diurnal pulse of power flows in a microcosm with P (gross photosynthesis) and R (respiratory consumption) processes graphed. (a) Storage of organic matter during periods of charge in the light and discharged in the dark. (b) Rates of photosynthesis and respiration. (c) Inflowing light energy. (Odum, H. T.: *Environment, Power, and Society.* 14. 1971. Copyright Wiley-VCH Verlag GmbH & Co. KGaA. Reproduced with permission.)

Landscape as *Energy* Infrastructure

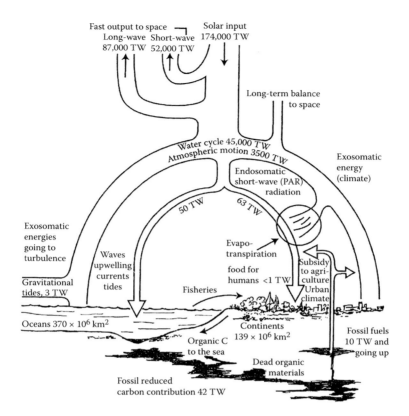

FIGURE 4.2 Distinction between endosomatic and exosomatic energies in the global budget of energy in the biosphere. Energy flows (average) are expressed in terawatts, 1 TW = 10^{12} W. (From Margalef, R., *Our Biosphere*, p. 56. International Ecology Institute, Oldendorf/Luhe, 1997. With permission.)

Hence, the production and respiration processes are closely linked energetically with the earth's physical processes (heat, wind, and water cycle). In fact, these processes provide not only the biochemical energy that is fundamental for the biological metabolism of species in the ecosystems ("endosomatic"), but also an external energy ("exosomatic"*), which helps maintain life and the organization of ecosystems (Margalef 1997); through the combustion process (for fossil sources) or other forms of recovery (for wind, solar, and hydropower), it responds to the extra-(solar) energetic needs of human activities[†] (Figure 4.2).

---

* The terms "endosomatic" and "exosomatic" were introduced by the biophysicist Alfred Lotka (1925) and reemployed by the mathematical economist Nicholas Georgescu-Roegen (1971). In this context, they were used specifically with regard to the organic and extraorganic instruments of the human species.
† According to the architect and urban designer Fernández-Galiano (2000, 20), architecture also can be considered as an "exosomatic artifact of man, existing outside of the body"; thus, the energy used in buildings and for the maintenance of the environment should be considered a form of external energy.

Indeed, what ecology and, more generally, natural sciences show us when we approach the field of landscape infrastructure is the fundamental fact that energy may no longer be considered as an external component of landscapes. Landscape infrastructure itself may be seen as a complex system of networks and exchanges that provides energy in the form of food, fuel, and heat for the biological metabolism of organisms. As an *infra*-structure,* it also sustains and supports human activities and those of other species in the biosphere by deploying the energy necessary for the transformation processes of natural resources and the construction of viable living environments.

4.3 HOW CAN DESIGN CAPITALIZE ON ENERGY IN THE LANDSCAPE INFRASTRUCTURE?

4.3.1 LANDSCAPE INFRASTRUCTURE DESIGN AS A PROJECT OF ENERGY EXCHANGES AND CONNECTIONS

The acknowledgment of landscape as energy infrastructure paves the way for renewing the role of landscape architecture practice in establishing new agencies, scales, and scopes for design that would be more "sustainable" (in its literal sense) in terms of energy use. Beyond the massive use of energy-efficient technologies, new strategies based on landscape architecture–specific methods, tools, and techniques may emerge with the goal of capitalizing those energy potentials that, as described by ecology, are embedded in the heterogeneous components of environmental systems.

The question that arises concerns the way in which landscape architecture can have recourse to and capitalize on the material and potential energies embedded in living systems as a fundamental resource for its practice.

An increasing number of sources in the contemporary literature on sustainable landscape design stress the need for practitioners to rethink the output of design (from building green facades to trans-boundary green infrastructures)—not simply as an *object* but rather as an *organism* permeated by flows and equipped with storages of energy and materials. The basic condition for establishing a truly sustainable approach to landscape architecture, urbanism, and planning would then be to design "within the limits of natural resources and natural laws" (Williams 2007). This purpose would entail the recourse and mobilization of natural resources as a commitment beyond the energy-efficiency standards outlined and required by public policies.

An energy-oriented reinterpretation of the approach *Design with Nature* (McHarg 1969) would result in taking account of and capitalizing on the considerably amount of energy in nature, hence designing by means of the available *site energies*. It is precisely in this sense that, in the introduction of his text *Sustainable Design, Ecology, Architecture and Planning*, the architect and urban planner Daniel Williams (2007), a former Howard T. Odum scholar, emphasizes the way of approaching landscape

---

* See the etymological meaning in the second note of p. 73.

Landscape as *Energy* Infrastructure

design practice through the adoption of the ecological lens.* For Williams, integrating design into the "ecology of the place" would result mainly in facilitating connections between the "sustainable energies and the materials" of the site and region in order to capture, concentrate, and store energies and materials. Moreover, this "ecological design of connections"[†] (i.e., designing "holistically and connectedly"; Williams 2007, 2) would also play a key role in increasing the human capacity to understand how ecological systems provide sustainable flows and storages of materials and energies from the site and how they auto-organize themselves in the function of their internal energy distribution. In fact, forms in nature represent a purely direct response to the way in which organisms capture the flows of energy and materials existing in their bioregions. Thus, as a result of biological processes, a form derives from the need for organisms to maximize the use and storage of energy and materials to implement their functions, according to their position in the ecological networks and food web.

Following Williams's viewpoint, we may assert that the fundamental lesson learned from ecology is that energy is a common denominator to all flows, processes, and materials within the biosphere. As Howard T. Odum stresses (1971, 21), "since forces [in ecosystems] are generated from energetic storages, their lines of action may also be represented by the same lines that indicate energy delivery." This ecological principle may be a source of inspiration for landscape designers to construct the spatial pattern of *lines of forces* of the landscape by means of consideration of these lines of energy delivery (i.e., the flows of energy and matter which are deployed within the ecosystem boundary or connect different subsystems). Through this method, the pathway of the successive levels of energy use and storage could be viewed as the basic armature of landscape infrastructure. This ecological pattern could be used as a support for spatial analysis of the project site and also as a fundamental tool to design the incremental development of the landscape infrastructure over time.

A comparable approach has been adopted by one of the Australian permaculture pioneers, Bill Mollison, as illustrated in the schema from his book *Permaculture: A Designer's Manual* in Figure 4.3. This drawing represents a schematic version of the energy pathways in the valley system. Mollison (1988, 13) stresses the need for designers to set up an "intersection net" from the source to the sink. This net is a compound web of life and technologies, which is designed to catch and store as much as energy as possible on its way to increase entropy. Hence, the designer's work consists in setting up useful energy storages in a landscape, proceeding from state A to state B as illustrated in the drawing. Ultimately, such storages, which are available for increasing yields, create what is normally called a "resource."

---

* See, for example, William's *Bioregionalism* model for site analysis at the regional scale with its three-dimensional characteristics based on air, water, and geological shades and their interactions with land use patterns in urban and agricultural systems (Williams 2007, 10).

† Because ecology is understood in terms of the study of spatial connectivity between organisms and their environment (Williams 2007).

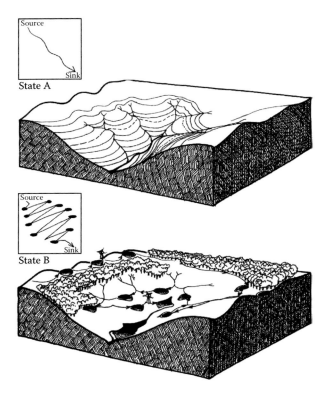

FIGURE 4.3 Designing to catch and store energy. Schematic version of energy pathway through the valley system. (From Mollison, B., *Permaculture: A Designer's Manual*, p. 14. Tyalgum, AU, Tagari Publications, 1988. With permission.)

4.3.2 Gardening with and Not against the Biological Engine of Living Systems

Having acknowledged "ecological design" as the design of the connections between the sustainable forces and energies available on the site, how can this design take into account the structural relationships between the flows of local renewable energies and the spatial organization of life within environmental systems?

The work of the landscape architect Gilles Clément represents a significant approach to landscape design through the capitalization and use of the flows of local renewable energies. In his practice, they represent the main determinants of the forms and functions of nature in the project site. These energies are intended to be mobilized not through the use of technological devices for energy capture and exploitation, but rather through landscape design and gardening traditional *savoir-faire* (e.g., species' associations) or new proactive strategies.

In his text of his 1999 *Le Jardin en Mouvement*, Clément emphases that, when it is not constrained by unsustainable and invasive landscape practices, nature is able to manage its own energy "system" and organization by itself through the deployment

Landscape as *Energy* Infrastructure

of its own metabolic and external energies. Thus, in the spaces that he calls "gardens in movement" (Figure 4.4), landscape practice results in a *laisser-faire* approach to natural energies "in presence" (growth, development, displacement, exchange), which ultimately enhances the very close and primordial link existing between energy and nature (Clément 1999).

FIGURE 4.4 Gilles Clément, *Jardin en Mouvement*, Parc André Citroën, Paris, February 23, 2014. (a) Description of the main species growing and their distribution in the garden; views of the two pieces of "domesticated wilderness" on both sides of the entrance path from the Park André Citroën. (b) Two views of the laissez-faire approach to vegetation adopted by Gilles Clément in the area on the left of the entrance path from the street and in the bosque. (Courtesy of Daniela Perrotti, 2014.)

80 Revising Green Infrastructure

Furthermore, Clément's work is an exhortation to minimize as much as possible the use of what he defines as the "contrary energy." As mentioned in his renowned text of 2004, *Le Manifeste du Tiers Paysage* (*The Third Landscape Manifesto*), this energy works *in opposition* to the energy cycles in the biosphere. Deployed under the form of fossil energy (but also human metabolic energy) to feed and operate machinery and technical equipment for site construction and maintenance, the contrary energy is used by those landscape practices that aim at constraining nature to a set of predetermined forms and functions, conceived in order to respond to standardized principles, such as geometry or cleanness. These unsustainable, "planetary" practices impact severely the biological features of the ecosystem substrates (soil, biomass, water, air); thus, they reduce the performances of the "biological engine" of the landscape, according to the ratio of "contrary energy" deployed. "The contrary energy is opposed to the energy which every living organism has at its disposal for its growth and development" (Clément 2004, 11, translation by the author).

4.4 WHAT AESTHETIC QUALITIES OF LANDSCAPE ARE UNDERSTOOD AS ENERGY INFRASTRUCTURE?

4.4.1 Experiencing Energy Exchanges and Connections within Landscape Infrastructure

As argued previously, the acknowledgement of the energy exchanges in living systems—which contribute additionally to the conception of landscape infrastructure as "ecology"—is a fundamental component of design practice. However, new research questions arise if we identify a new challenge for designers whose goal is to achieve an effectively "sustainable" process of landscape design: making inhabitants and landscape "users" aware of the material energies embedded in their living environment.

Design would thus represent a way of increasing human awareness of the fundamental energy exchanges that ensure the development, organization, and maintenance of life in heterogeneous forms within the landscape infrastructure. In this sense, landscape designers would be able to enact through their practice a form of aesthetic *mediation* between *on-site* energies and landscape inhabitants or, more generally, between energy and society. This mediation would initiate a significant change in the relationships between inhabitants and their living environment from a basic *physiological* involvement to an active "aesthetic engagement," following the definition adopted by contemporary philosophers in the field of environmental aesthetics (Berleant 1992, 1997; Blanc 2008; Brady 2003).

Our way of addressing the aesthetic qualities of energy landscapes refers clearly and directly to the etymological meaning of the term "aesthetic," derived from the Greek verb *aisthanomai* ("to perceive, feel, sense") as emphasized by the German philosopher Alexander Baumgarten in his work *Aesthetica* (1750). In fact, it is primarily through their senses that human beings are able to rearticulate their relationships to energy and perceive it as being primarily a natural phenomenon. Through a renewed aesthetic approach to energy, human societies may find a way to reestablish a balanced relationship with it and reformulate the choice of new, effective "natural proportions" to overcome the disproportionate relationship between contemporary societies and an

Landscape as *Energy* Infrastructure

"abstract," "generic" apprehension of technology.* Moreover, this relationship results in the current immoderate consumption of fossil energy and natural resources.

Therefore, renewing the link between human beings and the energies deployed in living systems represents a basic challenge for *ecological design*, especially if we intend to carry out this renewal through a deep re-questioning of the interplay between societies and technological innovations. This challenge raises the issue of energy as a socio-technical object and calls upon the adoption of an aesthetic approach to renewable energies in nature by landscape design practice.

4.4.2 Closing the Energy-Nature Loop

The abundant literature on the so-called issue of "social acceptance" of renewable energy projects has highlighted a large set of different, often antithetical, viewpoints (Aitken 2010; Brittan 2001; Cowell 2010; Pasqualetti et al. 2002), several questions arise when we focus on inhabitant perceptions and aesthetic representations of landscape as an integrated energy infrastructure. An original issue that we deem worthy to be addressed concerns the possibility to *experience* energies deployed or stored in living systems and the way in which this experience may be undertaken by *inhabiting* said energy landscapes. In other words, how can landscape design practice take into account the aesthetic implications of what we can define an "energetic dimension" of landscape infrastructure? How can landscape design contribute to the construction of new aesthetics of *landscape as energy infrastructures* beyond a merely consumerist approach to aesthetics?[†] How can a design approach be grounded on the aesthetic relationships that compose the landscape *infra*-structure?

Environmental art is an interesting source of inspiration for landscape architects in this field. Since the experimental work of Robert Smithson in the 1970s, it has challenged the issue of illustrating the aesthetic potential of living systems as a supply and storage system for different forms of energy. More recent works have addressed the aesthetic values inherent to the processes of conversion of energy and biological matter from one form or state into another, e.g., photosynthesis and evapotranspiration[‡]

---

* Nadaï (2012) refers to technologies as "assemblages made up of heterogeneous elements" (human and nonhumans) and, quoting the works of Law (1992), as "heterogeneous networks" that exist only by being embedded in the social realm. Thus, this definition of technologies aims at distinguishing a "situated technology," which the author defines as "embedded in heterogeneous networks at the local level" (e.g., wind farms, solar farms) from an abstract "generic technology" (e.g., wind power or solar energy). This latter is associated with an authoritative notion of technological potential (see *, p. 73). "It is because technological potential is conceived and approached as being 'technological' in a non-social sense that social interactions end up being posited as a barrier to its realization, saddling planning with the impossible mission of matching social processes with non-social targets" (Nadaï 2012, 110).

† In their reflection on the aesthetic value of "landscape machines," Paul A. Roncken, Sven Stremke, and Maurice P.C.P. Paulissen (2011, 70) warn against the risk of misunderstanding aesthetics as "a set of special effects that are supposed to entertain people and perhaps cure their bad habits."

‡ On photosynthesis, see, for example, the experimental work *Trees of Photosynthesis* (since 2006) of the Japanese artist Shigeko Hirakawa (Figure 4.5), the exposition of the work of the Danish artist Tue Greenfort *Photosynthesis* at the gallery Witte de With of Rotterdam (2006), or the "living portraits and landscapes" of the British duo Ackroyd & Harvey (since 2000). On evapotranspiration, see the automaton *La cuisse du Général Grenouille* by French artist Jean-Luc Brisson for his exposition at the Modern Art Museum of Villeneuve d'Ascq in 1985 (Brisson 1999).

(Figure 4.5). Through this experimentation, land and environmental artists have endeavored to increase social awareness of the significant presence of the flows of energy and matter processes in the biosphere.

Returning to our main scientific focus, how may landscape architects contribute to the transformation of the close, primordial link between energy and nature into a sensory experience? Through their practice, how can they experiment with new aesthetic languages in order to translate the living presence of energy in nature?

Looking at the edge between landscape and architecture, a significant example is the practice of the "meteorological architecture" designer Philippe Rahm, which represents a new way of approaching energy in the project field. This practice is led by a significant shift from the *visible* toward the *invisible*, and it is founded on the design of what is commonly defined "space and energy management," i.e., lighting, ventilation, and heating (Rahm 2006, 152). Rahm's approach entails the need for architecture to turn its practice toward the microscopic and the atmospheric and to "stretch" between the biological or physiological and the meteorological, in other words, between the infinitely large and the infinitely small. This design process results in the achievement of a new spatial organization, in which the project forms and functions "follow climate" and emerge *spontaneously* in response to it (Rahm 2006, 152, 155). If the density of the air and the intensity of the light (translated into standardized parameters,

FIGURE 4.5 Shigeko Hirakawa, *Tree of Photosynthesis*, Exhibition "Appel d'air," Jardins des Plantes of Rouen, 2010. A 22-meter-high oak tree on which the artist attached plastic discs containing photochromic pigments. They change color according to the intensity of solar radiations and their position on the tree; they reproduce photosynthesis process "artificially." (© Shigeko Hirakawa, 2013, courtesy of Daniela Perrotti.)

e.g., air temperature, relative humidity, sunray exposure) are seen as "the very matter of the space," the design practice may result in the production of a *new geography*, "an open-ended, shifting whether system embracing different climates and atmospheric qualities to be occupied and used according to our need and desires, the time of day, and the season" (Rahm 2006, 155). Hence, Rahm's practice represents an invitation for designers (not only architects, but, in a more holistic perspective, also urban planners and landscape designers) to establish their own practices on the *sensational* and the *phenomenal*; this would result in renewing their idea and way of generating the forms and uses of their project space. Through their interaction with body respiration, perspiration, and metabolism, phenomena of conduction and convection (e.g., the raising of the air or surface temperature, the evaporation of water, the renewal of air masses) become new paradigms for design; they inaugurate a shift from metric to thermal composition, from structural to climatic and meteorological thinking. "Space becomes electromagnetic, chemical, sensorial and atmospheric with thermal, olfactory and cutaneous dimensions with which we are immersed. The very act of inhabiting these spaces with the breath, perspiration and thermal radiation of our bodies in turn combines with this materiality; the physical environment of our surroundings" (Rahm 2009b, 32). Hence, the aesthetic, sensorial experience of Philippe Rahm's meteorological architectures becomes, at the same time, the main principle that drives the design process and the aim toward which it is oriented. The wide range of "sensual" exchanges between the body and the space (thermal exchanges and air flows, through the skin, the breath, and other sensorial interfaces) are the basic components of this renewed way of designing living space, not only on the architectural level, but also on the one of cities and landscapes. The act of inhabiting these spaces results in the aesthetic experience of the material energy embedded in the project site (climatic and physiological conditions). Hence, the design language results in an "aesthetic language" as it addresses directly the sensorial experience of the space. Ultimately, a new "open" landscape* (Rahm 2009b, 32, Figure 4.6) emerges from the wide range of physiological and climatic relationships established between the inhabitant and the space.

Another relevant example is the work of the North American WEATHERS office; it is based on the acknowledgement of material energies that make up the field of architectural, urban, and landscape design (electromagnetic, thermodynamic, acoustic, and chemical) as an essential building material. When the design process engages with climatic conditions as anything more than "a condition to be

---

* See, for example, the 2006 project "House Dilation" for an artist residence in the British countryside near the town of Grizedale (Figure 4.6). Here the act of living and moving through the different exterior and interior spaces that compose the house gives the inhabitants the possibility of experiencing distinct and opposite climate conditions. This results from the fact that the project "dilates" the functions in different places; climates; and calibers of light, temperature, and humidity, according to the time of day and the season. These specific climate characteristics are related to the shadow and humidity created by the surrounding vegetation (their process of photosynthesis and evapotranspiration), and each of them is associated to a particular activity and use. Indeed, the environment becomes the "last skin" of architecture. It filters the light, contains or repels moisture, and heats or cools each space differently according to its position in the project site (Rahm 2006, 141; Rahm 2009a, 24–25). Specifically on the relationship between water vapor, humidity, and space and the consequent relationships established between the occupant's thermal comfort and the presence of this flow of material energy in the space, see the 2005 project for the Mollier holiday homes on Lake Vassivière, France (Rahm 2009a, 112–114; Rahm 2009b, 36–37).

84 Revising Green Infrastructure

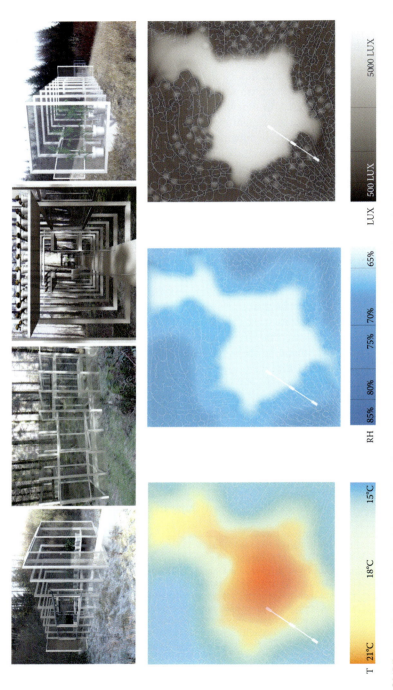

FIGURE 4.6 Philippe Rahm Architectes, House Dilation, Grizedale Arts, 2006. The project dilates the domestic functions through various qualities of temperature, relative humidity, and light, which correspond to a different place and climate for each part of the house. Three sites are selected for the house dilation: the meadow, the forest, and at the boundary between the two. Domestic activities are distributed with regards to the particular climate they require: the heat of the night forest, the warmth of the field in the winter during the day, the freshness of the forest edge in the spring. (© Philippe Rahm Architectes, 2013, courtesy of Daniela Perrotti.)

Landscape as *Energy* Infrastructure

tempered or mediated" (Lally 2009, 56), energy provides not only new forms and spatial strategies for designers, but also new aesthetic qualities associated with the multiple experiences that can be driven in the designed space. In this perspective, the design of material energies becomes a design of "thresholds"; if seen as thermal gradients, they engage a broader vocabulary within the human sensory system, which extends much beyond the boundary of the visible. This design is intended to counterbalance the binary contrast between conditioned and unconditioned spaces, which is very *unsustainable* in terms of its impact on global energy consumption. In contemporary cities, this binary contrast is often coincident with the "private versus public" dichotomy constructed by architectural surfaces and geometries.

The material energies that shape the landscape of the project site are seen as an "actionable medium," which designers should activate to design the climatic environment in coherence with the whole design process (Lally 2009, 63, Figure 4.7). For

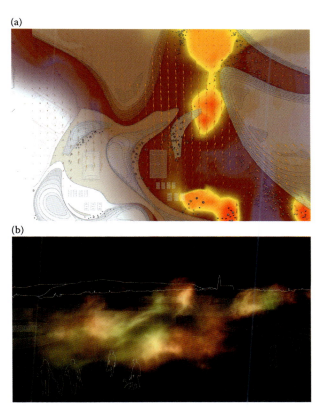

FIGURE 4.7 Sean Lally/WEATHERS, Vatnsmyri Urban Planning Competition Reykjavik, Iceland, 2007. (a) Climatic "wash" producing an "artificial" microclimate, controlling winds and permitting the extension of the seasonal activities and the period of usable time outdoors. (b) Urban plan defined by gradient thresholds, creating variable microclimates by worming the air, sculpted earth mounds, and vegetative soil by geothermal resources. (© Sean Lally/WEATHERS, 2013, courtesy of Daniela Perrotti.)

example, Sean Lally (2009, 2013), founder and principal of WEATHERS, raises the question of the way in which a design of material energies should approach the issue of urban heat islands—whether to continue facing them as a byproduct of anthropic activities or to operate upon these microclimatic conditions, bringing them into the design decisions. With the heat island example, Lally emphasizes the issue of transforming climatic variables into design materials. More precisely, he questions the way in which designers should "challenge" these material energies as potential generators of new spatial typologies and behaviors instead of systematically "capping" and "preventing them from operating to their fullest potentials or embracing their inherent proclivities as possible thresholds, circulating strategies or physical boundaries" (Lally 2009, 56). In other words, how could designers give these on-site energies the chance to influence social organizations of private and community life on the domestic, urban, and regional scales?

According to Lally, this experimental approach to material energies as (spatial and social) organization principles would also stimulate the imagination of the public and renew their aesthetic relationships with their living environment as a landscape of energy flows. Most of the WEATHERS projects are founded on the designer's ability to artificially increase the human body's sensory perceptions that detect these energy flows as, for example, in the 2011 installation proposal *Sirenuse* (with Thomas Kelley) for the city of Chicago (Lally 2013, 124–125, 178–179). This project was an engagement of the body that transcended the need for walls and canopies and a creator of space and outdoor activities over the different seasons of the year; here the borderline between landscape and architecture is blurred through the use of material energies as building blocks for defining spatial boundaries and shape.*

4.5 CONCLUSION

Reestablishing an aesthetic relationship between man and energy represents a very significant challenge for landscape design in view of achieving the energy transition trough a more holistic pathway. Indeed, in virtue of its fundamental socio-ecological functions, design practice in the field of landscape infrastructure can substantially contribute to redefining inhabitants' understanding of the role played by material energies in their physiological and social existence. It would also contribute to

---

* See also the project for the Vatnsmyri Urban Planning Competition of 2007 for the city of Reykjavik (Figure 4.7). Here the same geothermal resources that aliment the well-known Iceland thermal pools are used to affect the city's climatic conditions, including air and soil temperature for the growth of vegetation. A climatic "wash" is "designed" to connect each programmed landform in the site, controlling winds and permitting the extension of the seasonal activities and the period of usable time outdoors. The wash produces an "artificial" microclimate and acts as a connecting tissue between the different functions and activities of the project site (Lally 2009, 57; Lally 2013, 128–129, 231). For another example of what Lally defines as "neither landscapes nor building strategies, but climatic strategies," blurring the edge between architecture and its environment, see the 2008 proposal for the Tamula Lakeside Planning for the Estonian city of Võru (in collaboration with Morris Architects). Here, gradient climate zones are created on the ground floor of each of the buildings that compose the new urban development along the lakeshore. The gradient parks shift in size and intensity, growing and connecting with each other in warm or milder seasons or shrinking and acting as separated entities in cold winter time (Lally 2009, 60–62; Lally, 2013, 76–79).

Landscape as *Energy* Infrastructure

counterbalancing the dematerialized and abstract apprehension that humans have of energy in their daily lives, which results in a progressively dematerialized and "anesthetic" relationship between men and energy.

More concretely, this way of approaching the design of energy landscapes could overcome the strictly functionalistic logic that dominates a wide range of contemporary design practices and focuses on the mere achievement of energy efficiency or renewable energy targets. This is still the main logic behind most of the current zoning strategies for planning and developing "green" power plants. However, these strategies do not always result in the construction of effectively "sustainable energy landscapes," following the definition of Sven Stremke and Andy van den Dobbelsteen (2013), but rather in the production or reproduction of *renewable* energy landscapes. In other words, they do not draw effective sustainable scenarios for energy transition in different geographic and sociocultural contexts. On the contrary, they have more often produced off-limit enclaves under a system of strict surveillance, which may represent "a work of destruction or the reflection of a guilty conscience" rather than "a new achievement in the evolution of civilization" (Schöbel and Dittrich 2010, 60).

In an attempt to reach a more harmonious integration of the renewable energy plants into *their* landscape, some recent planning experiences have envisioned these decentralized renewable energy infrastructures as the "re-composition of sociotechnical links between landscape and energy" (Nadaï and van der Horst 2010, 144). They have been mainly oriented toward the establishment of planning processes that would be more "open" to the *logic* of the specific project site. A recent case study within the Narbonnaise Regional Natural Park in southwestern France (Nadaï and Labussière 2010, 2013, Figure 4.8) has shown how the process of planning and development of a wind power plant may pull an existing landscape into "a new existence" by adopting a "micro-siting" approach* focused on the "relational interplay" between landscape and renewable power (Nadaï and Labussière 2013, 124). Situated in a wildlife migration corridor, the planning process of this wind power development has been established through converting birds' intelligence in interacting with the wind into a "readable" quality of the landscape. Indeed, the process was led by the *aesthetic of the movement*, evoking the living presence of the birds within this landscape. This aesthetic allows the new wind power landscape to emerge from a "net of relations" between the birds, the wind, the turbines, the project developers, the bird-watchers, and the site. The resulting wind power development is envisioned as a way of "converting relations into other relations" (Nadaï and Labussière 2010, 218), and the *sensory* experience of the new energy landscape is driven through the specific "wind-related-kinetics" (Nadaï and Labussière 2010, 227) that characterizes the different living entities on the project site.

Hence, in the search for new pathways toward energy transition, the field of landscape infrastructure has a significant role to play in increasing social awareness of

---

* This was an experimental method of bird-watching, which was adopted in 2001 by the French Bird Protection Organization for siting new wind envelopes (repowering) in a preexisting wind power station within the perimeter of the Regional Park. It aimed at understanding and mapping how birds behaved in and reacted to a specific site where turbines were made to coexist with a migration corridor; "the moving presence of birds" was translated into textual and visual representations.

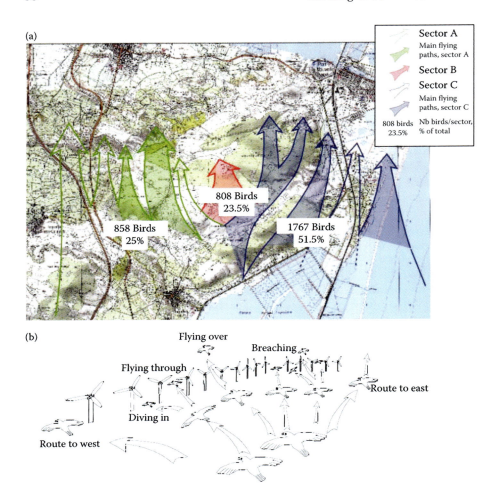

FIGURE 4.8 Ligue de Protection des Oiseaux (French Bird Protection Organization). Project of repowering of a wind power station in the Narbonnaise Regional Natural Park (Aude, France), located in a diffuse migration corridor, 2002–2007. (a) Micro-flying ways per sector. (b) Birds' reaction to the existing park. (From Ligue de Protection des Oiseaux délégation Aude, ABIES Bureau d'études énergie et environnement, *Suivi ornithologique des Parcs éoliens du plateau de Garrigue Haute [Aude], Rapport Final*, Etude financée par l'ADEME [Agence de l'Environnement et de la Maîtrise de l'Energie], 2001, pp. 16, 64. English translation by Alain Nadaï and Oliver Labussière. *Landscape Research*, 35, 2, pp. 221, 223, 2010. © Ligue de Protection des Oiseaux, 2013. With permission.)

the living presence of energy in our landscapes as this field encompasses the stratified web of aesthetic relationships established between man and place. Following this perspective, design practice could effectively contribute to reestablishing the primordial biophysical and aesthetic interconnections existing between energy and society through the mediation of "nature."

Landscape as *Energy* Infrastructure

REFERENCES

Aitken, M. (2010). Why we still don't understand the social aspects of wind power: A critique of key assumptions within the literature. *Energy Policy* 38(4): 1834–1841.

Allen, S. (1999). Infrastructural urbanism. In *Points + Lines: Diagrams and Projects for the City*, 48–57. New York: Princeton Architectural Press.

Aquino, G. (2011). Preface to *Landscape Infrastructure. Case Studies by SWA*, ed. The Infrastructure Research Initiative at SWA, 7. Basel: Birkhäuser Verlag.

Bélanger, P. (2009). Landscape as infrastructure. *The Landscape Journal* 28(1): 79–95.

Bélanger, P. (2012). Landscape infrastructure: Urbanism beyond engineering. In *Infrastructure Sustainability and Design*, ed. S. Pollalis, A. Georgoulias, S. Ramos, and D. Schodek, 276–315. London: Routledge.

Berleant, A. (1992). *The Aesthetics of Environment*. Philadelphia: Temple University Press.

Berleant, A. (1997). *Living in the Landscape: Toward an Aesthetics of Environment*. Lawrence, KS: University Press of Kansas.

Blanc, N. (2008). *Vers une Esthétique Environnementale*. Versailles: Editions Quæ.

Brady, E. (2003). *Aesthetics of the Natural Environment*. Edinburgh: Edinburgh University Press.

Brisson, J. (1999). *L'évaporation Motrice*. Paris: Ed. Actes Sud.

Brittan, G. G., Jr. (2001). Wind, energy, landscape: Reconciling nature and technology. *Philosophy and Geography* 4(2): 169–184.

Clément, G. (1999). *Le Jardin en Mouvement*. Paris: Pandora.

Clément, G. (2004). *Manifeste du Tiers Paysage*. Paris: Sujet/Objet.

Cowell, R. (2010). Wind power, landscape and strategic, spatial planning. The construction of "acceptable locations" in Wales. *Land Use Policy* 27: 222–232.

Fernández-Galiano, L. (2000). *Fire and Memory: On Architecture and Energy*. Translated by Gina Cariño, Boston: MIT Press.

Georgescu-Roegen, N. (1971). *The Entropy Law and the Economic Process*. Cambridge, Massachusetts: Harvard University Press.

Lotka, A. J. (1925). *Elements of Physical Biology*. Baltimore: Williams and Wilkins.

Lally, S. (2009). When cold air sleeps. *Architectural Design* 79(3): 54–63.

Lally, S. (2013). *The Air from Other Planets: A Brief History of Architecture to Come*. Zürich: Lars Müller Publishers.

Law, J. (1992). Notes on the theory of the actor network: Ordering, strategy and heterogeneity. *Systems Practice* 5(4): 379–393.

Margalef, R. (1997). *Our Biosphere*. Oldendorf/Luhe: Ecology Institute.

McHarg, I. L. (1969). *Design with Nature*. New York: The Natural History Press.

Mollison, B. (1988). *Permaculture: A Designer's Manual*. Tyalgum AU: Tagari.

Nadaï, A. (2012). Planning with the missing masses: Innovative wind power planning in France. In *Learning from Wind Power: Governance, Societal and Policy Perspectives on Sustainable Energy*, eds. J. Szarka, R. Cowell, G. Ellis, P. Strachan, and C. Warren, 108–132. Palgrave: MacMillan.

Nadaï, A. and O. Labussière (2010). Birds, turbines and the making of wind power landscape in South France (Aude). *Landscape Research* 35(2): 209–233.

Nadaï, A. and O. Labussière. (2013). Playing with the line, channelling multiplicity: Wind power planning in the Narbonnaise (Aude, France). *Environment and Planning D: Society and Space* 31(1): 116–139.

Nadaï, A. and D. van der Horst. (2010). Introduction to *Landscape Research* 35(2): 143–155.

Odum, H. T. (1971). *Environment, Power and Society*. New York: John Wiley & Sons.

Pasqualetti, M. J., P. Gipe, and R. W. Righter. (2002). A landscape of power. In *Wind Power in View. Energy Landscapes in a Crowded World*, eds. M. J. Pasqualetti, G. Paul, and R. W. Righter, 3–16. San Diego: Academic Press.

Perrotti, D. (2011). *Paysage-infrastructure ou de la dimension infrastructurelle du paysage*, PhD diss., Politecnico di Milano, Ecole Nationale Supérieure d'Architecture de Paris-la Villette.

Perrotti, D. (2012). Conceiving the (everyday) landscape of energy as a transcalar infrastructural device: The progressive construction of a work hypothesis. *Projets de Paysage*, 7.

Portoghesi, P. ed. (1969). *Dizionario Enciclopedico di Architettura e Urbanistica*. Roma: Istituto Editoriale Romano.

Rahm, P. (2006). Form and function follow climate. In *Environ(ne)ment*, G. Clemente et P. Rahm, ed. G. Borasi, 152–159. Montreal: Centre Canadien d'Architecture/Milan: Skira.

Rahm, P. (2009a). *Architecture Météorologique*. Paris: Archibooks.

Rahm, P. (2009b). Meteorological architecture. *Architectural Design* 79: 30–41.

Roncken, P. A., S. Stremke, and M. P. C. P. Paulissen. (2011). Landscape machines: Productive nature and the future sublime. *Journal of Landscape Architecture* 6: 6–19.

Schöbel, S. and A. Dittrich (2010). Renewable energies—Landscape of reconciliation? *Topos* 70: 56–61.

Strang, G. L. (1992). Infrastructure as landscape, landscape as infrastructure. In *CELA 1992: Design + Values. Conference Proceedings*, ed. E. Rosemberg, 8–15. Charlottesville: University of Virginia.

Stremke, S. and A. van den Dobbelsteen. (2013). Introduction to *Sustainable Energy Landscapes: Designing, Planning, and Development*, eds. S. Stremke and A. van den Dobbelsteen, 3–10. Boca Raton: CRC Press.

Williams, D. E. (2007). *Sustainable Design, Ecology, Architecture and Planning*. Hoboken, NJ: J. Wiley.

5 Landscape Machines
Designerly Concept and Framework for an Evolving Discourse on Living System Design

Paul A. Roncken, Sven Stremke, and Riccardo Maria Pulselli

CONTENTS

5.1 Introduction 91
5.2 Design and Ecology 92
5.3 Landscape Machines Explained 93
5.4 Aesthetic Aspect Considered 95
5.5 Metabolism of *Landscape Machines* 96
5.6 An Example: Ems Full Hybrid 101
5.7 Conclusions 103
Acknowledgments 111
References 111

… the "environmental problem" is not a problem of man's relation to his natural surroundings, but first and foremost a problem of man's relation to himself. It is not enough, then, to talk about—and work on—environmentally clean technologies, or to further enhance our environmental education efforts, or even to better understand the dynamics of natural processes from a complexity perspective. Fine efforts as they are, they will fall short if at the same time we do not place before us as the central problem the cultural limits of that predator man provoking the environmental damages.

Reframing Complexity, Juarrero, Sotolongo, van Uden, and Capra 2007, ix

5.1 INTRODUCTION

In the past decade or two, the functional value of landscapes has (re)gained interest. This affects landscape design in several manners, foremost by redefining the meaning and scope of "design." Functional design may not be the exclusive expertise of landscape architects, but also of ecologists, planners, eco-tech firms, and those responsible for land management. Terms such as ecosystem services, climate-proof design, green investments, and landscape infrastructure mark a paradigm shift from *beautification and preservation* to landscape-related *production*. The purpose, it appears, is to argue that natural processes can contribute to economic and geopolitical benefits as part of a more general turn to sustainable development. Ecosystems and biodiversity should no longer be isolated aims, but part of a larger framework to synthesize natural and cultural expressions and benefit both. This is reflecting a staggering ambition. It is both related to an unimaginably large and diverse set of data and an equally large complexity of interrelationships.

Our concern in this chapter is that this development could also be interpreted as a shift away from *aesthetics* in general, which would be rather counterintuitive as there are also manifold incentives for an increasingly experience-based society. Ecological theories, planners' scenarios, eco-tech products, and management pragmatism breathe a more confident aptitude regarding scientific methods such as monitoring, system modeling, quantification, and extrapolation. At times, the general scientific demand for clear methodology, theory, and repeatable results seem to be undermining the methods of conventional design. There is a fraction within the community of landscape architects that intends to serve as advertisers of what serious scientists have determined, providing for a fashionable beautification by an occasionally (eroticizing) style intervention (see Meyer 2008; Selman 2008). Another fraction, mainly consisting of academics, intends to upgrade design arguments and methods to match the results of co-academics. According to us, it is not yet determined what best serves the development of living system design. The indispensable position of design has yet to be found and nurtured.

Admittedly, a strict notion of a method matches one of the lowest ambitions of practicing designers perhaps because of an implicit fear that repetition is what designers must avoid. Nevertheless, their "designerly ways of knowing" (Cross 2007) and, specifically, their aesthetic explorations are enthusiastically appraised when designers are part of a research team. In this chapter, we therefore regard the increase of interest in functional landscapes as a momentum to not only develop a theoretical and methodological framework to synthesize common ground for both designers and scientists, but also as an incentive to specify unique design theories and methods. The same momentum was present when the discourse on Landscape Urbanism was reflected (Thompson 2012; Waldheim 2006) and will thus be included in our references.

5.2 DESIGN AND ECOLOGY

The integration of ecological system knowledge into landscape design is not a new subject (Hough 1995; McHarg 1969). It is, in fact, one of the key components for the "jump over the garden fence," the upscaling and the integration of complexity, long-term planning, and design. By the recollections of Dirk Sijmons (2012), one of the most experienced landscape architects in synthesizing ecological knowledge and designerly intentions, it is possible to trace what conditions have determined

Landscape Machines 93

past attempts. This historical development is, of course, not all logical or reasonable. On the contrary, the development of landscape design has been strongly related to pragmatic, entrepreneurial and political conditions instead of fundamental scientific ambitions (Sijmons 2012, 285–287). Landscape design is, above all, a practiced specialism, called for when all conditions are clear enough to start the construction phase. A principle lesson of recent history is that landscape architects are still too romantic in their perception of natural processes. Many landscape architects would rather satisfy their anonymous and broad audiences with accessible clichés of matured landscapes as if succession will always resolve in the ultimate (picturesque) scene (Crandell 1993). This may be causing a pitfall for any profound collaboration between frontline science and practiced design (Moore 2010; Waldheim 2006). A related lesson is that one should be wary of creating either a dichotomy between picturesque craftsmanship and scientific analysis or denying this binary to exist (Thompson 2012, 16). The development of living system design is a mere specialism within the widening applications of landscape architecture. Not all of the landscape architectural paradigms and traditions have to be converted to match this specific ambition.

Only very recently, the many trials and errors of past design experiments in large-scale housing development, nature development, and environmental policies have all been stopped at a "grinding hold," either failing to "reach the hearts and minds of the people" or "by the sheer complexity of the org-ware trying to mimic the complexity of society" (Sijmons 2012, 281–290). The current momentum for academic reflection on the methods and theories of landscape architecture will benefit from this (temporary) hold.

To avoid abstract discussion in this chapter, we therefore thought it important to introduce an academic design concept, capturing the extremely large ambition to design a living system. These conditions we defined by a paradoxical term: *landscape machines* (Roncken 2011; Roncken et al. 2011). The concept of *landscape machines* differs from those of McHarg or Hough because it is not the current or past landscape system that is used as a foundation for future development, but a new, initially even artificial landscape system that will nevertheless develop into a self-sustaining system. The premise is that it is not possible to reverse the history of a landscape system and go back in time (Sijmons 2012, 290). This is underlined by the particularity of a *landscape machine*; it is a design that allows the processing of material and resources that are not endemic, but imported (*Fremdkörper*). How should this be understood?

5.3 LANDSCAPE MACHINES EXPLAINED

Landscapes are, to most people, the opposite of machines because the first are related to nature and the second to mankind. This is an understandable cultural bias, yet it is also an unnecessary obstruction to envision the integrative character of nature, technique, and mankind. For instance, all garden designers know that gardens are, in fact, very similar to machines because they have to "deliver" with great regularity an anticipated show of colors, structures, and multisensory tasks and thus allow experiences to arise within the human perception. If the show fails to perform, than the machinery has to be tweaked: the soil improved, the plants exchanged, the watering regulated. The paradox of this design is that nothing resembles this clocklike machinery; all appears in natural bewilderment. The presence of this paradox is essential, both for the efficiency of the designers' craft and the bewilderment of the audience.

The definition of a *landscape machine* can principally be explained in threefold (see Figure 7.4):

1. It is a productive landscape that, by a design intervention, will resolve an existent malfunction in the physical environment. This malfunction may already be explicitly present by negatively affected ecological, societal, and economic development. The malfunction may also be artificially introduced in a landscape because the environmental interactions are expected to be resilient against an introduced stress, i.e., to respond with a beneficial processing. The design effort lies in the determination of the components, scale, position, time, and set of human/animal interactions by which a landscape could adapt to a desired functional situation. The malfunction (or induced stress) needs to be quantified to predict the material interactions, and it also needs to be qualified to understand the type of interactions to facilitate new routines of human/animal involvement.

2. The machine aspect consists of ecologically described processes that are either enlarged or stimulated to perform. These will continuously interact with each other, affecting the shape, scale, and position of components within the landscape. There is a dynamic exchange, a continuous shift of ecological interactions because there is a continuous disturbance of the system by large-scale harvesting of crops, fresh water, cleaned soil, or animal stock. There is need for a bookkeeping model of all the input and output that runs through the system.

3. The design and evaluation of the functionality is made explicit by an input-output ratio, i.e., metabolism of the system. This can be monitored both quantitatively (e.g., amounts of water retention and waste decomposition) and qualitatively (e.g., human and animal responses and well-being). The overall development can be simplified by four stages: an *initial stage*, a *growth stage*, a *yield stage*, and a *steady-state stage*. During the *initial stage*, an intervention is made in the landscape and the related societal/(a) biotic types of engagement. The *growth stage* is transitional due to various parallel successions that interact. During the *yield stage*, the *landscape machine* entirely regulates itself, is powered by renewable resources, and will provide a maximum amount of ecosystem services and goods. The *steady-state stage* would be the ideal state of the *landscape machine* because it indicates that the continuous harvesting of products can coincide with continuous shifts within the landscape, maintaining an abundance of biodiversity (ad. 2), preferably developing into a dynamic and dissipative ecosystem, such as mangrove forests, wetland systems, or highland peats. Yet it could also evolve into a *steady state* that is no longer productive. This would mean a failure according to the intended design but a success to a newly introduced ecological state.

These three rather simple rules allow a great variety of *landscape machines* to be developed. A fundamental and paradoxical aspect needs to be discussed here. What is very striking in the stated rules for development is the inclusion of a possible alternative ending of the intended design. The first three stages work toward a gradual development of some expected outcome. It takes time to change vast hectares of land by use of periodical occurrences, such as erosion and natural dispersion of plants and animals. This gradual development can be calculated and predicted by diverse system models. There may even be managerial calibrations if occurrences do not match the intended results. It is however feasible that managerial interventions will not suffice to change the course of events. In that case "a vital serendipitous relationship between formal design and chance" (Sijmons 2012, 298) might be the only option left.

What about human experience? Human experience and judgment change as environments change. Suppose that there is suddenly a hole in the ground. Some will neglect it and walk around, some will decide to fill it, some will reflect its meaning, and yet others will complain about it to the owners. It is people's response that will make landscapes alive beyond the liveliness of pristine naturalness. Human experiences are explicitly adaptive by their serendipitous interests and their various responses to explore new possibilities that arise. Living systems are not only changing by physical means, but also by experiential involvement of the people involved. Even if systems develop as planned in accordance with predicted models, humans may, due to an accumulation of experiences during the whole process, not accord with the expected results. It is this accumulative aspect of experiences that adds a specific addition to the success of living system design. Initially unwanted results may turn out to become desirable products. For instance, the relationship between landscape development and a human sense for serendipity offers the opportunity to first regard exotic planting a nuisance for endemic development and next discover the incredible qualities of that exotic planting for certain cancer treatments.

Our argument is to incorporate such serendipity into the design and modeling of living system design. How can this be done? Landscape philosopher Ian Thompson points out a similar nuance in the discourse on landscape urbanism that he aligns with Daoistic belief (2012, 14). Daoism is not against action, but only seeks "to act when the time is ripe and the occasion demands it." A relevant research question would be how to include the idea of the serendipity as a functional part of system modeling. *Landscape machines* can thus be regarded as environments in which all humans, even including the general public, are deliberately estranged, teased into new performances by the unfamiliar conditions of developing interactions, and, more importantly, are allowed to intervene—by some type of regulation. This is mainly possible because landscapes evolve rather gradually in pace with changing human aptitudes and reinterpretations (see Selman 2008).

5.4 AESTHETIC ASPECT CONSIDERED

So human experience and judgment change as environments change, and human experiences are explicitly adaptive by their serendipitous interests and their enthusiasm to explore new possibilities that arise. Through this notion, we explicitly refer to a different interpretation of aesthetics than is conventionally used by stylistic designers. Our interpretation of aesthetics is more in line with the idea of affordances (Gibson 1977).

It is, by now, increasingly acclaimed that we partly "read" the landscape besides being "involved" in landscapes according to biological instincts. Walking in the landscape using muscular power and hand–eye coordination are as much a part of the aesthetic experience as looking and daydreaming. They cannot be separated (Abram 2012; Kaplan and Kaplan 1989; Varela et al. 1993). Their combined effect is what constitutes an aesthetic interaction, which is referred to by James Gibson as "affordances." These are situations that are provoked by natural features, such as topography or objects, and cause humans (and animals) to grasp possibilities for interaction (Heft 2005, 123), for example, the discovery of a natural staircase formed by the roots of trees that hold on tight on the side of the mountain. Besides opportunities for interaction, limitations are a possibility to block certain types of interaction, such as swimming in dredge or moving through thick snow. Affordances are not completely deductible from the formal aspects of objects or topography, and neither can they be predicted by the human mind alone. They apparently arise contextually when circumstances are present. Affordances only exist if they are discovered by the act of engagement, something that was referred to as an essentially creative "antropotechnique" by the contemporary German philosopher Peter Sloterdijk (2011, 18). Aesthetics is, by the idea of affordances, a functional part of (ecological) system theories. The current momentum to explore functional landscapes therefore makes way for an equally functional definition of aesthetics. The quest to define the creative (Sijmons 2012, 289), both in living systems and in human response, may, in large part, depend upon it.

5.5 METABOLISM OF *LANDSCAPE MACHINES*

If we mention the metabolism of *landscape machines*, we mean more than physical interactions (matter). Cultural meanings, incentives for affordances including serendipity, and social processes should somehow be incorporated. This ambition is currently developed as a theoretical option only. *Landscape machines* are only developed within academia and are purely speculative; they are paper realities only. The visualization of experiential scenarios is thus an important means to simulate any aesthetic (affordance-related) responses by human interaction. These visualizations, preferably multisensory and cinematographic, have to conjure a high degree of empathy. Besides this, the design follows a procedure that is only partly dependent upon local circumstances; roughly it follows these points:

Examine (four points):

- Examine the confinement of the *landscape machine*.
- Examine potential ecosystem services.
- Examine historic systemics of the site and past/present social engagement (e.g., cultural embedding).

Landscape Machines 97

- Examine external and internal metabolic relationships and mark by what they can be measured.

Define (four points):

- Define desirable nutrient cycles and feedback systems (recycling).
- Define nutrient cycles geographically and describe what has to be connected/isolated.
- Define desirable human, animal, and plant life involvement (affordances and landscape ecology).
- Define what type of yield is possible over what timespan (strive for abundance and diversity).

A rather pragmatic part of the procedure is to administrate an input–output scheme of the metabolism. This scheme, together with accompanying cross-sections that show the dimensions in the landscape, indicate what types of interactions may take place. We argue, and have witnessed, that such schemes can serve as the neutral ground for both the designer and the involved specialists to foster the research and design process. One of the possible schematic representations can be found in thermodynamics and especially within the evolutionary thermodynamics contained in the dissipative model by Enzo Tiezzi (2011).

Evolutionary thermodynamics is a branch of science that predominantly allows for the explanation, in general terms, of processes that regulate the functions of living systems. Ilya Prigogine, Nobel Prize awarded 1977, formulated the concept of *dissipative structures* to define these systems based on their ability to let coherent forms/structures emerge and maintain in time (i.e., *steady state*). Dissipative structures are thermodynamic systems, open to energy and matter, that self-organize toward higher complexity and organization (Prigogine and Stengers 1984). Living systems, such as cells, plants, animals, and human beings, as well as ecosystems, cities, landscapes, and the planet Earth all belong to this category. In a wider sense, social, cultural, and economic systems also imply the emergence of ordered structures and thereby resemble thermodynamic systems (see Barnett 2013; Capra 1996).

Referring to the Second Law of Thermodynamics, every energy transformation leads to an increase of entropy, that is, heat flow toward colder sinks, loss of work capacity, disorder of matter. We can affirm that dissipative structures behave like intermediate systems in between sources (e.g., the sun) and sinks (e.g., for heat). They are fed by energy and material inflows from an external source and provide heat and waste outflows (entropy) to an external sink. It is the continuous flow through the system that enables any intermediate system (for example, a landscape) to exist, develop, and increase in complexity. In more recent years, Enzo Tiezzi (2011) proposed advances that he synthetically expressed with the acronym COOS,* which stands for "Confined Ontic Open

---

* Referring to the Latin name of the Greek island that was the home of Hippocrates.

System." According to Tiezzi, all living organisms, whether simple or complex, share similar thermodynamic properties: "They are *open systems* with their own evolutionary autonomy ... they are *confined* inside a bounded space in which they develop their processes, they are *ontic*, maintaining their internal evolutionary memory which cannot be deleted because it obeys the arrow of the time" (Tiezzi 2011, 2901).

It is, specifically, the concept of "onticness" that allows the inclusion of changing judgment and appreciation for the functionality of the *landscape machine*. The *onticness* of a system results from a coevolutionary process in which systems have developed through a succession of "choices" and chances. The state of a system at a given time and its future development follow a pathway that depends on events and thus takes memory of history as embedded information. Systems will not evolve backward; they will always respond progressively, according to the complex whole of new circumstances.

The concept of *openness* refers to the exchanges between the living system and its external environment as expected in dissipative systems. Because living systems are open, they are exposed to perturbations inducted by external changes. The presence of a border, far from determining a condition of isolation, allows the system to be consistent and recognized as a singular identity. This *confinement* works as a permeable membrane, an interface that allows for the modulation of the relationships and exchanges between the living system and the external environment and thus conditions its evolution. Finally, within the whole, these concepts operate *systemically*, indicating a dissipative, living system. As a framework to collect both the sublimely large set of mathematical data and dynamic exchanges, the simplicity of COOS is instructive.

As a means to visualize COOS, the diagrams by H. T. Odum are used (Odum 1994; Odum and Odum 2001). Odum's diagrams show how systems behave coherently to the Second Law of Thermodynamics. The COOS concept helps to determine where and why there are necessary connections even if data is missing. In Figure 5.1, we report three diagrams corresponding to the first three evolutionary steps of human-natural integrated systems within *landscape machines* (based on Bastianoni et al. 2009). The diagrams visualize a standard development, starting with the *initial phase* (Figure 5.2b) in which there are still processes present to install and speed up the new landscape development (e.g., pumping water, importing resources by means of transportation, planting mechanization, etc.). Every intervention of any *landscape machine* should somehow improve the current situation in landscape developments (Figure 5.2a), representing the growing trend of the world population and human consumption that determine a feedback-enforcing factor (ρ_1, ν_1). Quantities (ρ_0, ν_0) of resources are exploited progressively. Considering that, within a "business as usual" perspective, a progressive exhaustion of N is expected in the medium-long run; our capacity to exploit R, instead of N, should be guarded and even highly improved in the future.

The diagram in Figure 5.2b shows how a designed system-asset, including technological and natural processes, can increase the capacity to assimilate renewable flows even though an initial investment of nonrenewables is necessary. In other words, within a "quasi-sustainability" scenario, one can use flows withdrawn from

Landscape Machines 99

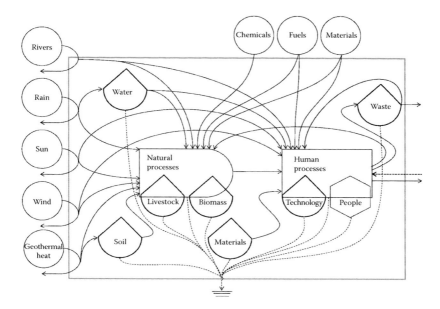

FIGURE 5.1 A visualization of COOS, using the Odum-type of diagram. A boxed confinement can be marked, with a metabolism within and external (re)sources as well as a sink (below). Central is the dual core of both natural processes and human processes, separate and at the same time interrelated within the whole of the *landscape machine*. The visualization represents a set of interacting processes, energy, and material flows that resemble functions and services performed. Processes, systems, and connections among them have a spatial and temporal dimension, each of which is the potential object of a deep investigation and design.

N (ν_g), such as fossil fuels and extracted materials, to create assets and boost living systems (E) to assimilate a growing amount of renewables from R (ρ_{1+e}).

The diagram in Figure 5.2c is a hypothetical scenario in which a desirable natural-human symbiosis has been achieved giving rise to a sustainable self-sufficient system based on renewable sources exclusively. In this case, a self-reinforcing productive system-process based on renewables and a decreasing adjustment of human demand merge and achieve a long-term steady state within a "sustainable" scenario.

The theoretical framework expressed through these diagrams leaves much room for imagination. Let's hypothesize that system E is a *landscape machine*. This would involve both natural and human processes and be potentially productive and not critically depending on nonrenewables. There is an initial direction from a medium-term quasi-sustainability and a long-term sustainable perspective afterward. Can we imagine *landscape machines* to grow in size and numbers, develop in quality, and, eventually, perform autonomously in the long run? The following list indicates all the specifications according to the acronym of COOS. It also consists of specifications when developing toward the *steady-state phase*, in which developments could turn out to be different than expected or even consciously changed due to progressive insights and/or changed aesthetic engagement (through affordances). The aesthetic engagement is discussed in both the *ontic* and the *systemic* aspects of the future landscape.

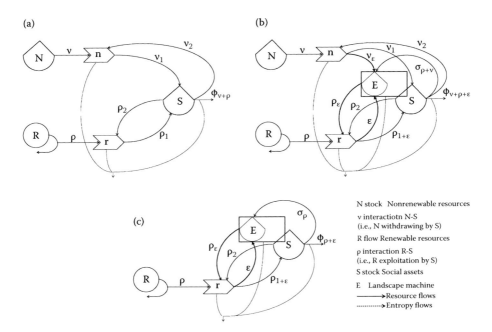

FIGURE 5.2 N represents a (limited) stock of nonrenewable resources; R represents a (unlimited) source of "renewables" entering the system; S represents social assets as a proxy of the whole of behaviors, organization, muscle power, structures, and settlements that belong to human life; symbols n and r represent processes through which nonrenewable and renewable resources are withdrawn and assimilated, respectively. Resource flows are represented by ν for nonrenewable and ρ for renewable. The flow φ represents outputs. Dotted arrows down to the heat sink represent the flow of entropy (not useful energy). (a) Current (business as usual) scenario: exploitation of both renewable (R) and nonrenewable (N) resources by a population (stock S). (b) "Quasi-sustainability" scenario: An investment of nonrenewable (N) resources to create assets (E), such as *landscape machines*, that allow for an augmented renewable capacity ($\rho_{1+\varepsilon}$). (c) "Sustainable scenario": Exploitation of renewable resources (R) within a sustainable rationale.

Key concept	Specification
Confined	Boundaries: Permeable-nonpermeable "barriers" perform functions of *control*, *link*, and *communication*, such as those performed by membranes, filters, and interfaces in order to regulate-constrain-monitor relationships and exchanges with other systems and the external environment. This refers to both consistency-perception of in-out limits and specific processes-functions of filter-control (e.g., noise, in-out flowing water, wind, animal and people movement, and connections to the local-global market).
Open	Connectivity: In-out flows allow the system to grow and develop, fed by energy, materials, water, and other resources (but also people as workers or users, imported goods, information) and then discharging waste, emissions, water outflows (but also exporting food, materials, and goods to the market). This implies the system behaves like a *node* (and plays a role) in a wider, even global, network of processes with a proper physical consistency and space-time rhythms.

Ontic	Historical dimension: Combination of choice and chance, (eco)systems coevolution, emergence of novelties. Events occurred in the past may be indicators of the present state of the system. Structures go along with availability of local resources as well as the historical-climate-cultural-social-economic context.
	Steady-state future: New dynamic equilibrium will influence the settlement of plants, animals, and human involvement indicated by monitoring developing affordances.
Systemic	Self-consistency: The system has a coherent configuration and recognizable functions-services. A cohesion of actors-elements not only derives from an identifiable physical structure (within given boundaries), but also from the whole organization that implies the sharing of intentions, aims, future perspectives, and the outcoming of a unique identity.
	Multifunctionality: The system is made of interacting elements-processes that perform different functions-services (with different space-time rhythms and different aims and users) but cooperate and coevolve in an integrated whole. The organization-configuration mirrors this bio-cultural diversity and heterogeneity—instead of monofunctional homogeneity—as a combination of structures-functions in the same place, resulting in an enhanced land-use intensity and augmented chance for combinations, self-organization, and emergence of novelties.
	Continuity: Interactions among elements in a network of processes in which, for example, outputs from one side become inputs to another (e.g., energy-matter exchanges, health-related phenomenon), enhancing both diversity-specialization of elements and their cohesion-unity-continuity in space and time. Production chains as well as networks of resources (e.g., energy, water, community services), communication (e.g., wired and wireless), discharged energy-matter collection and treatment (e.g., grey water, waste, heat), people, and goods transportation.

5.6   AN EXAMPLE: EMS FULL HYBRID

Over the past five years, several types of *landscape machines* have been developed at Wageningen University (see http://www.landscapemachines.com). One particular example may serve to exemplify the *landscape machines*, COOS, and the more conventional designerly visualizations. The project is titled "The Ems, full hybrid" and is situated in the northern delta region of the Netherlands, in line with similar German estuaries, such as the river Elbe, connecting to Hamburg (van der Togt and Papenborg 2012).

As stated previously, due to the propositional reality of the research (paper products only), these visualizations serve an important role in the examination of past, present, and future aesthetic engagement (through affordances). Human processes are based and highly dependent on primary processes and, at the same time, actively contribute to their development (feedback loops). These include technological assets as well as human (aesthetic) engagement, affecting a sense of identity and health.

The main challenge in this project is to vitalize the estuary system because it has been immobilized into a non-dynamic system that needs continuous dredging to maintain only one particular shipping route. A similar situation can be found in many deltas around the globe. Due to human creations, such as deep-sea harbors, most dynamic systems need continuous and costly maintenance. Imbalanced input-output ratios are maintained by relentless human intervention that continuously effect the potentially beneficial emergence of natural balance. The design effort is to rejuvenate foundations for local and sustainable socio-economical prospects. "The Ems, full hybrid" reveals that a sea delta can be restored to a natural balance of the width, depth, and shape of the whole delta while adding deconstructed bits and pieces of (retired) oil platforms as effective breeding grounds for, for example, mussels (Figure 5.3a and b).

Within the context of the project, waste, by the presence of hundreds of abandoned oil rigs, is a growing stock. It is, however, also exploitable as a resource (the part not treated elsewhere) to provide feedback to both human (e.g., mechanical waste treatment) and natural processes (e.g., new affordances for plants and animals and human tourism). The oil rigs can be shipped to the Ems delta as non-endemic elements (*Fremdkörper*) and become part of a new landscape system (Figure 5.3b).

FIGURE 5.3 (a) Project area encircled. Position and ownership of all the gas- and oil-rigs in the North Sea. (b) Disassembled oil rigs create new habitats. The introduction of "Fremdkörper" rig structures in the dynamic estuarine system result in an intriguing "hybrid" landscape wherein nature is productive and industry produces nature. (From Togt, R. v. der and Papenborg, J., *The Ems Full Hybrid*. [MSc], Wageningen University, Wageningen, 2012.)

Landscape Machines

Besides a newly introduced recycling industry for the region, the oil rig groins can be used as rigs to gradually reshape the estuary into a new dissipative system.* In Figure 5.1 is visualized that a part of the "waste" can be exploited internally as feedback flows. The exceeding part is exported. The dashed arrow to the right represents economic flows as an expected income. Among the set of possible products and services, creative engagement and welfare are either feeding internal flows and/or exportable products (e.g., tourism). This type of systemic interaction also needs to be visualized using empathic images and experiential scenarios (Figures 5.4 through 5.9).

An important part of the design research was dedicated to the understanding of the estuary system in upper, middle, and lower parts. A complicating factor in this (and other) delta redesigns is that these parts operate as different *landscape machines*, containing different specifications according to the COOS acronyms. The upper part, in the estuary mouth, is characterized by a different *confinement*, *openness*, *onticness*, and *systemic* set of characteristics than the middle and lower parts. These differences cannot be understood by the thermodynamic diagrams alone. The COOS specifications help to study such variances.

5.7 CONCLUSIONS

As a neutral ground amid the various experts, designing living systems, a framework for input-output monitoring can serve to collect (missing) data and describe dynamic interactions. We believe the simplicity of COOS offers a feasible common ground to discuss the intentions and methods for designed landscapes. A designerly benefit of COOS is that it explicitly includes human interaction in its most apocryphal and inconsequent state (*Onticness*). Such a model is needed because both human and ecological development do not follow logical systems. The state of human experience is changing due to changing landscapes that offer yet unfamiliar affordances. Such changes, among other dynamic interactions, are the shared concern of designers and scientists and can partly be modeled according to evolutionary thermodynamics. What ecological and thermodynamic scientists can learn from designers is to continuously reframe in what way experiential behavior could modify existing ecological routines. The COOS acronyms offer a simple and elusive overview of the main properties involved. This allows designers to further integrate their role within the discourses on living system design.

Much more than to simply "clarify the message" of ecological experts or to "gain the hearts of the audience," designers can claim their data and dynamic interactions, including aesthetics by means of affordances. They may still lag behind in terms of empirical evidence, yet their boldness by simplicity has the capacity to compensate for that. To sometimes cut right through the mesmerizing complexity of lacking data or unknown types of interaction by providing a prototype of

---

* Based on data from "rigs to reefs" projects in the United States, the placement of "oil rig" groins will cost $15,000 per meter, and conventional dams will cost almost €200,000 per meter. This means a cost difference of roughly factor 10, taking into consideration the less required length using conventional dams.

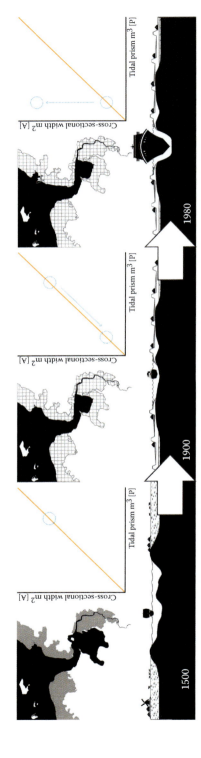

FIGURE 5.4 The adaptation of the Ems estuary. Over the course of centuries, the Ems estuary has been adapted to better the needs of men and increase safety and prosperity. These adaptions have resulted in an unbalanced system, which requires continuous dredging and maintenance to prevent the system from returning to its balanced but shallow state. (From Togt, R. v. der and Papenborg, J., *The Ems Full Hybrid*. [MSc], Wageningen University, Wageningen, 2012.)

Landscape Machines

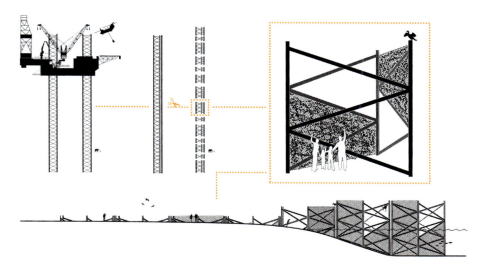

FIGURE 5.5 Decommissioned oil rigs as building blocks. To steer the system back toward a balanced state and conserve navigable depth, superfluous oil rigs are towed to the estuary. The top part is recycled, and the jackets are disassembled and placed on shallow sandbanks to steer sedimentation. (From Togt, R. v. der and Papenborg, J., *The Ems Full Hybrid*. [MSc], Wageningen University, Wageningen, 2012.)

the intended landscape experiences, to sometimes do as is expected and visualize information stunningly simply so that people gain awareness of what they are talking about. At all times, to deny the merely beautified ecological ideologies that do not include humans as part of—yet to be experienced—future landscape affordances.

For the time being, the most pragmatic agenda would be to continue to develop academic designs such as *landscape machines*. The momentum to strengthen the profile of such designerly research is present and will serve to improve a type of modeling that is an effective means to communicate in between disciplines. Such *landscape machines* cannot be simplified to purely technocratic schemes; they most likely contain a multitude of *confined* parts, smaller *landscape machines* within one coherent *landscape machine*, that operate by diverse specifications on *onticness* and *openness*. In the end, this geographical diversity provides a sense of identity that, if understood, can help to alter reality in unforeseen ways, aspiring serendipity. By focusing on the feedback loops within the input-output schemes of such living systems, designers will learn to see where certain systemic relationships are still lacking and what elements (*Fremdkörper*) must be imported into existing systems to provide exactly the right friction to create a (re)new(ed) and thriving dissipative system.

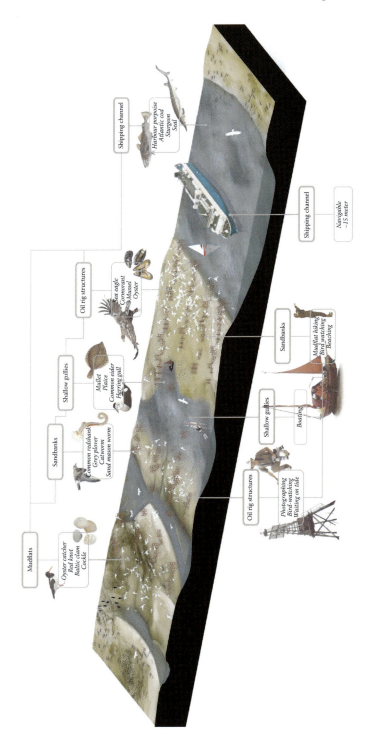

FIGURE 5.6 The Ems estuary, a growing landscape. With the groins placed, sediment flows are altered and gullies will divert their course. This step is repeated several times until the system is self-maintaining and able to "breathe" on its own. This system approach creates numerous positive side effects with a booming biodiversity and new economic uses. (From Togt, R. v. der and Papenborg, J., *The Ems Full Hybrid*. [MSc], Wageningen University, Wageningen, 2012.)

FIGURE 5.7 Disassembled oil rigs create new industries. The oil rigs are deployed to heal the estuary, restore lost habitats, and reconnect former fishing villages to the Ems. At the same time, it introduces an entirely new habitat in which mussels and shellfish will thrive, in turn, attracting other wildlife and creating a new industry for the reconnected villages. (From Togt, R. v. der and Papenborg, J., *The Ems Full Hybrid*. [MSc], Wageningen University, Wageningen, 2012.)

108 Revising Green Infrastructure

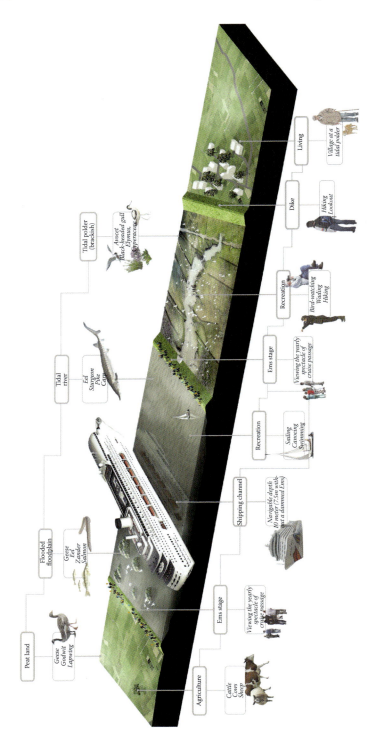

FIGURE 5.8 *Aerial cross-section of the Ems and a tidal polder.* The Ems' water level is pushed up during a vessel transport on the Ems of a large cruise ship. When the ship has passed, the water is used to flush out excess sediment and keep the polders functional and dissipative. (From Togt, R. v. der and Papenborg, J., *The Ems Full Hybrid.* [MSc], Wageningen University, Wageningen, 2012.)

Landscape Machines

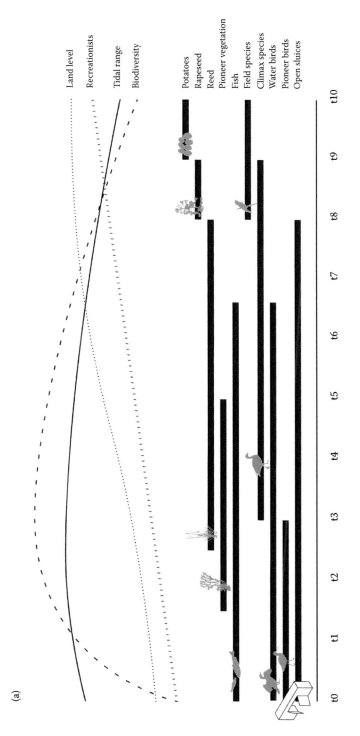

FIGURE 5.9 (a) The dynamics of a tidal polder. When left in standard operation, the sediment level in a tidal polder will increase, the biodiversity will decrease, and climax species will gradually take over.

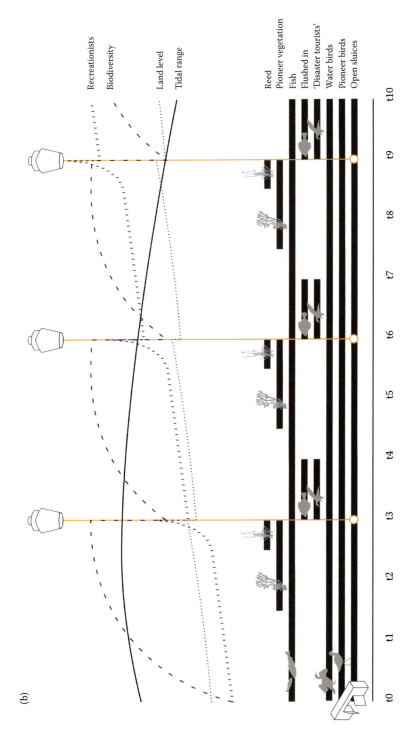

FIGURE 5.9 (Continued) (b) Dynamic development of a "resetting" polder. When a polder is reset, it is brought back to a pioneer stage, keeping constant change and high biodiversity. (From Togt, R. v. der and Papenborg, J., *The Ems Full Hybrid*. [MSc], Wageningen University, Wageningen, 2012.)

Landscape Machines

ACKNOWLEDGMENTS

The research presented in this paper was partly conducted during a visit of Riccardo Maria Pulselli to the Netherlands. The WIMEK research school in Wageningen financed his visiting fellowship.

REFERENCES

Abram, D. (2012). Becoming animal. *Green Letters: Studies in Ecocriticism, 13*(1): 7–21.
Barnett, R. (2013). *Emergence in Landscape Architecture*. London: Routledge.
Bastianoni, S., R. M. Pulselli, and F. M. Pulselli. (2009). Model of withdrawing renewable and nonrenewable resources based on Odum's energy systems theory and Daly's quasi-sustainability principle. *Ecological Modelling, 220*: 1926–1930.
Capra, F. (1996). *Web of Life: A New Scientific Understanding of Living Systems*. New York: Anchor Books, Doubleday.
Crandell, G. (1993). *Nature Pictorialized*. London: The Johns Hopkins University Press.
Cross, N. (2007). *Designerly Ways of Knowing*. Basel: Birkhauser.
Gibson, J. J. (1977). The theory of affordances. In *Perceiving, Acting and Knowing*, R. Shaw and J. Bransford (Eds.). Hillsdale, NJ: Erlbaum.
Heft, H. (2005). *Ecological Psychology in Context: James Gibson, Roger Barker, and the Legacy of William James's Radical Empiricism*. London: Routledge.
Hough, M. H. (1995). *Cities and Natural Process*. London: Routledge Press.
Juarrero, A., P. Sotolongo, J. v. Uden, and F. Capra. (2007). *Reframing Complexity: Perspectives from the North and South*. Paper presented at the Exploring Complexity, Havana, Cuba.
Kaplan, S., and R. Kaplan. (1989). *The Experience of Nature: A Psychological Perspective*. CUP Archive.
McHarg, I. L. (1969). *Design with Nature*. New York: The Natural History Press.
Meyer, E. K. (2008). Sustaining beauty—The performance of appearance: Can landscape architects insert aesthetics into our discussions of sustainability? *Journal of Landscape Architecture, 98*(10): 6–23.
Moore, K. (2010). *Overlooking the Visual, Demystifying the Art of Design*. Abingdon: Routledge.
Odum, H. T. (1994). *Ecological and General Systems. Introduction to Systems Ecology*. Niwot, CO: Colorado University Press.
Odum, H. T., and E. C. Odum. (2001). *A Prosperous Way Down. Principles and Policies*. Boulder, CO: University Press of Colorado.
Prigogine, I., and I. Stengers. (1984). *Order Out of Chaos: Man's New Dialogue with Nature*. Flamingo Press.
Roncken, P. (2011). Agrarian rituals and the future sublime. In *Images of Farming*, W. Feenstra and A. Schiffers (Eds.), 102–113. Heijningen: Jap Sam Books.
Roncken, P., S. Stremke, and M. Paulissen. (2011). Landscape machines, productive nature and the future sublime. *Journal of Landscape Architecture, Spring 2011*(11): 68–81.
Selman, P. (2008). What do we mean by "sustainable landscape?" *Sustainability: Science, Practice and Policy, 4*(2): 23–28.
Sijmons, D. (2012). Simple rules: Emerging order? A designer's curiosity about complexity theories. In *Complexity Theories of Cities Have Come of Age*, J. Portugali, H. Meyer, E. Stolk, and E. Tan (Eds.), 281–309. Berlin, Heidelberg: Springer.
Sloterdijk, P. (2011). *Je Moet je Leven Veranderen, Over Antropotechniek*. Amsterdam: Uitgeverij Boom.
Thompson, I. H. (2012). Ten tenets and six questions for landscape urbanism. *Landscape Research, 37*(1): 7–26.

Tiezzi, E. (2011). Ecodynamics: Towards an evolutionary thermodynamics of ecosystems. *Ecological Modelling* (222): 2897–2902.

Togt, R. v. der and J. Papenborg. (2012). *The Ems Full Hybrid*. (MSc), Wageningen: Wageningen University.

Varela, F. J., E. Thompson, and E. Rosch. (1993). *The Embodied Mind: Cognitive Science and Human Experience*. Massachusetts: Massachusetts Institute of Technology.

Waldheim, C. ed. (2006). *The Landscape Urbanism Reader*. New York: Princeton Architectural Press.

6 Problems of the Odumian Theory of Ecosystems

Georg Hausladen

CONTENTS

6.1 Introduction 113
6.2 Eugene Odum's Organismic Understanding of Ecosystems 114
6.3 Why Ecosystems Are Not Organisms 116
    6.3.1 Boundaries 116
    6.3.2 Development 116
    6.3.3 Self-Organization 117
    6.3.4 Homeostasis 117
    6.3.5 Functional Units 118
6.4 Howard Odum's Physicalistic Understanding of Ecosystems 119
6.5 Limits of Howard Odum's Understanding of Ecosystems 121
    6.5.1 Biological Systems as Dissipative Structures 121
    6.5.2 The Concept of Information Used by Howard Odum 122
6.6 Naturalisms of the Odum Brothers 124
6.7 Critique of Odumian Naturalism 125
6.8 Conclusion 126
Acknowledgments 127
Notes 127
References 131

> The landscape is not just a supply depot but is also the *oikos*—the home—in which we must live.
>
> **E. P. Odum**
> *1971, 267*

6.1 INTRODUCTION

It is a banal statement that ecological knowledge is relevant for landscape architecture. The landscape architect designs with organisms (mostly plants), and if he wants to design successfully, he must know something about organisms and their relationships to their abiotic and biotic environments. Hence, the objects he designs are somehow always ecological systems. To focus on entire *landscapes* as ecological systems *explicitly* is not as self-evident. It is accompanied by a shift in the theoretical point of reference from autecology to synecology or from the *individual* organism to the ecological *community*. Among the theories of synecological systems, there is especially one that is very popular *beyond* ecology, e.g., in landscape architecture[1]— the so-called "Odumian" ecosystem theory.

The brothers Eugene and Howard Odum are among the most famous ecologists of the second half of the 20th century. Their popularity is largely owing to one work, Eugene Odums' *Fundamentals of Ecology* (1953).[2] It is considered one of the standard textbooks on ecology and has made ecology popular to a large audience beyond the field. Along with the book, one concept became especially famous: that of the *ecosystem*. Eugene himself did not contribute much to the progress of the theory of ecosystems because his book is mainly a summary of theoretical concepts that were developed by others.[3] Moreover, it is exceedingly problematic to talk about *the* Odumian theory because the brothers' views differ fundamentally (in spite of all their commonalities).[4] While Eugene Odum views ecosystems as "organisms," his brother Howard, in contrast, understands them as (physically describable) "machines" (Hagen 1992, Chapter 7). Beyond ecology, both concepts, meaning the so-called (holistic) *organicism* and the so-called (reductionistic) *physicalism*, play decisive roles as cores of certain antagonistic ideologies that are very influential in the debates on environmentalism and nature conservation.[5] Within ecology, both concepts were heavily criticized and rejected. I will focus on this point in the following. That is, I will show why Eugene Odum's organicism is insupportable, and I will focus on the limits of Howard Odum's physical theory of ecosystems, especially the concept of *information* he uses.

What I want to demonstrate may appear banal to expert scientists because it does not go beyond knowledge that can be found in good textbooks. But it has relevance beyond science because scientific knowledge is often interpreted falsely, twisted and transformed into ideological statements. This often leads to assertions and adoptions that are absurd from a scientific point of view and highly problematic and even dangerous from a philosophical perspective. Basically, the problem arises from the metaphysical idealization of scientific knowledge, in other words, from the aspiration to explain not only natural phenomena but the *whole world* through science. This idealization is called (scientistic) *naturalism*. The writings of the Odum brothers are paragons of that. I will go into that point in the Sections 6.6 and 6.7. The overall purpose of this chapter is to reach some clarification concerning the relevance and applicability of certain ecological knowledge in general and the "Odumian" theories in particular within a practical context, e.g., landscape architecture.

6.2   EUGENE ODUM'S ORGANISMIC UNDERSTANDING OF ECOSYSTEMS

Eugene Odum's theory is said to be a "lineal descendant of the organismic tradition in ecology" (McIntosh 1980, 229; Bergandi 1995). Frederic Clements' theory of ecological communities is regarded as an epitome of that tradition (Clements 1916, 1936).[6] After all, Eugene Odum's view of the *development* of ecosystems corresponds especially to the Clementian understanding (Kirchhoff 2007, 230; Voigt 2009, 216). This is remarkable in so far as the concept of the ecosystem contradicted

Problems of the Odumian Theory of Ecosystems

this understanding in the beginning. Arthur Tansley introduced it in 1935 within the framework of his fundamental critique on Clements and his scholar John Phillips (Phillips 1934, 1935a, 1935b).[7] Through the textbook of Eugene Odum, the "organismic" holism found its way back into the debate. However, this had little influence on ecology. The idea was and still is generally rejected there. Those ecosystem theoreticians who stand in the tradition of the Odum brothers form an exception.[8]

That Eugene Odum considers ecosystems as "organisms" can be indicated mainly by three points: (1) For him, the relationship between an ecosystem and its components (i.e., between the whole and its parts) corresponds to the relationship between an organism and its organs. (2) Thus, the organization of ecosystems equates to that of organisms. And (3) the development of the ecosystem conforms to the ontogenesis of the individual organism.

On (1), for Odum, ecosystems constitute one "level of organization" in the range of what he calls "biosystems." (E. P. Odum 1971, 4f). For him, all biosystems (from the cell to the biosphere) feature the same kind of relationship between the parts of the system and the whole system: Systems of one level (e.g., an individual) are related to systems of the next level (the population) in a way that is equivalent to the relationship between an organ and an organism. This also pertains to the ecosystem and its components (the community and its abiotic environment): "From the standpoint of interdependence, interrelations and survival, there can be no sharp break anywhere along the line. The individual organism, for example, cannot survive long without its population any more than the organ would be able to survive for long as a self-perpetuating unit without its organism. Similarly, the community cannot exist without the cycling of materials and the flow of energy in the ecosystem" (E. P. Odum 1971, 5). Systems of one level are thus *existential* for systems of the next level and vice versa. For Odum, the main function of the ecosystem concept is "to emphasize obligatory relationships, interdependence, and causal relationships, that is, the coupling of components to form functional units" (E. P. Odum 1971, 9). Therefore, ecosystems are *functional* units.

On (2), for Odum, "Ecosystems are capable of self-maintenance and self-reproduction as are their component populations and organisms" (E. P. Odum 1971, 33). The processes in ecosystems lead to a certain stabilizing mechanism: the *homeostasis*:[9] *"The interplay of material cycles and energy flows in large ecosystems generate a self-correcting homeostasis with no outside control or set-point required"* (E. P. Odum 1971, 35). I will show why the concept of homeostasis is not applicable to ecosystems.

On (3), Odum defines "ecological succession" as "development of ecosystems" (E. P. Odum 1971, 251). For him, "The development of ecosystems has many parallels in the developmental biology of organisms" (E. P. Odum 1971, 251). Odum defines three parameters of succession: "(1) It is an orderly process of community development that is reasonably directional and, therefore, predictable. (2) It results from modification of the physical environment by the community; that is, succession is community-controlled even though the physical environment determines the pattern, the rate of change, and often sets limits as to how far development can go. (3) It culminates in a stabilized ecosystem in which maximum biomass (or high information content) and symbiotic functions between the organisms are maintained per unit of available energy flow" (E. P. Odum 1971, 251). "The final or stable community ... is the climax community; it is self-perpetuating and in equilibrium with the physical habitat" (E. P. Odum 1971, 264). Thus, the development of the ecosystem (like the ontogenesis of an individual organism) has a *final state* (the climax), meaning it is a *teleological* process. This process "is directed towards achieving as large and diverse an organic structure as is possible" (E. P. Odum 1971, 257), meaning the ecosystem *differentiates* over the course of its development (like the organism when developing from a juvenile to an adult stage).[10] Like Clements, Odum talks about "adult" ecosystems; he is interested in the question of, if ecosystems "age" like organisms and, in the course of that, become more vulnerable to "diseases" and "perturbations" (E. P. Odum 1971, 257).

Those three points should clarify that, for Odum, the "organism" is more than just a metaphor[11] or a more or less suitable analogy. For him, ecosystems *are* organisms. He regards them as *ontological units*.

6.3 WHY ECOSYSTEMS ARE NOT ORGANISMS

6.3.1 Boundaries

The *spatial* boundaries of an organism, likewise, are its *functional* boundaries. In other words, within those boundaries, there can be found all components that depend essentially on the organism *and* which the organism essentially depends on (organs). That means that the boundaries of the organism are *independent* of a definition and must be accepted by the researcher.[12] What belongs to the organism is "defined" by the organism *itself*. However, the situation is different with ecosystems: insofar as ecosystems show obvious spatial boundaries (e.g., the lakeshore or the edge of a forest), those boundaries (usually) are the result of *external* factors (geomorphology, land use, etc.). But, principally, they depend on a definition: The researcher determines (with his definition) what the ecosystem *is*.[13] Whether a certain component belongs to it depends only on whether it is included by this definition. An example: the bear that catches salmon can be regarded as a component (consumer) of the ecosystem "river" although it does not live *in* the river. But if the bear is included, the definition results in a system that should also include the berries the bear eats or the cave it sleeps in, i.e., something quite different from the river. Thus, ecosystems are *not ontological* units.

6.3.2 Development

The *ontogenesis* of an organism is determined by its genetic program. In this case, something *unfolds* (differentiates) that is encoded (inherent) in this program. Here the fundamental difference between the development of an organism and changes within an ecosystem becomes apparent because we do not know of a "program" that determines the "development" of an ecosystem. In what way an ecosystem "develops" depends on which species (can) exists in a certain area at a certain time. Some species are components of the system only for a limited time period. An example: an early stage of the succession of a forest is dominated by birch. In the course of succession, the birch trees create shady conditions that allow the beech to germinate and develop. Over time, however, the birch trees disappear because they are suppressed

Problems of the Odumian Theory of Ecosystems   **117**

by the beech. Thus, the "development" of ecosystems is a *sequence* (precisely, a succession) of certain combinations of species (associations).[14] Thus, the question arises: Which argument justifies the *identity* of the ecosystem "forest" in view of the fact that the composition of components (species) gradually changes.[15]

6.3.3 SELF-ORGANIZATION

Ecosystems can be referred to as *self-organizing systems*. This kind of self-organization, however, differs fundamentally from the self-organization of the organism that is, nowadays, called *autopoiesis* (Maturana 1975; Maturana and Varela 1987). A unit produces and reproduces *itself*, i.e., the whole (organism) produces its parts (organs) while the parts, on the other hand, produce the whole. Therefore, the features and the existence of the parts and the whole *depend* on each other, alternately. In contrast, the self-organization of an ecosystem is more similar to what can be observed when a crystal or a cyclone evolves: Certain units (organisms, molecules, etc.) organize according to their individual properties, which are *independent* from the whole. This is called "self-assembly" (Trepl 2005, 85; Keller 2008, 2009).[16] The fact is that in biological systems (also in the case of the individual organism) both kinds of self-organization play a role and that the "organic" concept of self-organization also makes sense (e.g., Sendova-Franks and Franks 1999) for higher biological units (e.g., social insects as ants and bees) does not matter here. Furthermore, it is not relevant whether the system that emerges in the course of the self-assembly is in thermodynamic equilibrium with its surrounding (like the crystal that is thus a *static* system) or not (like the so-called dissipative structures, such as cyclones, which show a *dynamic* organization).[17] Crucial is just the relationship between the whole and its parts.

6.3.4 HOMEOSTASIS

The difference in the organization of organisms and ecosystems also becomes obvious in the concept of *homeostasis*. The concept denotes a particular kind of system *equilibrium* or *resistance* to fluctuating environmental conditions, in which the organization of the system is maintained against external fluctuations *actively*. That means the system *regulates itself*; it *reacts* to the fluctuations (and thereby receives its identity). The classic example is the regulation of temperature by an endothermic organism. But regulation is only possible within certain limits. If those limits are exceeded, the organism *dies*. Homeostasis characterizes an *organic* equilibrium.[18] In contrast, ecosystems are not able to do the same. Admittedly, in their case, it also makes sense to say that they resist environmental fluctuations. However, this resistance is not an active but rather a *passive* process. The ecosystem does not *react* to changing environmental conditions. It is not in an organic equilibrium, but in a *flow* equilibrium. An example: a lake has a particular amount of different kinds of nutrients (e.g., phosphate). The nutrient level depends on the input of nutrients, on the bonding and release by certain system components (the organisms but also abiotic processes, e.g., the bonding and release of phosphate in the sediment of the lake) as well as on the output of nutrients. If the concentration increases (e.g., by increased eutrophication), this can be compensated for by biotic and abiotic processes to a certain extent.[19] But if the concentration exceeds a certain limit, the system (its species composition as well as its material flows) *changes* fundamentally from an aerobic (oligotrophic or eutrophic) to an anaerobic (hypertrophic) organization. In physics, one also speaks of a *phase transition* of the system.[20] Nevertheless, the lake remains a lake (although the system "lake" is now organized totally differently). In this context, the concept of *death* does (obviously) not make sense. The system simply *changed*.

6.3.5 FUNCTIONAL UNITS

For Odum, the components of the ecosystem are functional to the extent that they are *necessary for the existence* of the whole and that the whole, on the other hand, is necessary (functional) for the existence of the components. As Kirchhoff and Voigt show, Odum thereby uses a certain concept of function that can be referred to as "etiological" (Kirchhoff 2007, 218; Voigt 2009, 192). This concept not only describes the contribution of a function bearer to the performance (an arbitrary defined target state) of a system (this is called *dispositional* function), but also *why it exists* (Wright 1973).[21] The classic example is that it is *the* function of the heart to pump blood through the organism. Or it is *the* function of a fuel pump to pump fuel to the engine. This not only describes the effect of the component (delivery rate of the heart or fuel pump) but, at the same time, points to a *predefined* target state of the system: the survival of the organism or the roadworthiness of the car. In contrast, the (freely) selectable function of cracking a nut can be attributed to the fuel pump, or the function of being used as nourishment or trophy can be attributed to the heart. The etiological concept can only make sense in the case of organisms or artifacts (although in different ways) (McLaughlin 2001, 2005). The difference is that, in the case of the artifact, the fact that the function bearer (the fuel pump) is there, i.e., how and why it was produced, can only be explained by the assumption of an *external* purpose (a rational being): The car does not *produce* the pump, and the pump, on the other hand, does not produce the car. However, the situation is (obviously) different with organisms: The function bearer (the heart) not only contributes to the (re)production (and the existence) of the whole but, furthermore (mediated through the whole), to its own production. With regard to organisms (and in contrast to cars), thus, one can speak of a *self-producing* (and self-reproducing) system (McLaughlin 2001, 2005). In the *scientific* context, the etiological concept of function can only be used meaningful for organisms (as, *by definition*, an artifact is no scientific issue). But, here, one important point should be noticed.

The etiological concept of function implies a *teleological* judgment (i.e., the assumption of a goal or a purpose). In biology, this only has *heuristic* relevance, i.e., it serves to help us understand the organism as a certain scientific, namely *biological*, item. It is the basis for differentiating the living from the nonliving in principle, i.e., *it constructs the subject area of biology*.[22] Only through this we can speak about the "development" of an organism meaningfully, attribute a "program" to it, say that something is "good" or "bad" for it, that it organizes "itself," that it is in a "homeostatic" equilibrium, and that its components (organs) have "functions" for the whole

Problems of the Odumian Theory of Ecosystems 119

(and vice versa). The organism *itself* "defines" the reference state to which we refer with all those statements. For proving that ecosystems can be regarded as organisms, one has to show that also in their case such a teleological judgment (although only as regulative idea) is necessary. But this has not been shown yet and is not expected either (Weil 2004; Keller 2008, 2009; Trepl and Voigt 2011).

Admittedly, ecosystems are not organisms, but we can ascribe functions to them and their components: The latter have certain functions, e.g., for the materials cycle or transformation and/or storage of energy. But then one leaves, in a way, the level of science and enters the level of technology. The "ecosystem" is determined by external purposes (and is a *means* for those). The "materials cycle" and/or the "flow of energy" are purposes *that are determined by us*, i.e., the ecosystem is an *artifact* (Voigt 2009, 235; Trepl and Voigt 2011, 72)[23]—a technological system. This point is fundamental to understand what we mean when we talk about "*ecosystem services.*" What these "services" and thereby the "ecosystems" *are* depends on our *interests* and, thus, is not something natural but at least something *political*.

6.4 HOWARD ODUM'S PHYSICALISTIC UNDERSTANDING OF ECOSYSTEMS

For Howard Odum, ecosystems are *physical* systems. Thus, he follows Tansley's intention, which describes ecosystems as "one category of the multitudinous physical systems of the universe" (Tansley 1935, 299). Odum's understanding of ecosystems (resp. of biological systems in general) is greatly influenced by the work of Alfred Lotka (primarily by his 1925 book *Elements of Physical Biology*).[24] Odum's works on the thermodynamics of ecosystems as well as his representation of energy flows in ecosystems (the so-called *energese*) were very influential for the ecology of ecosystems.[25] In contrast to other ecologists who worked on physical descriptions of ecological systems as well, Howard Odum converted everything to "energy" (Taylor 1988, 230; Hagen 1992, Chapter 7; Golley 1993, 107; Mansson and McGlade 1993).[26] Insofar as he not only describes biological systems physically, but, furthermore, claims that this description represents an all-encompassing (sufficient) explanation of such systems (see the following), his view can be declared to be *physicalistic*. In the following, I will give a short summary of some focal aspects of his theory. Thereupon I will especially go into two points: the limits of a physical description of biological systems as dissipative structures and the limits of the concept of *information* Odum uses.

1. Odum starts his remarks concerning the "energy in ecological systems" with the classic (physical) definition of "energy": "Energy is defined as the ability to do work" (E. P. Odum 1971, 37). After that, he points out the importance of the first and the second laws of thermodynamics for the understanding of biological systems (E. P. Odum 1971, 37). He states that "[o]rganisms, ecosystems and the entire biosphere possess the essential thermodynamic characteristic of being able to create and maintain a high state of internal order, or a condition of low entropy … Low entropy is achieved by a continual dissipation of energy of high utility (light or food,

for example) to energy of low utility (heat, for example). In the ecosystem, 'order' in terms of complex biomass structure is maintained by the total community respiration which continually 'pumps out disorder' " (E. P. Odum 1971, 37). Thus, Odum regards ecosystems (and biological systems in general) as a certain kind of physical system: as dissipative structure. I will explain what is meant by this in Section 6.5.1.

2. For Odum, "[t]he variety of manifestations of life are all accompanied by energy changes … The essence of life is the progression of such changes as growth, self-duplication, and synthesis of complex relationships of matter" (E. P. Odum 1971, 37). On the other hand, he does not differentiate between living and nonliving systems as for both the same (thermodynamic) laws apply.[27] Obviously, at this point, Odum's statement is quite inconsistent: On the one hand, thermodynamic laws seem to be essential for the understanding of living systems; on the other hand, the difference between living and nonliving systems is negligible. I will show below that, within the framework of a thermodynamic theory, nothing can be said about the difference between living and nonliving systems.

3. For Odum, the main feature of an ecosystem is its *productivity*, i.e., the transformation of energy into *biomass* and its storage in that form (E. P. Odum 1971, 43 pp.). He uses the relationship between *production* and *respiration* for classifying ecosystems as well as their states of succession (E. P. Odum 1971, 252). Furthermore, he classifies the biotic components with regard to their *function* within the "trophic chain" and associates them with different trophic levels: *producers*, *consumers*, and *decomposers*. Several times, he emphasizes "that *this trophic classification is one of function and not of species as such*" (E. P. Odum 1971, 63).[28] Thus, Odum abstracts from the various properties and the specific quality of the biotic components of the ecosystem (individuals of certain species) and *reduces* them to their "trophic," i.e., energetic function.[29] I will show that species, as such, not only do not matter in his theory, but cannot even be articulated by it.

4. With reference to the writings of Alfred Lotka, Odum suggested a law that describes the "development" of ecosystems thermodynamically: the "maximum power principle" (Odum and Pinkerton 1955). "The theory of maximum power control of self-organization suggests survival of those combinations of components that contribute most to the collective power of their system" (Odum 1994, 451).[30] Thus Odum introduces a different teleological principle than his brother Eugene: the development of ecosystems does not result in a state of maximum diversity (i.e., inner differentiation) but in a state of maximum power, i.e., a state in which the system performs as much work per unit of time as possible.

5. To measure the "complexity" ("order") of biological systems, Howard Odum uses the concept of entropy and information, respectively.[31] In so doing, he relates two different concepts: the thermodynamic concept of entropy (by Ludwig Boltzmann) and the concept of information (by Claude Shannon) used in communication theory (Odum 1994, Chapter 17). For him, "energy" and "information" are related fundamentally.[32] In that point,

Problems of the Odumian Theory of Ecosystems

he follows a course that is common in system theory; however, it is quite problematic as I will show. Apart from this, Odum summarizes fairly (actually categorically) different things with the concept of information: "Flows of genes, books, television communications, computer programs, human culture, art, political interactions, and religious communications are examples of information" (Odum 1994, 19; Odum 2007, 87).

6.5 LIMITS OF HOWARD ODUM'S UNDERSTANDING OF ECOSYSTEMS

6.5.1 BIOLOGICAL SYSTEMS AS DISSIPATIVE STRUCTURES

Odum views ecosystems and biological systems in general as *dissipative structures*. I will briefly explain what is meant by this: the first law of thermodynamics deals (to put it simply) with the *quantity* of energy within a system, the second law with the *quality* of different forms of energy. According to the first law, energy can "neither be created nor destroyed." However, it is possible to *transform* energy. According to the second law, through each transformation, "useful" energy (meaning energy that can be used to do work) is converted to "useless" energy (which cannot be used to do work anymore), e.g., heat. In other words, energy is *dissipated* due to each transformation. The measure of the amount of the "useless" energy in a system is its *entropy*.[33] Entropy is produced *inevitably* as long as a system is not in thermic equilibrium with its surroundings (that is, a state of maximum entropy).

For us, the following is relevant: The increase of entropy in a system can only be compensated for by adding (continuously) energy to the system that is poor in entropy and by emitting energy that is rich in entropy (e.g., heat) to the surroundings of the system. Energy that is poor in entropy is also called *exergy* (it is defined as energy by which work can be done). In other words, a state of low entropy, i.e., high exergy (a "distance" to the thermodynamic equilibrium) can be maintained in a system only by consuming entropy-poor energy, meaning by the (continuous) *dissipation of exergy* (and by the simultaneous output of entropy-rich energy to its surroundings). Systems that are capable of doing that (and that emerge in the course of self-organizing processes) are called *dissipative structures*.[34] The classic example of dissipative structures are the so-called Bénard cells, but also tornados, hurricanes, and cyclones, or simply the vortex emerging in the bathtub can be counted among them. During the emergence of such structures, a system shifts from an "unorganized" to an "organized" stage. In physics, this is regarded as a certain kind of *phase transition*.[35]

Organisms exhibit some characteristic features as dissipative structures. Bénard cells or cyclones evolve due to an energy gradient (differences of temperature and/or pressure). Thereby they show a specific organization (convection cells, vortex, etc.), which is kept as long as the energy gradient does not exceed or go below a critical value. They contain a *certain* content of exergy. Because of the phase transitions, the content changes *discontinuously*. The exergy of an organism (i.e., its biomass), in contrast, increases *continuously* in the course of its ontogenesis until it reaches a maximum in the adult stage. But principally, for the emergence of an organism, not only an external energy gradient is necessary, but a *germ* (in which the program of its development is inherent) is needed. The emergence and existence of the germ can in no way be explained by the energy gradient. Thus, using the theory of dissipative structures, one can point to a *necessary condition* of living creatures: that they must take up energy that is poor in entropy and release energy that is rich in entropy. However, this is not a *sufficient explanation of their mode of existence* (as is often suggested).[36] The peculiarity of living systems cannot be described within the framework of a physical theory because the concept of function or purpose (see Section 6.3) cannot be used meaningfully for physical objects.[37] Thus, one cannot say that it is the "purpose" (or the "function") of water molecules to build a convection cell or a cyclone or that it is their purpose to transfer energy more efficiently (unless one *defines* the structure or the effect to be the purpose).

6.5.2 THE CONCEPT OF INFORMATION USED BY HOWARD ODUM

Besides the physicalistic reduction of biological systems to dissipative structures, another problem of Howard Odum's theory is his use of the concepts of *entropy* and *information*. The problem, however, did not occur with him but rather with the "fathers" of the information theory: Claude Shannon and Warren Weaver. It is not a specific problem of the ecosystem theory but of system theory in general. Even today, it pervades the literature and experts in the subject of information and complexity theory seem to be confused by it. The problem has two aspects: (1) *The equalization of the thermodynamic concept of entropy* (introduced by Boltzmann) *with the concept of information used in communication theory* (introduced by Shannon) and (2) *the equalization of the syntactic and semantic dimension of the concept of information*. Before going into these points, I want to outline, briefly, what Boltzmann and Shannon meant by "entropy" and "information," respectively.

Boltzmann's entropy: $S = k \ln W \, [\text{J/K}]$

W is a measure of the number of microstates (distribution of elements) that realize a certain macrostate of a system; k denotes the Boltzmann constant.[38] The entropy S is proportional to W, which means S is higher if a certain macrostate is realized through many different microstates. The classical example is a gas-filled box: The number of microstates that realize the macrostate "all molecules in one half of the box" is smaller than that realizing the macrostate "all molecules distributed evenly."[39] Thus, in the second case, the entropy is higher.

Shannon's information: $I = \log_2 W \, [\text{bit}]$

Here too, W measures the possible states of a system. However, Shannon's concept did not refer to physical systems but to texts and codes, i.e., to the sequence and distribution of *signs* (e.g., letters, numbers, etc.). He was interested in the capacity of certain communication channels (e.g., a telegraph). The information of a system that can be found in two states (e.g., a coin) is I = 1 bit. If it can be found in just one state, then one has $I = 0$ bit (meaning that no information can be transported by the system). Shannon, therefore, introduced a measure for distinguishable states of a system, i.e., for its capacity to transport information (respectively for the number of distinctions that are needed to describe the system). Considering a system of n elements that can be assigned to i distinct symbols (e.g., the letters of the alphabet or different biological species) that occur with the frequency p_i; the specific content of information per symbol is[40]

$$H = -\sum_{i=1}^{n} p_i \log_2 p_i$$

Here we come to the first aspect of the problem mentioned above. Shannon also named this measure H "entropy" (Shannon and Weaver 1963, 51) and, therefore, suggested not only a *formal* but a *substantial* relationship to Boltzmann's "entropy." This "fundamental" relationship between thermodynamics and information theory was propagated above all by Leon Brillouin (1956). Odum and his followers (but also a vast number of system theorists) have adopted it, which has been a serious mistake. Shannon mentioned the formal similarity of his concept to Boltzmann's entropy (Shannon and Weaver 1963, 51). To call it the same, however, is not justifiable from a scientific point of view because no substantial similarity exists between the two concepts.[41] Shannon's naming is only based on the formal similarity *and* a strategic piece of advice by John von Neumann who counseled, "Nobody knows what entropy really is, so in a debate you will always have the advantage" (Janich 2006, 57; Thims 2012, 2). Thus, Shannon opened a "Pandora's box of intellectual confusion" (Danchin 2001, cited in Thims 2012, 2). Peter Janich called it a "coup de main" ("Husarenstück"), a "usurpation" and "slyness" ("Schlitzohrigkeit") (Janich 2006, 57). Being unaware of the real story, even reputable researchers have attempted to give this "fraud" ("Etikettenschwindel") a scientific fundament (Janich 2006, 57; Thims 2012). The question why this is all so attractive to such a huge mass of scientists is mysterious. Perhaps the reason is that mind ("information") and matter ("energy") seem to be linked by the equalization, and therefore, the old Descartian dualism between *res cogitans* and *res extensa* seems to be solved once more. But it is not that simple.

Thereupon, we come to the second equalization, that of the syntactic and the semantic dimension of "information." In information theory, it is common to distinguish between the *syntactic*, the *semantic*, and the *pragmatic* dimensions of "information."[42] On the syntactic level (where Shannon's "entropy" is located), the concept refers to the sequence, the frequency, and the distribution of (distinguishable) signs (e.g., of letters in a text or of individuals of different species in an ecosystem). On this level, "information" is *quantifiable*, i.e., it is possible to say that an information bearer contains a certain *amount* of information. The *meaning* of the signs is, however, meaningless on this level. No statement is possible about what a text *is about*—about its *sense*. Therefore, the text must be *interpreted*, meaning one has to shift to the *semantic* level of the words and sentences (Capurro 2000, III). Concerning semantic problems, Shannon's concept of information has no validity.[43] The pragmatic dimension of information should be mentioned just for the sake of completeness. On this level, information is not only interpreted with respect to its (semantic) meaning but, furthermore, as *a means to an end*, i.e., with regard to possible *actions*. An example: on

124   Revising Green Infrastructure

the syntactical level, the statement "here is an apple" is just a sequence of letters; on the semantic level, it can be a plain statement about the fruit or the brand; and on the pragmatic level, it can be advice to pick or eat the apple in order to appease hunger or throw it at someone.

For our subject, that has the following meaning: the syntactical concept of information introduced by Shannon and used by Odum and his followers can be utilized to quantify the amount of "information" in a certain system and, thus, to make systems comparable in this respect. However, a difference between two texts, e.g., a cookbook and Kant's *Critique of Pure Reason*, cannot be made if the amount, distribution, and frequency of signs (e.g., letters) are the same in both. The same applies to different species or ecosystems: a mouse cannot be distinguished from a frog by means of the syntactical concept of information as long as both exhibit the same number and distribution of distinct elements (e.g., different genes or cells).[44] The same is true, e.g., for a coral reef and a tropical rainforest as long as both contain the same number of individuals belonging to the same number of distinct species with similar distributions. And if the distribution, etc., of elements is the same in the frog and Kant's *Critique,* no difference can be made between them either. Shannon's "information" measures a *quantity*, and that is its *only quality* (meaning). Moreover, that means as well that certain ecological objects are not only not distinguishable in Howard Odum's theory, but they are not *comprehensible* at all because such objects must be understood as *historic singularities*. Biological species as well as specific combinations of them ("communities") count among this class of objects.[45]

6.6   NATURALISMS OF THE ODUM BROTHERS

Both Eugene and Howard Odum not only describe ecological systems with their theories but also make statements about "economy," "politics," "art," and "religion," that is, about "culture." Both transfer their conception of the ecosystem to human society but thereby follow different ideologies.

Eugene Odum (1971, 270) states that the development of ecosystems "has many parallels in the development of human society itself." For him, "civil rights," "education," and "culture" are symbiotic relationships that emerge as a result of society's (organic) development.[46] "The principles of ecosystem development bear importantly on the relationships between man and nature" although the "goals" of the respective developments "often conflict" (E. P. Odum 1971, 267). Nevertheless, "recognition of the ecological basis for this conflict between man and nature is a first step in establishing rational land-use policies" (E. P. Odum 1971, 267). Odum emphasizes that the "populations of men like other populations, are part of biotic communities and ecosystems" (E. P. Odum 1971, 513). For him, this point is "basic to modern ecology" (E. P. Odum 1971, 35). Indeed "men" are "more dominant" (E. P. Odum 1971, 514); the difference from other species is, however, just a gradual one: "his ability to control his immediate surroundings, and his tendency to develop culture … are greater than those of other organisms" (E. P. Odum 1971, 512). For Odum, recognition of the principles that govern ecosystems as well as human society results in a political claim: "*The time has come for man to manage his own population as well as the resources on which he depends*" (E. P. Odum 1971, 510). The ecological principles he suggested should

Problems of the Odumian Theory of Ecosystems

serve therefore (E. P. Odum 1971, 510). Yet "management" is limited: "We cannot take over management of everything!" (E. P. Odum 1971, 35). "Complete domination of nature is probably not possible and would be very precarious, or unstable" (E. P. Odum 1971, 514). Therefore, his relationship to technology is ambivalent: "Technology is, of course, a two edged sword; it can be the means of understanding the wholeness of man and nature or of destroying it" (E. P. Odum 1971, 6).

Howard Odum exceeds the realm of ecology with his theory as well. "Where humans comprise a major part [of ecosystems], new kinds of systems evolve with human culture at the hierarchical center. Information processing, social structure, symbolism, money, political power, and war become important components along with the vegetation, consumer organisms, and the inanimate work of the biosphere" (Odum 1994, 508). For him, "culture" and "humans" are "programs" that are "programmed": "Apparently, humans evolved from earlier animal stages with a fundamentally greater ability to be reprogrammed by environmental conditions. This ability allowed humans to become the ecosystem's computer program ... Humans become the main programming entities for new kinds of ecosystem[s] such as systems of hunting ... Emerging with the new ability to program were new mechanisms of controls, such as economic motivations, family, government, and religion. The social structure and programs are called *culture*" (Odum 1994, 508). Thus, for Odum, "humanity and nature" and even "religion" can be studied using his energy theory (H. T. Odum 1971).[47] His relationship to technology is much more optimistic. In contrast to his brother, who emphasizes the necessity of conserving natural ecosystems (which he regards as models for the organization of human society), Howard Odum states, "new kinds of landscapes and interface ecosystems need to evolve with the help of ecological engineering" (Odum 2007, 9). These "new" kinds of systems he calls "technoecosystems."[48]

6.7 CRITIQUE OF ODUMIAN NATURALISM

For the Odum brothers, ecology is a special kind of science because it is not only concerned with statements about certain *scientific* objects (ecosystems) but obviously also with *political* claims and *technological* interventions, which are deduced from the former. Such a deduction of *normative* instructions from *descriptive* statements is called the *naturalistic fallacy* in philosophy (Moore 1922). When theories about nature are used to state something about human society, with which the assumption of *actions*, *purposes,* and, thus, *subjects* is indispensable for an understanding, one speaks of *naturalism*. If the theory is a scientific one, the naturalism is called *scientistic* (Keil 2010, 155). On this point, another difference between the two brothers becomes apparent because Eugene uses a scientifically untenable model (that of the superorganism) to describe societies, while his brother invokes a physical theory (*the* physical model—the "mechanism") to explain biological and, furthermore, social systems.[49]

For Hubig and Luckner (2008, 53), the main problem of such approaches is "that the viewpoint from which the naturalistic determinations of man are taken remains itself undeclared."[50] These approaches are "monistic" as Hubig (2011, 202) states elsewhere, i.e., they "are seeking for *the one and only* reason."[51] All approaches of that kind, Hubig (2011, 203) continues, have the problem of explaining "how the relationship of the mind that models such relationships to the world, while being itself part of that world, can be conceived as part of the model."[52] That means the philosophical problem is, to put it simply, that naturalists (such as the Odum brothers) do not reflect on their own (epistemic) position. But what does that mean? And what exactly is the problem?

At this point, something must be said about the relationship of descriptive (scientific) and normative (e.g., political) statements (i.e., values). Generally, science and technology are regarded as "value-neutral." That is right in a certain respect, but nevertheless, they are not free of values at all. Hubig (2007, 61) states, "Value-neutral is technology [and science] as far as it can be regarded as a necessary condition for disposal in relation to values in general."[53] However, this "value-neutrality" itself is a "higher-level value" (Hubig 2007, 61). This fundamental value of science is that it is directed toward the production of so-called *instrumental* knowledge (Habermas 1969), i.e., toward *regulation* and *control* (Heidegger 1962; Hubig 2006, 19 f.) of systems. That kind of knowledge enables the "mastery of nature," that is, obtaining *power* over it (Hubig 2007, 60 ff., 205 ff.). Exactly that and nothing else can be accomplished by science (and technology).

Apart from this general value, scientific theories (and, therefore, the objects they refer to) are also structured by *specific* cultural values. These are part of the theory's "paradigm" (Kuhn 1976) and belong to the "hard core" (Lakatos 1970) of the research program, meaning that they can neither be verified nor falsified (Lakatos 1970). They constitute "implicit philosophies," "the unconscious of science" (Foucault 1974, 11). For example, it can be shown that theories of synecological systems are structured by certain political values (Voigt 2009). The values that structure, for instance, organismic theories (such as Eugene Odum's) derive from a conservative ideology that can be traced back to the philosophy of Gottfried Wilhelm Leibniz (especially his "Monadology") (Eisel 1991; Cheung 2000; Eisel 2005; Kirchhoff 2007). By contrast, it can be shown that thermodynamics (and physics in general) are structured by *economic* values (more precisely, by those of the industrial capitalism), meaning that through such theories (as Howard Odum's) nature is regarded *inherently* from an economic point of view (Eisel 2009).

One can say that, through the scientific method, cultural values are "projected" into nature and, thus, placed at the disposal of technology. The cultural constitution of scientific theories is inevitable. The problem of naturalistic (scientistic) theories is that they are not reflected concerning this constitution ("their viewpoint") but neglect this relationship by declaring "culture" as a natural matter (Eisel 2009). And precisely this point is dangerous because, apart from the fact that "reason" is thereby lost, naturalistic approaches can serve as a basis for ideologies that take their implicit values for granted and thus will (or can) not understand or accept a different position.

6.8 CONCLUSION

What is the use of the Odum brothers' theories for landscape architecture? First, it should be summarized that (1) an ecosystem (and thereby a landscape as a physical system) cannot be regarded as an organism. What it *is* depends on how we define it. The organism, in contrast, "defines" itself. (2) By the theory of dissipative structures, one can refer to a necessary condition of living creatures but not give a sufficient

Problems of the Odumian Theory of Ecosystems

explanation of their mode of existence. (3) When talking about "information," one must distinguish between (a) the thermodynamic concept of entropy and the concept of information used in communication theory and (b) the syntactic and semantic (and pragmatic) dimensions of information. (4) The Odum brothers take up a naturalistic point of view that is quite problematic in a philosophical respect. Apart from these theoretical and metaphysical shortcomings, the theories of the Odum brothers can serve (if anything) as a basis for the production of so-called *instrumental* (technological) knowledge (as it is valid for science in general). However, for landscape architecture, such knowledge is undoubtedly *useful* but utterly not sufficient because architecture, as such, must go beyond science and technology. Therefore, another kind of knowledge is required, that is, the so-called *reflexive* knowledge (Trepl 1997; Hubig 2013). When Eugene Odum states, "The landscape is not just a supply depot but is also the *oikos*—the home—in which we must live" (E. P. Odum 1971, 267), he cuts right to the chase of the matter. But, being scientists (and engineers), the Odum brothers can indeed say something about the "supply depot" but unfortunately nothing essential about that "home" (Heidegger 1951, 1962; Hubig 2013).

ACKNOWLEDGMENTS

I want to thank Julia Steil, Ludwig Trepl, Sören Schöbel, Daniel Czechowski, Thomas Hauck, and Tina Schrier for their constructive comments and further patient help.

NOTES

1. See, for example, Chapter 5 by Roncken et al. and Chapter 4 by Perrotti (in this book), Stremke and Koh (2010, 2011), and Stremke and van den Dobbelsteen (2011).
2. The book was written by Eugene Odum except for the chapter "Energy in Ecological Systems" that was written by his brother. In the following, I will refer to the third edition of the book (E. P. Odum 1971).
3. Arthur Tansley's work on the ecosystem concept (Tansley 1935), Raymond Lindeman's on the trophic dynamic viewpoint in ecology (Lindeman 1942), George Hutchinson's theory of biogeochemical cycles (Hutchinson 1947), and Howard Odum's work on thermodynamics of ecosystems (e.g., Odum and Pinkerton 1955; Odum 1956) should be mentioned here.
4. Practically, the ideas of the brothers cannot be distinguished easily because the ideas of Howard influenced ecology mainly indirectly through the writings of his brother (Hagen 1992, 124).
5. Recently the antagonism can be seen in the "energy transition" (in Germany) that shows the conflict between progressive "eco-technocrats" and conservative "species conservationists" (Trepl 2013).
6. Clements (1916, 2) wrote, "As an organism the formation arises, grows, matures, and dies. Its response to the habitat is shown in processes or functions and in structures which are the record as well as the result of these functions. Furthermore, each climax formation is able to reproduce itself, repeating with essential fidelity the stages of its development. The life history of a formation is a complex but definite process, comparable in its chief features with the life history of an individual plant." Kirchhoff (2007) and Voigt (2009) showed that the Clementian theory is no typical example for the organismic view of synecological systems.

7. The organismic holism that Phillips presents in his writings reminds Tansley "irresistibly of the exposition of a creed—of a closed system of religious or philosophical dogma. Clements appears as the major prophet and Phillips as the chief apostle" (Tansley 1935, 285).
8. For example, Sven E. Jørgensen, Bernard C. Patten, William J. Mitsch, and Enzo Tiezzi (just to name a few) (Mitsch and Jørgensen 1989; Jørgensen et al. 1992; Jørgensen 2006; Jørgensen et al. 2007; Tiezzi 2011).
9. "*Homeostasis* ... is the term generally applied to the tendency for biological systems to resist change and to remain in a state of equilibrium" (E. P. Odum 1971, 34).
10. Elsewhere, Odum emphasizes (with reference to Clements) that the development of ecosystems is not only a "succession of species," but a "development process" (Odum 1999, 304).
11. Odum himself contradicts the accusation that he views ecosystems as superorganisms: "The analogues of goals in teleological systems can be discovered indeterminate ecosystems if it would be useful, as it is for organisms, to regard ecosystems as purposeful. In general, however, this is not the case, and teleological metaphors that are sometimes applied to ecosystems (e.g., Odum 1969) are only that. The feedback control model ... can be fit to ecosystems, but it is not really needed, and the determinate model ... is more consistent with a strictly mechanistic view of nature which science favors at the present time." (Patten and Odum 1981, 889).
12. "To liken an ecosystem to a Clementsian (super)organism is fundamentally wrong and misleading, because an organism has an outside boundary (skin, epidermis, cell wall etc.) produced by itself and accepted by the researcher" (Haber 2011, 219).
13. Tansley already pointed this out. For him, ecosystems are "mental isolates for the purposes of study" (Tansley 1935, 300).
14. In complexity theories, one also speaks about systems moving along a *trajectory* (e.g., Ashby 1956, Chapter 2).
15. Gleason already invoked these arguments against Clements (Gleason 1917, 1939)·
16. In this regard, Ashby (1962) is also quite revealing. A very good overview of what "self-organization" can mean with regard to ecosystems is given by Solé and Bascompte (2006).
17. Concerning this difference, see, e.g., Halley and Winkler (2008).
18. For the original definition of homeostasis, see Cannon (1929). To receive more detailed information on this context, see Weil (1999) and Trepl (2005, 423f, 470).
19. One should notice that it is not the ecosystem but its components that control the concentration.
20. See, e.g., Haken (1980). Especially concerning ecosystems, see, e.g., Solé and Bascompte (2006, 144).
21. Simplified, one could say the dispositional concept of function is about *one* function, the etiological concept about *the* function of a function bearer.
22. By the teleological judgment, however, nothing can be *explained* for the purpose of a causal explanation. It only has *regulative* relevance. Being a science, biology also has to aim at such a causal explanation. Fundamental in this context is Kant's concept of the "natural purpose" ("Naturzweck") (Kant 1974, § 65): According to Kant, we must assume that the organism does not have an *external* purpose (like an artifact) but an end in itself (McLaughlin 1989).
23. At this point, one will argue that *natural* ecosystems are not artifacts because they were not constructed by us. But this does not play a role in this context. In fact, it is crucial that we mentally make them artifacts by attributing functions to them and therefor define them as a means to an end (McLaughlin 2001, Chapter 3).
24. Lotka's writings are viewed to be classic in regard to a physical description of biological systems.

Problems of the Odumian Theory of Ecosystems 129

25. "H. T. Odum's monographic study of Silver Springs in 1957 replaced Lindeman's Cedar Bog Lake as the paradigmatic ecosystem report and his diagrams of energy flow through the spring were widely republished in textbooks and research papers" (Golley 1991, 132). In this regard, the following writings of Odum are counted among the most important: Odum and Odum (1955), Odum and Pinkerton (1955), and Odum (1956, 1957).

26. Odum himself denies this accusation (Odum 1995).

27. "The relationships between producer plants and consumer animals, between predator and prey, not to mention the number and kinds of organisms in a given environment, are all limited and controlled by the same basic laws which govern nonliving systems, such as electric motors or automobiles" (E. P. Odum 1971, 37).

28. "It is important to emphasize again that the *concept of trophic level is not intended for categorizing species*" (E. P. Odum 1971, 66).

29. Ditto for the combination of species in ecological "communities": "Communities not only have a definite functional unit with characteristic trophic structures and patterns of energy flow but they also have compositional unity in that there is a certain probability that certain species will occur together. However, species are to a large extent replaceable in time and space so that functionally similar communities may have different species compositions" (E. P. Odum 1971, 140).

30. Odum considers his law to be "a better and earlier, if reverse but equivalent statement, to Prigogine's *building structure by maximizing dissipation*" (Odum 1995: 521). Odum's law has been reformulated by his scholars in different variations (for example, by Jørgensen 2000).

31. "Complexity … is measured by the information content …, and this measure is entropy" (Odum 1994, 320).

32. "The amount of complexity and information was found to be a logarithmic function of the energy" (Odum 1994, 320).

33. It should be noticed that the entropy itself is *not a form of energy,* but rather describes the amount of energy in a system that cannot be used for doing work at a certain temperature. One only has to regard the unit to understand that it is not indicated as "Joule" [J] but as "Joule/Kelvin" [J/K].

34. See, for example, Prigogine (1978, 1988).

35. Bénard cells can be produced easily: Take a petri dish, fill it with water, and heat it from below. Beyond a critical temperature, an alveolate structure emerges inside the dish (the Bénard cells). The emergence of the structure is linked to a more efficient heat transfer: Below the critical temperature, the heat is transferred by *conduction*, above by *convection*. The evolving structures are thus also called *convection cells.*

36. A very famous example is Erwin Schrödinger's *What Is Life?* (Schrödinger 1944). It is furthermore irrelevant here that, in the "genesis of life" (meaning in the emergence of self-reproducing molecules), an external gradient of energy was maybe an indispensable condition and that maybe it offers a sufficient explanation of the phenomena.

37. "For the fact remains that agency, function, and purpose are all conspicuously absent from the kinds of systems with which physics deals, and no one has succeeded in offering an account of how they might emerge from the dynamics of effectively homogeneous systems of simple elements, however, complex the dynamics of their interaction might be." (Keller 2007, 310).

38. It indicates the energy of a particle per degree Kelvin and has the value 1.38×10^{-23} J/K.

39. This has stochastic reasons and can be easily understood by a respective textbook.

40. This formula underlies the so-called Shannon Index that is used for calculating "biological diversity." The only difference is that, in the Shannon Index, the natural logarithm *ln* is used. The value of p_i corresponds to the frequency of the species i in a system of n species.

41. As Thims puts it, "*Information theory*—the mathematical study of the transmission of information in binary format and or the study of the probabilistic decoding of keys and cyphers in cryptograms—is *not* thermodynamics! This point cannot be overemphasized enough, nor restated in various ways enough. Information theory is not statistical mechanics—information theory is not statistical thermodynamics—information theory is not a branch of physics. Claude Shannon is not a thermodynamicist. Heat is not a binary digit. A telegraph operator is not a Maxwell's demon. The statistical theory of radiography and electrical signal transmission is NOT gas theory. The equations used in communication theory have **absolutely nothing** to do with the equations used in thermodynamics, statistical mechanics, or statistical thermodynamics. The logarithm is not a panacea. Boltzmann was not thinking of entropy in 1894 as 'missing information.' Concisely …: thermodynamics ≠ information theory" (Thims 2012, 1; emphasis in original). See also Janich (2006).
42. The differentiation is attributed to Charles W. Morris who is considered a founder of semiotics (c.f. Capurro 2000, III.2).
43. "Shannon's mathematical theory was only a theory on the syntactical level … but with no reference to the semantic … and pragmatic … levels" (Wersing 1996, 221 cited in Capurro and Hjørland 2003, 291). See also Janich (2006, Chapter 4).
44. Jørgensen uses the number of genes for calculation of the information content of ecosystems (Jørgensen and Svirezhev 2004).
45. Odum himself claimed this not explicitly (but implicitly). He rather emphasizes (as already shown) that he is only interested in "functions," not in species or communities as such (i.e., as historic singularities). His followers, however, claim that the singularity of ecological entities is of crucial importance, e.g., Enzo Tiezzi: "The singularity of an event also becomes of particular importance: if a certain quantity of energy is spent to kill a caterpillar, we lose the information embodied in the caterpillar. But were this the last caterpillar, we should lose its unique genetic information forever. The last caterpillar is different from the n-th caterpillar" (Tiezzi 2011, 2900).
46. "In the pioneer society, as in the pioneer ecosystem, high birth rates, rapid growth, high economic profits, and exploitation of accessible and unused recourses are advantageous, but as the saturation level is approached, these drives must be shifted to considerations of symbiosis (that is, 'civil rights,' 'law and order,' 'education,' and 'culture'), birth control, and the recycling of recourses" (E. P. Odum 1971, 270).
47. Resp. Odum 2007: "A study of humanity and nature is thus a study of systems of energy, materials, money, and information. Therefore, we approach nature and people by studying energy system networks" (Odum 2007, 1). "This chapter shows the systems nature of religion, its place in the energy hierarchy, relationships of religion and science, global pluralism, adaptive God, alternative views of the cosmos, and the kind of religion needed for times ahead" (Odum 2007, 313).
48. "Some formerly wild components of ecosystems may be incorporated into technological systems as hybrids of living units and hardware homeostatically coupled. These might be called *technoecosystems*" (Odum 1994: 526).
49. Thus, basically just the latter can be regarded as a *scientistic* naturalist.
50. Translated by G. H. In the original: "dass die Warte, von der aus die naturhaften Bestimmungen des Menschen getroffen werden, selbst unausgewiesen bleibt."
51. Translated by G. H. In the original: "suchen nach dem *einen* Grund."
52. Translated by G. H. In the original: "Alle Versuche dieser Art stehen unter der Begründungshypothek, (...) zu erklären, wie das Verhältnis desjenigen Geistes, der diese Prozesse eines Weltverhältnisses als Teil der Welt modelliert, als Moment eben des derart Modellierten erfasst werden kann."
53. Translated by G. H. In the original: "Wertneutral' ist Technik, sofern sie als notwendige Bedingung des Disponierens unter Werten überhaupt gelten kann."

Problems of the Odumian Theory of Ecosystems

REFERENCES

Ashby, W. R. (1956). *An Introduction to Cybernetics*. London: Chapman & Hall.

Ashby, W. R. (1962). Principles of the self-organizing system. In *Principles of Self-Organization: Transactions of the University of Illinois Symposium*, ed. H. von Foerster and G. W. Zopf, 255–278. London: Pergamon Press.

Bergandi, D. (1995). "Reductionist holism": An oxymoron or a philosophical chimera of E. P. Odum's systems ecology. *Ludus Vitalis* 3: 145–180.

Brillouin, L. (1956). *Science and Information Theory*. New York: Academic Press.

Cannon, W. B. (1929). Organization for physiological homeostasis. *Physiology Review* 9: 399–431.

Capurro, R. (2000). Einführung in den Informationsbegriff. http://www.capurro.de/infovorl -index.htm (accessed February 21, 2014).

Capurro, R. and B. Hjørland (2003). The concept of information. *Annual Review of Information Science and Technology* 37: 343–411.

Cheung, T. (2000). *Die Organisation des Lebendigen—Die Entstehung des biologischen Organismusbegriffs bei Cuvier, Leibniz und Kant*. Frankfurt am Main: Campus.

Clements, F. E. (1916). *Plant Succession: An Analysis of the Development of Vegetation*. Washington: Carnegie Institution of Washington.

Clements, F. E. (1936). Nature and structure of the climax. *Journal of Ecology* 24: 252–284.

Danchin, A. (2001). *The Delphic Boat: What Genomes Tell Us*. Cambridge, MA: Harvard University Press.

Eisel, U. (1991). Warnung vor dem Leben—Gesellschaftstheorie als "Kritik der Politischen Biologie." In *Industrialismus und Ökoromantik—Geschichte und Perspektiven der Ökologisierung*, ed. D. Hassenpflug, 159–192. Wiesbaden: Deutscher Universitäts-Verlag.

Eisel, U. (1997). Unbestimmte Stimmungen und bestimmte Unstimmigkeiten: Über die guten Gründe der deutschen Landschaftsarchitektur für die Abwendung von der Wissenschaft und die schlechten Gründe für ihre intellektuelle Abstinenz—mit Folgerungen für die Ausbildung in diesem Fach. In *Vor der Tür—Aktuelle Landschaftsarchitektur aus Berlin*, ed. S. Bernard and P. Sattler, 17–33. München: Callwey.

Eisel, U. (2005). Das Leben im Raum und das politische Leben von Theorien in der Ökologie. In *Strukturierung von Raum und Landschaft—Konzepte in Ökologie und der Theorie gesel-schaftlicher Naturverhältnisse*, ed. M. Weingart, 27–42. Münster: Westfälische Dampfpost.

Eisel, U. (2009). Objektivismuskritik. In *Befreite Landschaft—Moderne Landschaftsarchitektur ohne arkadischen Balast*, ed. U. Eisel and S. Körner, 4–76. Freising: Lehrstuhl für Landschaftsökologie, Technische Universität München.

Foucault, M. (1974). *Die Ordnung der Dinge*. Frankfurt am Main: Suhrkamp.

Gleason, H. A. (1917). The structure and development of the plant association. *Bulletin of the Torrey Botanical Club* 44: 463–481.

Gleason, H. A. (1939). The individualistic concept of the plant association. *American Midland Naturalist* 21: 92–110.

Golley, F. B. (1991). The ecosystem concept: A search for order. *Ecological Research* 6: 129–138.

Golley, F. B. (1993). *A History of the Ecosystem Concept in Ecology: More than the Sum of the Parts*. New Haven, CT: Yale University Press.

Haber, W. (2011). An ecosystem view into the twenty-first century. In *Ecology Revisited—Reflecting on Concepts, Advancing Science*, ed. A. Schwarz and K. Jax, 215–227. Dordrecht, Heidelberg, London, New York: Springer.

Habermas, J. (1969). *Technik und Wissenschaft als "Ideologie."* Frankfurt am Main: Suhrkamp.

Hagen, J. B. (1992). *An Entangled Bank—The Origins of Ecosystem Ecology*. New Brunswick: Rutgers University Press.

Haken, H. (1980). Synergetics. *Naturwissenschaften* 67: 121–128.

Halley, J. D. and D. A. Winkler (2008). Consistent concepts of self-organization and self-assembly. *Complexity* 14: 10–17.

Heidegger, M. (1951). Bauen Wohnen Denken. In *Martin Heidegger—Gesamtausgabe, I. Abteilung, Band 7: Vorträge und Aufsätze*, ed. F.-W. v. Hermann, 138–156. Frankfurt am Main: Vittorio Klostermann.

Heidegger, M. (1962). *Die Technik und die Kehre.* Stuttgart: Klett-Cotta.

Hubig, C. (2006). *Die Kunst des Möglichen I—Technikphilosophie als Reflexion der Medialität.* Bielefeld: Transcript.

Hubig, C. (2007). *Die Kunst des Möglichen II—Ethik der Technik als provisorische Moral.* Bielefeld: Transcript.

Hubig, C. (2011). "Natur" und "Kultur": Von Inbegriffen zu Reflexionsbegriffen. In *Homo naturalis—Zur Stellung des Menschen innerhalb der Natur*, ed. A. M. Rabe and S. Rohmer, 202–226. Freiburg, München: Karl Alber.

Hubig, C. (2013). Dialektik des entwerfens. entwurfswissenschaften als reflexion. In *Wissenschaft Entwerfen*, ed. S. Ammon and E. M. Froschauer, 267–285. München: Wilhelm Fink.

Hubig, C. and A. Luckner (2008). Natur, Kultur und Technik als Reflexionsbegriffe. In *Naturalismus und Menschenbild*, ed. P. Janich, 52–66. Hamburg: Felix Meiner.

Hutchinson, G. E. (1947). Circular causal systems in ecology. *Annals of the New York Academy of Science* 50: 221–246.

Janich, P. (2006). *Was ist Information?* Frankfurt am Main: Suhrkamp.

Jørgensen, S. E. (2000). The tentative fourth law of thermodynamics. In *Handbook of Ecosystem Theories and Management*, ed. S. E. Jørgensen and F. Müller, Washington, DC: Lewis.

Jørgensen, S. E. (2006). An integrated ecosystem theory. *European Academy of Science.* Liège: EAS Publishing House: 19–33.

Jørgensen, S. E., B. D. Fath, S. Bastianoni et al. (2007). *A New Ecology—Systems Perspectives.* Oxford, Amsterdam: Elsevier.

Jørgensen, S. E., B. C. Patten, and M. Straškraba (1992). Ecosystems emerging: Toward an ecology of complex systems in a complex future. *Ecological Modelling* 62: 1–27.

Jørgensen, S. E. and Y. M. Svirezhev (2004). *Towards a Thermodynamic Theory for Ecological Systems.* Oxford: Elsevier.

Kant, I. (1974). *Kritik der Urteilskraft.* Frankfurt am Main: Suhrkamp.

Keil, G. (2010). Naturalismuskritik und Metaphorologie. In *Information und Menschenbild*, ed. M. Bölker, M. Gutman and W. Hesse, 155–171. Berlin, Heidelberg, New York: Springer.

Keller, E. F. (2007). The disappearence of function from "self-organizing systems." In *Systems Biology*, ed. F. C. Boogerd, F. J. Bruggeman, J.-H. S. Hofmeyr and H. V. Westerhoff, 303–317. Amsterdam, Oxford: Elsevier.

Keller, E. F. (2008). Organisms, machines, and thunderstorms: A history of self-organization, part one. *Historical Studies in the Natural Sciences* 38: 45–75.

Keller, E. F. (2009). Organisms, machines, and thunderstorms: A history of self-organization, part two. Complexity, emergence, and stable attractors. *Historical Studies in the Natural Sciences* 39: 1–31.

Kirchhoff, T. (2007). *Systemauffassungen und biologische Theorien: Zur Herkunft von Individualitätskonzeptionen und ihrer Bedeutung für die Theorie ökologischer Einheiten.* Freising: Lehrstuhl für Landschaftsökologie, Technische Universität München.

Kuhn, T. S. (1976). *Die Struktur wissenschaftlicher Revolution.* Frankfurt am Main: Suhrkamp.

Lakatos, I. (1970). Falsification and the methodology of scientific research programmes. In *Criticism and the Growth of Knowledge*, ed. I. Lakatos and A. Musgrave, 91–196. Cambridge: Cambridge University Press.

Lindeman, R. L. (1942). The trophic-dynamic aspect of ecology. *Ecology* 23: 399–418.

Lotka, A. J. (1925). *Elements of Physical Biology.* Baltimore, MD: Williams & Wilkins Company.

Problems of the Odumian Theory of Ecosystems 133

Mansson, B. A. and J. M. McGlade (1993). Ecology, thermodynamics and H. T. Odum's conjectures. *Oecologia* 93: 582–596.

Maturana, H. and F. Varela (1987). *Der Baum der Erkenntnis.* München: Goldmann.

Maturana, H. R. (1975). The organization of the living: A theory of the living organization. *International Journal of Man-Machine Studies* 7: 313–332.

McIntosh, R. P. (1980). The background and some current problems of theoretical ecology. *Synthese* 43: 195–255.

McLaughlin, P. (1989). *Kants Kritik der teleologischen Urteilskraft.* Bonn: Bouvier.

McLaughlin, P. (2001). *What Function Explains—Functional Explanations and Self-Reproducing Systems.* Cambridge University Press: Cambridge.

McLaughlin, P. (2005). Funktion. In *Philosophie der Biologie*, ed. U. Krohs and G. Toepfer, 19–35. Frankfurt am Main: Suhrkamp.

Mitsch, W. J. and S. E. Jørgensen (1989). *Ecological Engineering: An Introduction to Ecotechnology.* New York: Wiley.

Moore, G. E. (1922). *Principia Ethica.* Cambridge: Cambridge University Press.

Odum, E. P. (1953). *Fundamentals of Ecology.* Philadelphia, PA: Saunders.

Odum, E. P. (1971). *Fundamentals of Ecology.* Philadelphia, PA: Saunders.

Odum, E. P. (1999). *Ökologie—Grundlagen, Standorte, Anwendung.* Stuttgart, New York: Thieme.

Odum, H. T. (1956). Efficiencies, size of organisms, and community structure. *Ecology* 37: 592–597.

Odum, H. T. (1957). Trophic structure and productivity of Silver Springs, Florida. *Ecological Monographs* 27: 55–112.

Odum, H. T. (1971). *Environment, Power and Society.* New York: John Wiley & Sons.

Odum, H. T. (1994). *Ecological and General Systems—An Introduction to Systems Ecology.* Niwot, CO: University Press of Colorado.

Odum, H. T. (1995). Energy systems concepts and self-organization: A rebuttal. *Oecologia* 104: 518–522.

Odum, H. T. (2007). *Environment, Power and Society for the Twenty-First Century: The Hierarchy of Energy.* New York: Columbia University Press.

Odum, H. T. and E. P. Odum (1955). Trophic structure and productivity of a windward coral reef community on Eniwetok Atoll. *Ecological Monographs* 25: 291–320.

Odum, H. T. and R. C. Pinkerton (1955). Time's speed regulator—The optimum efficiency for maximum power output in physical and biological systems. *American Scientist* 43: 331–343.

Patten, B. C. and E. P. Odum (1981). The cybernetic nature of ecosystems. *The American Naturalist* 118: 886–895.

Phillips, J. (1934). Succession, development, the climax, and the complex organism: An analysis of concepts. part I. *Journal of Ecology* 22: 554–571.

Phillips, J. (1935a). Succession, development, the climax, and the complex organism: An analysis of concepts: part II. Development and the climax. *Journal of Ecology* 23: 210–246.

Phillips, J. (1935b). Succession, development, the climax, and the complex organism: An analysis of concepts: part III. The complex organism: Conclusions. *Journal of Ecology* 23: 488–508.

Prigogine, I. (1978). Zeit, Struktur und Fluktuationen. *Angewandte Chemie* 90: 704–715.

Prigogine, I. (1988). *Vom Sein zum Werden.* München: Piper.

Schrödinger, E. (1944). *What Is Life?* Cambridge, MA: Cambridge University Press.

Sendova-Franks, A. B. and N. R. Franks (1999). Self-assembly, self-organisation and division of labour. *Philosophical Transactions of the Royal Society B* 354: 1395–1405.

Shannon, C. E. and W. Weaver (1963). *The Mathematical Theory of Communication.* Champaign, IL: University of Illinois Press.

Solé, R. V. and J. Bascompte (2006). *Self-Organization in Complex Ecosystems.* Princeton, Oxford: Princeton University Press.

Stremke, S. and J. Koh (2010). Ecological concepts and strategies with relevance to energy-conscious spatial planning and design. *Environment and Planning B: Planning and Design* 37: 518–532.

Stremke, S. and J. Koh (2011). Integration of ecological and thermodynamic concepts in the design of sustainable energy landscapes. *Landscape Journal* 30: 2–11.

Stremke, S. and A. van den Dobbelsteen (2011). Exergy landscapes: Exploration of second-law thinking towards sustainable landscape design. *International Journal of Exergy* 8: 148–174.

Tansley, A. G. (1935). The use and abuse of vegetational concepts and terms. *Ecology* 16: 284–307.

Taylor, P. J. (1988). Technocratic optimism, H. T. Odum, and the partial transformation of ecological metaphor after World War II. *Journal of the History of Biology* 21: 213–244.

Thims, L. (2012). Thermodynamics ≠ information theory: Science's greatest Sokal affair. *Journal of Human Thermodynamics* 8: 1–120.

Tiezzi, E. (2011). Ecodynamics: Towards an evolutionary thermodynamics of ecosystems. *Ecological Modelling* 222: 2897–2902.

Trepl, L. (1997). Zum Verhältnis von Landschaftsplanung und Landschaftsarchitektur. In *40 Jahre Landschaftsarchitektur an der Technischen Universität München*, ed. 83–96. Freising: Lehrstühle für Landschaftsarchitektur TU München.

Trepl, L. (2005). *Allgemeine Ökologie Band 1—Organismus und Umwelt*. Frankfurt am Main: Peter Lang.

Trepl, L. (2013). Energiewende—Ende der Ökologiebewegung? *Landschaft und Ökologie*. Spektrum. (accessed January 26, 2014).

Trepl, L. and A. Voigt (2011). The classical holism-reductionism debate in ecology. In *Ecology Revisited—Reflecting on Concepts, Advancing Science*, ed. A. Schwarz and K. Jax, 45–83. Springer.

Voigt, A. (2009). *Die Konstruktion der Natur—Ökologische Theorien und politische Philosophien der Vergesellschaftung*. Stuttgart: Franz Steiner.

Weil, A. (1999). Über den Begriff des Gleichgewichts in der Ökologie. In *Landschaftsentwicklung und Umweltforschung*, ed. 7–97. Berlin: Technische Universität Berlin.

Weil, A. (2004). *Möglichkeiten und Grenzen der Beschreibung synökologischer Einheiten nach dem Modell des Organismus*. Dissertation. Department für Ökologie. Freising: Technische Universität München.

Wright, L. (1973). Functions. *The Philosophical Review* 82: 139–168.

Section II

Culture and Specificity

7 The Garden and the Machine

Thomas Juel Clemmensen

CONTENTS

7.1 Introduction 137
7.2 Modern Pastoralism and "The Machine in the Garden" 138
    7.2.1 Modern Pastoralism 139
    7.2.2 Aesthetic and Utilitarian Landscape 140
7.3 Landscape as Infrastructure and the Garden as Machine 141
    7.3.1 Landscape as Infrastructure 141
    7.3.2 The Garden as Machine... 143
7.4 Garden as a Third Nature 144
    7.4.1 Garden of Lancy 146
    7.4.2 Garden of River Aire 147
7.5 Conclusion: Surface versus Palimpsest 150
References.. 151

7.1 INTRODUCTION

When the American cultural historian Leo Marx in 1964 wrote *The Machine in the Garden* as a work of literary criticism, he used the concepts of "garden" and "machine" to explain how the American people through generations have tried to reconcile two very different views on their country: on the one hand, the uncomplicated rural paradise, and on the other hand, the industrialized and urbanized nation. By doing this, Marx brought attention to an underlying conflict between pastoral and progressive or utilitarian ideas in the American ideology of space* and observed how the machine was accommodated in the garden (see Figure 7.1). Although it is fair to say that today the machine is not so much in the garden as it is indistinguishable from the garden, they are inexorably intertwined (Strang 1996), and this set of concepts still seems to possess an explanatory power. Thus, the aim of this chapter is to explore how the concepts of garden and machine might inform our understanding of the complex relationship between infrastructure and nature.

---

* Marx calls his subject the ideology of "space" rather than "landscape" because he considers space to be an essential component of both architecture and landscape. And according to Marx, the term *space* also may remind us initially that a landscape is a physical entity whose meaning and value we construct and for which we have a variety of other names: land, topography, terrain, territory, environment, etc. (Marx 1991).

FIGURE 7.1 The Lackawanna Valley (oil on canvas) by Georges Inness, which Leo Marx uses on the cover of his book *The Machine in the Garden* (1964). The painting depicts the colonization of the American West. According to Marx, the train represents the machine in the garden. (Georges Inness, 1855.)

As we shall see, the concept of the garden is not only associated with pastoral ideals, but can also be regarded as a kind of a third nature, drawing on the theories of the English landscape historian John Dixon Hunt. In this context, the garden can be regarded as an important place of reflection on the relationship between the given nature (first) and the transformed nature (second), which involves an important distinction from the way Marx uses the concept of the garden. Rather than being something accommodating our machines, the garden becomes something that ultimately reveals how the machines are involved in the transformation processes.

In this chapter, this distinction is used to shed a critical light on the promotion of landscape *as* infrastructure, which has been developed in the current discourses on landscape urbanism and ecological urbanism. It will be argued that the emphasis of performance, functionality, and horizontality, which seems to follow the promotion of landscape *as* infrastructure, in some cases, could be counterproductive in relation to the environmental problems being addressed and that we need gardens of reflection, interrogation, and doubt in order to engage with the deeper complexities of territorial transformations. In order to substantiate this argument, we will visit two gardens designed by the Swiss landscape architect Georges Descombes.

7.2 MODERN PASTORALISM AND "THE MACHINE IN THE GARDEN"

According to Leo Marx, it is possible to identify three different versions of the myth of the origins of America, which correspond to three variants of the American ideology of space: the progressive, the primitive, and the pastoral. The progressive

The Garden and the Machine   139

variant relates to the idea of America as a vast uncivilized territory whose resources were waiting to be exploited. Nature was to be defeated, colonized, and turned into something useful. In contrast, the primitive variant relates to the idea of America as an unspoiled paradise where nature was identified with freedom, authenticity, spontaneity, and the opportunity to avoid civilization's many social downsides. Between these opposites exists the pastoral variant, which relates to the idea of America as the place where the old dream of achieving harmony between man and nature could be realized. The ideal was neither the old European civilizations nor the wild nature but a "middle landscape" where the best of both worlds could be combined (Marx 1964).

Despite these different variations, Marx argues that, in reality, there only exists one American ideology of space. According to Marx, primitivism never established the conceptual basis for an independent ideology, and only a very modest part of the American population has adopted it as a guide for their behavior. Apart from its motivation on various environmental movements, its actual consequences are easily overlooked. In contrast, both the progressive and pastoral ideals have had a great influence throughout American history, in which their mutual tensions repeatedly and persistently have manifested themselves. Although the two ideals, in principle, are ultimately irreconcilable, this logical contradistinction has, in practice, been relatively easy to conceal and ignore. In this way, Marx believes that a fusion between the progressive and pastoral ideals has emerged, a single ideology, which embraces their recurring contradistinctions both in discourse and practice (Marx 1964).

7.2.1 MODERN PASTORALISM

American urban scholar Peter G. Rowe uses the term "modern pastoralism" to capture the antagonistic nature of the American ideology of space, an idealization of rural life and the contact to nature merged with a modern technical orientation. Just as Marx, Rowe considers pastoralism to be a complex artistic and ideological motif, in which both the urban and the rural are at play. In pastoralism, along with the idealization of rural life and the contact to nature also exists an idea about resolving the conflict between the two different worlds. In contrast to pastoralism, the modern technical orientation is not associated with a particular type of environment but is ubiquitous and primarily concerned with a mode of acting on the world. According to Rowe, it is sustained by three complementary ideas: a technological way of making things, a technocratic way of managing things, and a scientific way of interpreting people and their environment (Rowe 1991).

Both Marx and Rowe have looked toward suburbia when describing the kind of "middle landscape" that seems to be the result of modern pastoralism (Rowe 1991) (see Figure 7.2). However, modern pastoralism is not a specifically suburban phenomenon, and the features associated with it can be identified in many different environments (Marx 1991). Equally, the explanatory power of modern pastoralism is not limited to North America as it addresses fundamental tensions between opposing values and perceptions of the natural environment, which can be identified throughout the history of the Western World.

FIGURE 7.2 Broadacre City (section of scale model) by Frank Lloyd Wright. According to Peter G. Rowe, Broadacre City has all the features of "modern pastoralism" in its use of technological and technocratic means within a clearly defined rural agrarian setting. (Rowe, P. G. Making a Middle Landscape. Massachusetts: MIT Press, 1991, 235; Frank Lloyd Wright, 1935.)

7.2.2 AESTHETIC AND UTILITARIAN LANDSCAPE

The Danish landscape scholar Ellen Braae has identified the same kind of tensions associated with modern pastoralism in relation to when Denmark in the 18th century was in the middle of an ecological crisis because of an excessive use of the country's natural resources. In this context, Braae uses a distinction between the "aesthetic" and the "utilitarian" to describe the emergent split between sense and sensibility, between science and art, which characterized the 18th century, a period in time marked by dawning industrialization, the advance of science, and the Romantic movement. On the one hand, nature was made an object of scientific studies and industrialization; on the other hand, nature was cultivated as a sensuous phenomenon by the arts (Braae 2011).

According to Braae, this split has subsequently created the basis for a gap between the realities in our everyday landscapes and the ideals, which has continued to influence landscape architecture. Although most of our natural environment has been subjected to a utilitarian approach and an industrial style cultivation with emphasis on optimization and maximization, landscape architecture, to a large extent, has rested on landscape as an aesthetic category characterized by a contemplative relationship to

The Garden and the Machine

nature. The aesthetic became a refuge and a compensation for the loss of connection to nature, which came along with the development of the natural sciences (Braae 2011).

With the description of a split between sense and sensibility, the inherent contradictions of modern pastoralism seem less dramatic as one aspect, the pastoral, seems more rooted in the aesthetic, and the other aspect, the progressive, seems firmly rooted in the utilitarian. In this way, it could be the old split between sense and sensibility, between science and art, which, in the first place, has made a hybrid ideology such as modern pastoralism possible. It explains how it has been possible to hold on to romantic ideas of living in harmony with nature and, at the same time, overexploit the natural resources.

According to Braae, the split between sense and sensibility, between science and art, is not only linked to the many ecological crises that have followed industrialization, but also to the inability of landscape architecture to address and engage with these problems. Consequently, Braae calls for new interpretations of the aesthetic and the utilitarian (Braae 2011), something she locates in relation to the current recovery of landscape architecture often associated with the discourses of landscape urbanism and ecological urbanism. In this context, Braae identifies the ecological imperative with a reinterpretation of the utilitarian and the emphasis on dynamic processes and performance with a reinterpretation of the aesthetic. According to Braae, these reinterpretations together constitute an important shift from *statis* to *dynamis* in relation to landscape architecture, a shift from static pictures to dynamic systems and relationships that appear as an ability to create frameworks for open systems (Braae 2011).

7.3 LANDSCAPE AS INFRASTRUCTURE AND THE GARDEN AS MACHINE

Seen in relation to the discussion in Section 7.2, one might wonder whether the current discourses on landscape urbanism and ecological urbanism and the promotion of landscape as infrastructure is just another and perhaps more sophisticated chapter in the history of modern pastoralism? The combination of the terms ecology and urbanism certainly evoke the kind of harmony between man and nature that can be associated with the pastoral ideal, and the idea of landscape as infrastructure implies some kind of utilitarian and technological approach to the natural environment. However, landscape as infrastructure might as well involve the kinds of new interpretations of the aesthetic and the utilitarian that Braae calls for. Whether the current promotion of landscape as infrastructure is an extension or a rejection of modern pastoralism seems an interesting and important question given the fact that the promotion of landscape as infrastructure is often linked to the idea of revealing and solving environmental problems caused by a predominantly utilitarian and technological approach to the natural environment.

7.3.1 LANDSCAPE AS INFRASTRUCTURE

According to Canadian landscape scholar Pierre Bélanger, who has contributed to the current discourses on landscape urbanism and ecological urbanism, the urbanregional landscape should be conceived as infrastructure (see Figure 7.3). The main

FIGURE 7.3 Transporting Columbus: A Proposal for a Contemporary Transportation Node in Columbus, Ohio (3-D rendering) by OPSYS, which is codirected by Miho Mazereeuw and Pierre Bélanger. The proposal was a competition finalist and received honorable mention for Re-Wired Design Competition, American Institute of Architects, 2007. "Drawing from disurbanist precedents in the Midwest such as Frank Lloyd Wright's Broadacre City and Ludwig Hilberseimer's Lafayette Park," the design case study "develops a land use pattern that synthesizes mass transportation systems with the imperatives of the urban economy and of regional ecology." (OPSYS, http://www.opsys.net/index.php?/projects/disurbanism/ [accessed October 18, 2013], 2007.)

argument is that economy and environment today have become inseparable and mutually codependent whereas, in the past, urban industrial economies were forced to contaminate and destroy the environment in service of the economy (Bélanger 2009). Infrastructures that, in the past, were only engineered to support urban economies should now simultaneously support the environment. As a consequence, infrastructure, once the sole purview of the profession of civil engineering, is taking on extreme relevance for landscape planning and design practices. Bélanger describes how landscape practice stands to gain momentum by widening its sphere of intervention to include the operative and logistical aspects of urbanization and talks of new paradigms of longevity and performance, which decisively break with the Old World pictorial, bucolic, and aesthetic tradition of landscape design (Bélanger 2009).

Another North American landscape scholar to promote landscape as infrastructure is Gary Strang, who has formulated similar thoughts about the relationship between infrastructure and the contemporary landscape. With reference to Marx's description of "the machine in the garden," Strang observes that current conditions create a situation in which the machine becomes inseparable from the garden or in which the garden and the machine are completely intertwined (Strang 1996). In doing so, Strang does not refer to the formal characteristics but to the functional integration between infrastructure and a constructed landscape, which relies on infrastructure for its preservation. In seeing infrastructure as landscape, Strang argues for

The Garden and the Machine

an approach that allows the natural landscape and the landscape of infrastructure, which occupy the same space, to coexist and preform multiple functions. He would like architects to be more like farmers who depend on the architecture of natural systems for their livelihood (Strang 1996).

7.3.2 THE GARDEN AS MACHINE

The emphasis on performability and functionality, which characterize the way Bélanger and Strang approach landscape as infrastructure, not only confirms the shift from *statis* to *dynamis*, which Braae describes, but also indicates that the utilitarian no longer is restricted to the performance of man-made systems in nature, but is expanded to include the performance of natural systems as well.

There seems to be no doubt that this shift in from *statis* to *dynamis* is a necessary and important step if landscape architecture is to challenge the static ideals of modern pastoralism and engage with the dynamic realities in our contemporary urban landscapes. As James Corner has explained, it allows us to see landscape as an active instrument for the enrichment of culture rather than as a passive product of culture (Corner 1999). The expansion of the utilitarian to include both man-made and natural systems is a bit more complicated. Seen in relation to the extent to which we already have altered the natural environment, it could seem only logical to make it our job to secure the performance of natural as well as man-made systems—as Bélanger reminds us, economy and ecology are not in opposition but codependent. In other words, it makes perfect sense for us to learn to intervene in ways that facilitate rather than disrupt natural systems and processes. However, given the many examples of how our attempts to control natural systems have seriously harmed natural environments and led to actual ecological crises, it also calls for caution.

Where the shift from *statis* to *dynamis* appears as a rejection of modern pastoralism and its inherent contradictions, the expansion of the utilitarian to include both man-made and natural systems appears more like an extension of modern pastoralism and the old dream of achieving harmony between man and nature. The machine is no longer accommodated in the garden, but the garden itself becomes part of an environmental machine that can be optimized (see Figure 7.4). Man-made and natural systems together are described as "surface systems" (Bélanger 2009, 92).

The current emphasis on performability and functionality, which follows the promotion of landscape as infrastructure, might prove valuable in addressing the massive environmental problems that characterize the current relationship with our natural basis. At the same time, there seems to be a danger that the environment, natural or man-made, becomes the object of crude instrumentality. Considering how modern industrial agriculture with all of its associated environmental downsides also relies upon landscape as infrastructure with an emphasis on performability and functionality, is seems a bit paradoxical that Strang wants architects to be more like farmers. In this respect, landscape as infrastructure might be one of the "mythic images" that, according to Rowe, help to mask inconsistencies and inadequacies in relation to modern pastoralism (Rowe 1991, 315).

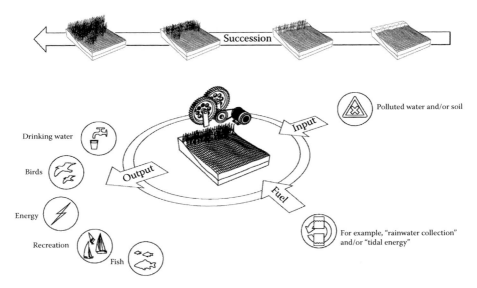

FIGURE 7.4 Conceptual visualization of the functional components of a "landscape machine" according to the "Landscape Machine Design Laboratory" founded at the Wageningen University in the Netherlands by landscape architect Paul Roncken. In this case, a landscape is quite literally considered a machine with a certain material input and output and driven by a critical amount of energy input (see Roncken et al.'s Chapter 5, "Landscape Machines: Designerly Concept and Framework for an Evolving Discourse on Living System Design" in this volume). (Alexander Herrebout, 2011.)

7.4 GARDEN AS A THIRD NATURE

Until now, the term garden has been used as a general metaphor for our cultivated landscapes covering everything from past preindustrial pastoral landscapes to present postindustrial urban landscapes. But in order to explore how to avoid the masking of inconsistencies and inadequacies in relation to the way we intervene in the natural environment, it makes sense to recall John Dixon Hunt's idea of the garden as a third nature, which he introduces in his book *Greater Perfections—The Practice of Garden Theory* (2000). According to Hunt's garden theory, the first nature represents the natural world and can be associated with the wild or the idea of wilderness. The second nature, in contrast, represents the natural world transformed by humans or the cultural landscape and can be associated with both agriculture and urban development—all the interventions related to human survival and habitation. *Terza natura* (the third nature), a phrase first coined by Italian Renaissance writer Jacopo Bonfadio (1508–1550), represents those human interventions that go beyond what is required by the necessities or practice of agriculture and urban settlement and can be associated with the garden (Hunt 2000). Apart from the clear schematic in this trichotomy, the garden's place within the three natures is complex, and one should not make the mistake of considering the three natures simply as different domains.

The Garden and the Machine

The third nature of gardens is best considered as existing in terms of the other two and will always be engaged in a dialogue with them (see Figure 7.5).

Seen in relation to the ideology of modern pastoralism and the discourses of landscape urbanism and ecological urbanism, which all appear to be preoccupied with the second nature, the idea of three natures provides a valuable and perhaps much needed framework for reflections on our relationship with our environment. First, it reminds us that no matter how synthetic our environment has become, the first nature will always be there and always play an important role, not as something original and uninfluenced by humans but as something beyond our total control. Second, it liberates the garden from its strong association with the second nature and the utilitarian approach to the natural environment. The Swiss architect Georges Descombes, who has been referring to Hunt's garden theory in relation to his own work, suggests that it also involves a liberation of the formal aesthetic that often clings to the idea of the garden. According to Descombes, the garden is an important place of reflection, interrogation, and doubt in regard to the relationship between the world given and the world transformed, a place that represents, simulates, and reveals what we are doing to the world given (Descombes 2012). In order to explore how this theoretical position might influence landscape design and planning in practice, we can visit the southwestern outskirts of Geneva to examine two of Descombes' projects in an area that once was his childhood landscape and since has become a sort of a open air laboratory to him.

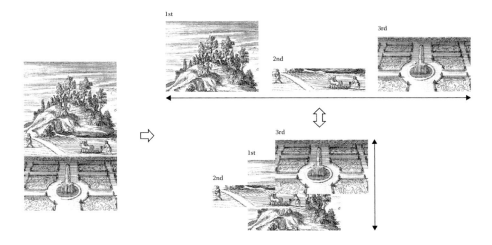

FIGURE 7.5 Frontispiece (section) Curiositez de la Nature et de l'Art (1705) by Abbé Pierre Le Lorrain de Vallemont, which, according to John Dixon Hunt, illustrates the idea of the garden as a third nature (Hunt 2000). Apart from the clear schematic in this trichotomy, the garden's place within the three natures is complex, and one should not make the mistake of considering the three natures simply as different domains (horizontal). The third nature of gardens is best considered as existing in terms of the other two and will always be engaged in a dialogue with them (vertical). (Thomas Juel Clemmensen, 2012.)

7.4.1 Garden of Lancy

The first project is situated in Lancy, an area that has undergone a quite dramatic transformation from rural hinterland to residential suburb to part of the larger urban landscape surrounding Geneva. As in so many similar cases, this transformation, which began early in the 20th century, involved a substantial modification of the existing topography. Not only the process of urbanization, but also the industrialization of the remaining agriculture has resulted in numerous landfills, regulated watercourses, and leveling of the terrain.

A majority of these transformations had already taken place when Descombes, in the early 1980s, was commissioned to develop a public park on a site next to a small tributary of the River Aire. Although this rivulet, called the Voiret, still maintained some of its meandering course and surrounding vale, its overall character was profoundly marked by the aforementioned transformations. In the project, Descombes exploits this in-between status to create a series of architectural interventions, which not only fulfill a recreational purpose, but also reveal something about the disrupted topography of the site (see Figures 7.6 and 7.7).

The most remarkable of these interventions is the so-called "tunnel bridge," which is situated where a new road was constructed across the Voiret in the 1940s. Instead of building a bridge, which, in Descombes opinion, would have been a more appropriate way of straddling the vale and rivulet, they used a "landslide" approach,

FIGURE 7.6 Plan of Parc de Sauvy (1980–1986) in Lancy, an area that has undergone a quite dramatic transformation from rural hinterland to residential suburb to part of the larger urban landscape surrounding Geneva. In this project for a public park, Georges Descombes and his team have designed a series of architectural interventions, which not only fulfill a recreational purpose, but also reveal something about the disrupted topography of the site. (From Descombes, G., *Shifting Sites*, Gangemi editore, Rome, 1988.)

The Garden and the Machine

FIGURE 7.7 Photo of the so-called "tunnel-bridge" in Parc de Sauvy, an architectural solution that is both an attempt to answer the brief directly and, at the same time, revealing something about the existing site conditions. According to Georges Descombes, the tunnel bridge exhibits the forces acting and interacting in the place, leading to its interpretation and reinterpretation. (Thomas Juel Clemmensen, 2013.)

based on the imperfect logic of drainage, and blocked the vale with an embankment and directed the rivulet through a culvert (Descombes 2009). The road later was to be widened, and the authorities requested a pedestrian underpass to create safe access to the new public park. Descombes saw this as an opportunity to address the specific history of the site. Crossing through the embankment suggested a tunnel, and crossing the rivulet suggested a bridge, so Descombes and his team proposed to make a tunnel bridge, a tunnel with a bridge inside, a solution that is both an attempt to answer the brief directly and, at the same time, revealing the existing site conditions (Descombes 2009). According to Descombes, the tunnel bridge exhibits the forces acting and interacting in the place, leading to its interpretation and reinterpretation (Descombes 1988).

It is precisely these kinds of architectural interventions that refer to the history of the place and question the consequences of its successive transformations, which turn the site into a place of reflection, interrogation, and doubt in relation to the way we transform our natural environment. Seen in this perspective, Descombes' project in Lancy is not so much a park as it is a garden.

7.4.2 GARDEN OF RIVER AIRE

The second project deals with the "re-naturalization" of the River Aire, which, from the 19th century to the middle of the 20th century, had undergone an extensive regulation and had been reconfigured as a canal system to control flooding and improve the agriculture of the valley (see Figure 7.8). In 2001, the authorities decided to restore the river to its original form and requested ideas for realizing the restoration through an invited competition, which Descombes and his team Superpositions won. One of the

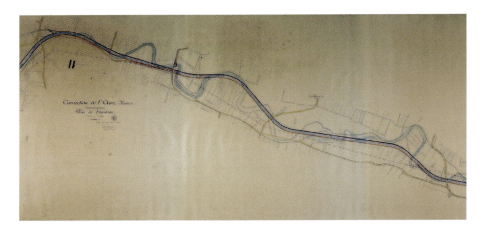

FIGURE 7.8 Nineteenth-century map of the River Aire (detail) depicting the project for a canal superimposed on the existing river. The map seems to illustrate Georges Descombes' idea about looking in the past to recall the future. (Municipal Archives, City of Geneva.)

key objectives was to secure the nearby neighborhood Praille-Acacias-Vernets against flooding, so it was once again flooding that initiated a transformation of the river.

Similar to the case with the tunnel bridge in Lancy, Descombes and his team not only wanted simply to answer the brief, but also saw the commission as an opportunity to redevelop or reframe the whole problem. Instead of pretending to come up with a perfect solution, they wanted to engage with the real complexity of the problem and pose an open question: How can we make a better relationship between the contradictory desires of leisure, natural environment, and agriculture? In this sense, Descombes and his team wanted to create a linear garden, a platform for reflection and discussion (Descombes 2012).

In accordance with this approach, the existing canal should be preserved as a canal, not only because it was an important part of the cultural heritage of the region, but also in order to reveal what has changed or, as Descombes has formulated it, to be able to look in the past to recall the future (Descombes 2012). Instead of simply returning the river to its former course and maintaining the canal as a remnant of the past, Descombes and his team proposed to experiment with a model, which it would be tempting to call a "canal river," where the water on one side follows the original bank of the canal and on the other side, in certain places, is allowed to meander more freely in a zone parallel to the canal (see Figure 7.9). Letting the river free will encourage an interaction between the water and banks that will support the restoration of natural systems of flow and vegetation. The strait part of the canal, in contrast, will become a kind of boulevard that, on one side, buffers the land from the water (see Figures 7.10 and 7.11). In this way, according to Descombes, the new canal river represents an in-between condition, in some ways like a battle zone with areas for leisure, natural environment, and agriculture, a laboratory for new relationships between the river and the surrounding landscape (Descombes 2009).

The Garden and the Machine

FIGURE 7.9 Photo of study model (detail) illustrating the new river parallel to the existing canal that is proposed to contain a series of water gardens. Along the canal runs a promenade. According to Georges Descombes, the existing canal should be preserved as a canal, not only because it was an important part of the cultural heritage of the region, but also in order to reveal what has changed. (Descombes, G. Re-naturalization of River Aire, Lecture at Aarhus School of Architecture, May 24, 2012; Superpositions, 2007.)

FIGURE 7.10 Photo from the new condition at the River Aire. View from pont de Lully looking down at the former canal. (Thomas Juel Clemmensen, 2013.)

FIGURE 7.11 Photo from the new condition at the River Aire. View from pont de Lully looking down at the new meandering river running parallel to the former canal revealing how the river (once again) has been transformed by human intervention. (Thomas Juel Clemmensen, 2013.)

7.5 CONCLUSION: SURFACE VERSUS PALIMPSEST

From the work of Descombes, the idea of the three natures seems to involve a kind of in-depth approach to the territories concerned, in which different layers referring to the different variants of nature can be uncovered. The French landscape scholar Sébastien Marot, who has made an elaborate reading of Descombes' garden in Lancy, supports this understanding. According to Marot, the innovation of the project stems from the fact that it not only takes into account the changes in the territory, but also responds to the palimpsest-like nature that the successive modifications have conferred upon it, inviting people to experience this landscape precisely as a palimpsest (Marot 2003). In my view, this reading could also be applied to the linear garden of River Aire.

In relation to the current discourses on landscape urbanism and ecological urbanism, Marot has used the palimpsest as a metaphor for the contemporary landscape to articulate the need of a deeper and more complex notion of what landscape means today (Marot 2011), an idea that draws on the thoughts of André Corboz, who has described how the land, so heavily charged with traces and with past readings, seems very similar to a palimpsest (Corboz 1983). In this context, Marot compares the palimpsest to the theater, two very different metaphors used for landscape. With reference to J. B. Jackson, Marot describes the theater as the strong operational metaphor through which the classical landscape was envisioned and produced, a metaphor that has proved difficult to escape although its representational matrix has ceased to be operative (Marot 2011). However, seen in relation to modern pastoralism, the theater as metaphor might indeed have been operative, not as instrument to recover the future but as a cover for the inconsistencies following the combination of pastoral and progressive ideals.

The Garden and the Machine

With the decisive challenges to and rejections of the scenic landscape that followed the discourses on, first, landscape urbanism and, later, ecological urbanism, the theater as metaphor finally seems to loosen its grip on landscape architecture, and in the wake, new metaphors or analogies, to be more precise, are entering the field of landscape architecture. However, it seems that infrastructure and not palimpsest is the preferred analog. Considering how infrastructure, as opposed to the theater, is associated with enabling and dynamic qualities, this may appear logical, but it is not necessarily unproblematic. First, with the analogy of infrastructure, the scope of the functionality of the landscape runs the risk of being narrowed down to functions of necessity, which only renders landscape as a second nature, lacking the dimension of reflection, interrogation, and doubt, which Descombes calls for. Second, thinking of landscape as infrastructure seems to invoke a flattening rather than a deepening of territories with an overwhelming focus on landscape as a horizontal surface. It becomes clear when Bélanger refers to landscape design as "design of surface systems" and, with reference to Corner, describes landscape architecture as a "horizontal discipline" (Bélanger 2009, 91) or when German urban scholar Alex Wall describes landscape as a "functioning matrix of connective tissue" or "active surface" (Wall 1999, 233). Given this emphasis on the surface and horizontality, one could fear that landscape as infrastructure simply cannot accommodate the full strata of a territory and, as a result, becomes another vehicle for hiding in relation to the real and deeper complexities of the landscape. This potential problem seems to be confirmed by Australian landscape scholar Peter Connolly (2004), who has made a thorough critique of the lack of "landscape" in landscape urbanism. According to Connolly, Wall's description of landscape as an active programmable surface rests on an idea of landscape as an abstract space, which can also be associated with modernism and the idea of *tabula rasa*. With Connolly's critique of the surface as an abstract space, it seems more clear that not only the theater, but also the surface as analogues for landscape have significant drawbacks compared to the palimpsest as they, on a conceptual level, disregard or hide other or former texts in the territory. Seen in this light, with reference to Strang, maybe architects should be less like farmers and more like gardeners? When "designing nature as infrastructure," we should not forget the palimpsestic qualities of landscapes by simply thinking them as programmable surfaces. Landscapes constitute important places of cultural identity and an opportunity to reflect upon our changing perceptions of nature. That is why we need gardens, not the ones designed to accommodate our machines or optimized "as" machines, but gardens designed to promote reflection, interrogation, and doubt in relation to our changing perceptions of nature.

REFERENCES

Bélanger, P. (2009). Landscape as Infrastructure. *Landscape Journal* 28(1): 79–95.

Braae, E. (2011). Det skønne og det nyttige landskab. [The aesthetic and the utilitarian landscape] In *Grænseløse byer—Nye perspektiver for by- og landskaabsarkitekturen* [*Cities Without Limits—New Perspectives for Urbanism and Landscape Architecture*], 106–122. Aarhus: Aarhus School of Architecture.

Connolly, P. (2004). Embracing openness: Making landscape urbanism landscape architectural: Part 1. In *The Mesh Book—Landscape/Infrastructure*, 76–103. Melbourne: RMIT University Press.

Corboz, A. (1983). The land as palimpsest. *Diogenes 31*: 12–34.

Corner, J. (1999). Recovering landscape as a critical cultural practice. In *Recovering Landscape—Essays in Contemporary Landscape Architecture*, 1–26. New York: Princeton Architectural Press.

Descombes, G. (1988). Notes for David Cooper. In *Shifting Sites*, 38–39. Rome: Gangemi Editore.

Descombes, G. (2009). Displacements: Canals, rivers, and flows. In *Spatial Recall: Memory in Architecture and Landscape*, 120–135. New York and London: Rutledge.

Descombes, G. (2012). *Re-naturalization of River Aire*. Lecture at The Aarhus School of Architecture, May 24 2012.

Hunt, J. D. (2000). *Greater Perfections—The Practice of Garden Theory*. London: Thames & Hudson.

Marot, S. (2003). *Sub-urbanism and the Art of Memory*. Architecture Landscape Urbanism Series No 8. London: AA Publications.

Marot, S. (2011). Ecology and urbanism: About the deepening of territories. In *The Eco-Urb Lectures*, 73–110. Copenhagen: The Royal Danish Academy of Fine Arts Schools of Architecture, Design and Conservation.

Marx, L. (1964). *The Machine in the Garden: Technology and the Pastoral Ideal in America*. New York: Oxford University Press.

Marx, L. (1991). The American ideology of space. In *Denatured Visions—Landscape and Culture in the Twentieth Century*, 62–78. New York: MOMA.

OPSYS, http://www.opsys.net/index.php?/projects/disurbanism/ (accessed October 18, 2013).

Rowe, P. G. (1991). *Making a Middle Landscape*. Massachusetts: The MIT Press.

Stang, G. (1996). Infrastructure as landscape. In *Theory in Landscape Architecture. A Reader*, 220–226. Philadelphia, PA: PENN.

Wall, A. (1999). Programming the Urban Surface. In *Recovering Landscape—Essays in Contemporary Landscape Architecture*, 233–249. New York: Princeton Architectural Press.

8 Infrastructure Design as a Catalyst for Landscape Transformation

Research by Design on the Structuring Potential of Regional Public Transport

Matthias Blondia and Erik De Deyn

CONTENTS

8.1 Introduction 153
8.2 The Scheldt and Rupel Rivers as Structuring Landscape Figures in the Test Case of Klein-Brabant 156
8.3 Five Configurations for a Landscape-Embedded Light-Rail Connection 160
    8.3.1 "Polder Island": In and around Bornem 160
    8.3.2 "Polder Edge": Kruibeke/Bazel/Rupelmonde 160
    8.3.3 "Waterfront": Rupelmonde/Steendorp/Temse 161
    8.3.4 "Industrial Patch Regeneration": Hemiksem/Schelle/Niel 162
    8.3.5 "Postproductive Landscape": Boom 164
8.4 Research beyond the Design: The Emergence of Recurrent Design Criteria 165
8.5 Conclusions: Comparing the Proposed Design Schemes with Different Scenarios and Applying the Design Lessons on the Scale of Flanders 168
References 169

8.1 INTRODUCTION

Within the Flemish context, existing train and bus networks function reasonably efficiently at the intercity and local levels, respectively. However, public transport is not as solid in the in-between territories, which are strongly urbanized as well; here, mobility demands are much bigger than in central cities (Boussauw 2011, 152), and the resulting travel patterns are often of an intermediary distance. Individual car traffic markedly dominates regional mobility: For home–work movements between 10 and 50 km, the modal share of the car exceeds 70%.[*]

All across Europe, regional public transport networks have been developed to cover medium-range distances and successfully compete with car traffic.[†] These projects, often based on light-rail technology, suggest that an efficient regional transport system not only serves, but also influences travel behavior, thus making a long-term impact on spatial structure and urbanization patterns. A number of regional light-rail projects are currently being prepared in Flanders, but their strategic potential as a de facto tool for urbanism in a changing spatial context is not being considered.

In the past, however, the urbanization of Flanders has always closely grafted itself on infrastructure networks, including public transport.[‡] This relationship between infrastructure and urbanization has mostly been researched after its occurrence (De Block, Smets, and De Meulder 2011; Grosjean 2010; Van Acker, Smets, and De Meulder 2011). In retrospect, it seems that infrastructure works stand out as a highly effective leverage for guiding spatial development. However, nowadays, the context has changed in a fundamental way; as argued by Smets (2001, 121), infrastructures within the *urban context* produced a variety of public spaces—thus positioning themselves as a tool for urbanism—while networks for the surrounding territories, the *rural landscape*, belonged to the field of engineering with the spatial project only implicitly present. This shifted through the gradual merging of town and countryside, "discovered" in the late '90s through concepts such as sprawl, rurbanization, ville territoire, and edge city.[§] As the dichotomy between urban and rural was being challenged, so was the distinction in infrastructure design for both contexts. Reed (2006, 270) states, "The reformulated context within which public works have evolved is now characterized by dispersion, decentralization, deregulation, privatization, mobility and flexibility." At the same time, "the tendency to engineer for a single purpose" (Strang 1996) has shifted toward an integration of technical, structural, architectural, and biological disciplines, resulting in emerging disciplines such as Landscape Urbanism (Waldheim 2006) and Infrastructure Urbanism (Shannon and Smets 2010).

Specifically with regards to public transport, design efforts have focused on the local scale level: strategic master plans for station surroundings, which has evolved into a spatial planning concept called *Transit-Oriented Development* (Curtis, Renne, and Bertolini 2009). However, the question as to how public transport infrastructure could influence spatial organization on the scale of the territory is rarely tackled.

---

[*] For more detailed data, see Verhetsel et al. (2007).

[†] To name a few: RandstadRail (NL), Glattalbahn (CH), Mulhouse (FR), and Freiburg (DE).

[‡] For example, with the construction the interurban tramways, the vicinal railway network, and the intercity rail connections.

[§] For more on this topic, see Delbaere, D. (2007), Possible standpoints on the diluted city/Mogelijke posities ten opzichte van de verdunde stad. *World Architecture Magazine*, 2007/03 Emerging Belgian Architecture, School of Architecture—Tsinghua University, Beijing, China.

Infrastructure Design as a Catalyst for Landscape Transformation   **155**

Contributing to the debate on the issue of regional public transport as a catalyst for urban development, a consortium of four universities and one private partner collaborate in the ORDERin'F project*—an acronym for Organizing Rhizomic Development along a Regional pilot network in Flanders. This multidisciplinary and transdisciplinary research project investigates the feasibility of regional public transport (such as light rail) in Flanders—not only in terms of traveler potential, but also as a strategic instrument to steer spatial developments—with the explicit aim of a transferability to planning practice.

When aligning spatial-development patterns and public transport networks with each other, the *corridor* emerges as a key figure, a linear structure organized around a transport infrastructure, built up from a sequence of nodal concentrations around stops. One of the main design questions not tackled by existing evaluation tools (such as traffic models or environmental assessments) is to determine which corridors are to be developed, which spatial structures offer a backbone for public transport corridors to be built upon. Methodologically, this was researched through a research-by-design process executed on a test case region within Flanders. Three different concepts were tested: (a) the heavy- to light-rail transformation of existing train connections, (b) the *tramification* of bus corridors, and (c) corridors that strengthen landscape-embedded urbanization patterns.[†] The third concept will be elaborated in this chapter. Contrary to the other two, it is not directly tied to the existing infrastructure networks or mobility patterns of the region.[‡] Instead, it uses the spatial logic of landscape structures as the framework for steering regional development and mobility.

---

* Collaborating on the ORDERin'F project are the research group OSA from the Department of Architecture, Urbanism and Spatial Planning of the KULeuven; the Institute for Mobility (IMOB) from the University of Hasselt; the research group MOSI-T from the VUB; Lab'URBA from the Université Paris-Est; and finally, the private partner BUUR, a design office that specializes in urbanism and urban design within the context of Flanders. This multidisciplinary and transdisciplinary research project investigates the feasibility of regional public transport, such as light rail, within the strongly fragmented spatial structure of Flanders. Methodologically, ORDERin'F relies on a research-by-design process. For more information, see the OSA, BUUR, IMOB, MOSI-T, and Lab'Urba ORDERin'F, Organizing Rhizomic Development along a Regional pilot network in Flanders, Research Proposal (2009) or http://www.orderinf.eu.

† For more on this, see Blondia, M., and De Deyn, E. (2012), Understanding the development potential of regional public transport in a peri-urban context through research by design: The case-study of Klein-Brabant (Belgium). BUFTOD 2012—Building the Urban Future and Transit Oriented Development. Paris, 16–17 April 2012 (art.nr. 25).

‡ The two other scenarios are, in fact, based on infrastructure and mobility patterns as the base for a network proposal. (1) A first scenario uses existing and derelict rail infrastructures as the basis for a network layout. The tracks are already there, and having a trajectory embedded in the morphology of the landscape and the tissue is a major advantage. The curves and longitudinal profile of the lines were once designed to obtain a high speed and degree of comfort, which makes them ideal for an efficient light-rail service. (2) The second incentive for a scenario uses displacement patterns as a base for a layout of the system. An overview of the displacements in the area shows a very diffuse pattern. The existing bus line network tries to cope in a pragmatic way with this diffusion, responding to policies of basic mobility for the scattered spatial distribution. In certain areas, multiple bus lines aggregate along one corridor, creating a line along which many public transport trips take place. A transformation of these bus corridors into light-rail lines lays the groundwork for the second scenario.

8.2 THE SCHELDT AND RUPEL RIVERS AS STRUCTURING LANDSCAPE FIGURES IN THE TEST CASE OF KLEIN-BRABANT

The test case for the research-by-design process is the area of Klein-Brabant, a region that stands out for its spatial complexity and is, thus, exemplary for the rural-urban continuum. It is situated in the "void" between the major cities of Antwerp, Brussels, and Ghent at the heart of the Flemish Diamond.* Klein-Brabant is characterized by a diverse juxtaposition of open landscape, different types of housing tissue, and areas of economical activities strung together in a network of towns connected by ribbon development. Moreover, as a clear hierarchy between settlements or a clear dominance of certain mobility patterns is lacking, there is no apparent structure from which a public transport network layout can be derived.

A major driving force behind the dispersed urbanization of Flanders has been the sequence of consecutive infrastructure networks projected on the territory. A historical analysis of Klein-Brabant (Figure 8.1) shows that urbanization patterns, today considered dispersed, unstructured, or directionless, are, in fact, linked with the chronological layers of the transport network. Throughout history, every time new transport technology brought about new mobility networks, spatial developments occurred in the vicinity of nodes or lines along these networks, often with distinguishable morphological characteristics. For example, De Block and Polasky (2011) recognize a recurrent pattern in railway development of a dual centrality between the "old" town center and the "modern" 19th-century railway station connected by the *Stationstraat* (station street), which becomes an attractor for new functions. Similarly, the construction of highways during the second half of the 20th century led to large-scale economic development near intersections and along the secondary feeder road systems (Ryckewaert and Jocelyne 2011).

Underneath these chronological layers of infrastructure, the initial urbanization emerges, a fine-grained network of villages, closely linked with the topography and hydrology of the region.[†] There are notably higher settlement concentrations along the edges of river valleys as these historically provided the framework for infrastructure trajectories, production opportunities, and buildable land. Although less distinct nowadays, spatial patterns linked to the natural landscape are still discernible and are increasingly appreciated. In the case of Klein-Brabant, the rivers Scheldt and Rupel are the main landscape structures. As argued by Nolf et al. (2011, 265), water structures not only have a recreational and ecological function, but they are also the carrier of a regional identity. Secchi and Viganò (2011, 28) refer to the structure of

---

* Translation of *De Vlaamse Ruit*, a planning concept proposed in the Spatial Structure Plan of Flanders. ARP (Afdeling Ruimtelijke Planning) (1998). Ruimtelijk Structuurplan Vlaanderen—integrale versie. Brussel: Ministerie van de Vlaamse Gemeenschap, Departement leefmilieu en infrastructuur, Administratie ruimtelijke ordening, huisvesting, monumenten en landschappen.

† A clear example of this has been analyzed by Omgeving. In 2010, a preliminary study was made for the transformation of the A8 primary road. An analysis was made of the surrounding "Pajottenland" landscape. It was shown that early infrastructures were constructed mainly in accordance with topographic structures, following ridges or valleys. Later infrastructures have their trajectory based on the fastest connection between possible destinations, cutting through landscape structures. The analysis shows a gradual loss through time in legibility of the infrastructural landscape. For more details on this project, http://www.omgeving.be.

Infrastructure Design as a Catalyst for Landscape Transformation

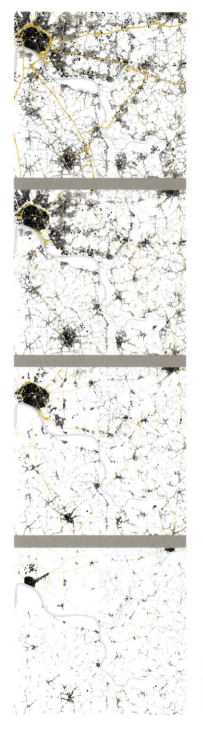

FIGURE 8.1 Analysis on the interdependency between different chronological layers of infrastructure networks. Based on a comparison with historical maps (1770 Ferraris map; 1892, 1969, and 2011 topographic maps), existing urbanization patterns were linked to the infrastructures around which they emerged in the past (from left to right): Until the 18th century: dispersed pattern of small settlements along main roads. 19th century: development of a dense railway network while bigger towns develop around stations. First half of the 20th century: automobility further stimulates dispersed urbanization, in particular, car-oriented ribbon development. Second half the 20th century: construction of highways, attracting large-scale developments. (Courtesy of NGI.)

the diffuse—or porous—city as a long-time construction of an articulated relationship with agriculture, production, and leisure, all three of which can be linked back to the structuring potential of the river valley.

Currently, there is a reappropriation of these landscape-based urbanization patterns, which could be described as "injecting a natural or rural quality into a urbanized territory" as shown in recent projects in Klein-Brabant such as *De Zaat* and *De Oude Scheepswerf*, in which old industrial areas are converted into water-oriented housing/mixed-use areas. At the same time, a reclamation of the historical waterfronts in town centers is also emerging. If the aim of a new public transport network is to strengthen the physical landscape by reconnecting it with a new infrastructure, the principles of these qualitative water-oriented developments could be expanded into a general strategy on the scale of the region; thus, public transport infrastructure becomes a catalyst for landscape transformation or, more precisely, for a landscape-embedded urbanization.

An analysis of the existing landscape (Figure 8.2) shows the rivers Scheldt and Rupel as shaping forces within Klein-Brabant. Urbanizations have developed along these rivers, but in different forms on each bank due to the dynamics between natural processes and human interaction with the landscape, according to specific local

FIGURE 8.2 Analysis of the landscape structure of Klein-Brabant.

Infrastructure Design as a Catalyst for Landscape Transformation 159

conditions. Thus, the landscape functions as a system of relationships between settlements or other developments and the underlying physical substrate, which, in itself, is also very diverse,* not only incorporating physical aspects, but also ecological, historical, and cultural components. They are the constraints, or points of reference, for the exploratory research by design. Section 8.3 analyzes the different riverside conditions and proposes regional public transport (light-rail) trajectories and spatial development schemes for each of them (Figure 8.3).

FIGURE 8.3 An overview of the five design configurations.

---

* Between and during the different Ice Ages, natural forces thoroughly reshaped the Flemish landscape as it was positioned in the transitional zone between temperate and cold conditions and thus subjected to continuous climate change dynamics. With each Ice Age, water levels decreased as ice packs were formed. During periods of thaw, large amounts of defrosted water ran back to the sea, enlarging the river's flow. At the same time, major changes in the topography occurred, caused by displacements of sand and loam, either through wind or water. As the growing river searched its way through the shifting landscape, the Scheldt changed its course and even its direction a number of times. The current trajectory of the Scheldt through Klein-Brabant and its confluence with the Rupel only occurred at the end of the last Ice Age. The shaping force of these dynamics has embedded itself in the structure of the current landscape and resulted in distinct differences on the different borders of the Scheldt and Rupel rivers.

8.3 FIVE CONFIGURATIONS FOR A LANDSCAPE-EMBEDDED LIGHT-RAIL CONNECTION

8.3.1 "Polder Island": In and around Bornem

Settlements on the south bank of the Scheldt river (Figure 8.4)—an inner curve prone to inundations—were founded on a sandy ridge, positioned slightly higher in the topography, thus protecting the villages from water risks. As no area-wide dyke system was developed, the spatial structure of urbanizations surrounded by lower polder and wetland landscapes still remains. This strongly limits expansion possibilities for urbanization around a new tramline. The potential for urban growth is mostly to be realized by a redevelopment or densification of the existing village tissue.

8.3.2 "Polder Edge": Kruibeke/Bazel/Rupelmonde

Downstream of the Scheldt-Rupel confluence, the river makes a curve, and the polder landscape switches banks (Figure 8.5). Here, poldering took place very differently: A single dyke along the waterfront was built to protect the entire area from flooding. Inland and parallel to this dyke runs the edge of a plateau, on which settlements grew. Over the last decade, as part of a flood risk–containment plan,* a second 8-m-high dyke was built near the edge of the plateau, and the old dyke was lowered, turning the polder into a floodable wetland. This new dyke offers a number of advantages as a structure for the new tramline trajectory to be bundled with; there are no crossing infrastructures or sharp curves, buildable land is available, and stops can be placed near town centers and directly next to new developable terrain—the strip of land between the edge of the plateau and the dyke, a part of the former polder that is now completely protected from flood risks. In the current situation, the secondary road connecting the towns is the main attractor for functions, resulting in car-oriented

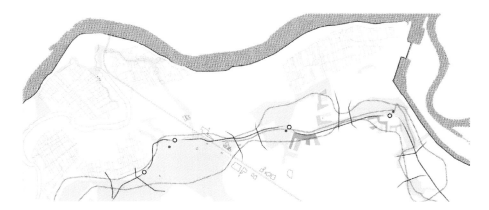

FIGURE 8.4 The "polder island" approach: densifying an elevated ridge in the polder landscape with a public transport connection as the backbone.

---

* Sigma plan.

Infrastructure Design as a Catalyst for Landscape Transformation 161

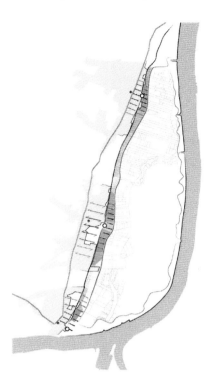

FIGURE 8.5 The "polder edge" approach: bundling the new line with the new dyke, connecting the developable areas. Transversal connections between the existing road and new tramline create dynamics in the existing fabric.

ribbon development. Adding a new mobility infrastructure that is not bundled with the existing road but instead runs parallel at a walkable distance creates new dynamics; transversal connections between both infrastructures are the framework for new public spaces and, near the tram stops, for new developments as well.

8.3.3 "Waterfront": Rupelmonde/Steendorp/Temse

On the outer curves of the river, developments followed different logics (Figure 8.6). For Rupelmonde and Temse (Figure 8.7) and, to a lesser degree, for Steendorp as well, an eroded cuesta front allowed a direct relationship with the river. This economical advantage is still legible in the current spatial structure; Temse and Rupelmonde developed as inner harbor settlements, and although the quays have lost their economic importance, they are still maintained as valuable and qualitative public spaces. The structure of this riverbank is an alternation between these strongly defined waterfronts and reclined relief slopes. The proposed tramline follows this dual structure. In the town centers, the tram is located directly on the waterfront, confirming its importance as a public space, making the water landscape very tangible from the tram. In between, the line retracts toward the edge of the plateau,

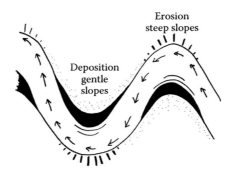

FIGURE 8.6 The difference between inner and outer curves of the river.

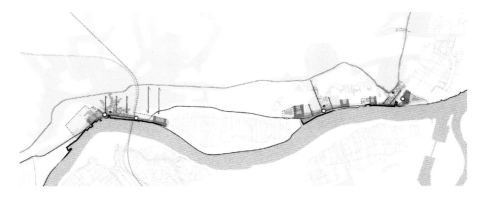

FIGURE 8.7 The "waterfront" approach: qualification of the historical waterfront developments combined with the reconversion of industrial sites at the edge of the towns.

leaving small patches of polder sites untouched. The development potential is mainly found in the reconversion of former shipyards and brick factories, which were built at the edge of the town centers; most of them have ceased their activities, and some of them have already been redeveloped.

8.3.4 "Industrial Patch Regeneration": Hemiksem/Schelle/Niel

South of Antwerp (Figure 8.8), the right bank of the Scheldt also has a clearly defined border. Here, the proximity to Antwerp and the easy connection to the North Sea created an ideal location for bigger industries. A secondary road and a railway track with branches toward the industrial plots run parallel to the riverbank, turning the entire strip along the Scheldt into a multimodal connected zone. Macroeconomical changes, mobility issues in the nearby settlements, the construction of the A12 highway just 3 km further inland, and the expansion of the port of Antwerp north of the city have put great pressure on the economic viability of these industries. There are different evolutions in the use of these sites; some of them have transformed into service-oriented companies, some sites are currently vacant, and others still remain active. As the strip

Infrastructure Design as a Catalyst for Landscape Transformation

FIGURE 8.8 The "patch regeneration" approach: light rail connecting existing centers along the corridor on the one side and specific sites for redevelopment altering with existing industrial activities on the other.

between the waterfront and the edge of the towns is quite wide, 500 m on average, there is a substantial redevelopment potential, which should be adapted to the large scale of the industrial fabric; individual patches can have independent campus-like developments as is already happening on Petroleum-Zuid.* In between the existing

---

* For more on Petroleum-Zuid, see http://www.agstadsplanning.be/kaai/kaai_petrol.html.

towns, the light rail claims a central location in the transversal section of the strip, claiming a spatially structuring role in specific redevelopments.

8.3.5 "Postproductive Landscape": Boom

Further south, along the right bank of the Rupel River, the industrial profile of the waterfront is more similar to that of Temse and Rupelmonde. Surfacing layers of clay provided the base material for brick production. Around Boom, the excavation occurred in a remarkably systematic way (Figure 8.9). Between the 13th and 18th centuries, a parcel-wise excavation was structured by a linear 400 × 400 m raster along the riverbank—400 m being the distance one could walk in five minutes. This "ladder" was stretched between the waterfront and a parallel service road. Along some of the "rungs" of this "ladder," one-street villages developed. At the start of the 19th century, the brick production industrialized, and a second excavation front was formed inland. Urbanization did not follow this new development. It remained within the old town centers and the raster structure. Also, brick factories stayed within the raster as they needed access to the water for transportation. However, during the 20th century, an increase in the scale of industries forced the brick factories inland beyond the initial grid, leaving the intricate small-scale "ladder" structure as an opportunity for qualitative redevelopment; small-scale infill projects can occur, creating a contemporary alternative to the existing one-street village typology with a transversal orientation toward the water rather than a linear waterfront for the "happy few." North of the service road, huge clay pits have transformed into valuable patches of reclaimed nature.

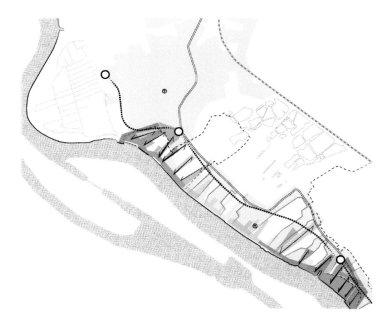

FIGURE 8.9 "Reinterpreting the productive landscape": positioning a public transport connection on the old service road between the two scales of development.

Infrastructure Design as a Catalyst for Landscape Transformation

The industrial areas in between them could benefit from a shift in mobility profile, which a new public transport line provides. Locating the light rail on the old service roads connects both scales of development while reusing existing infrastructure.

8.4 RESEARCH BEYOND THE DESIGN: THE EMERGENCE OF RECURRENT DESIGN CRITERIA

As the term "research by design" suggests, the "design" (*product*)—as a project and as an activity—described previously, is at the service of "research" (*process*) (Klaassen 2004; Schreurs and Kuhk 2011) as a means and a methodology. According to De Jong and Van der Voordt (2002), the subject of research by design can either be the *context* or the *object* of the design or, in some cases, both. In this case, the design was intended as an exploration of the *object* of regional public transport, not of the *context* of Klein-Brabant. To make the results of this exploration—in terms of *process*, not *project*—more tangible, a number of "design criteria" were defined. These are derived from the recurrence of certain questions or opportunities throughout the design. Making these explicit makes the design process communicable even without projecting them into a plan. It is also intended as a step toward the transferability of the research. Formulated differently, criteria are used to isolate the *object* of research from the *context* of research. Four of them are derived (Figures 8.10 and 8.11):

- The physical landscape as a structuring force. First, the degree to which the physical landscape has, in fact, been a structuring force for spatial developments has to do with scale (for example, the big scale of Scheldt river means it has a region-wide impact), with specific constraints (for example, soil conditions can limit development possibilities), or with opportunities (for example, in terms of resources). This criteria forms the basic foundation from which the other three are derived.
- The influence of morphological landscape characteristics. Second, the morphology of the landscape on the local scale can have both positive and negative features for the implementation of a new infrastructure. For example, the curvature of the river, efficient public transport requires fast connections with curves as smooth and as few as possible—in this case, a straight waterfront offers important advantages to compete with the congested car road in terms of travel time. Also, the alternation between inner and outer curves in the river-scape is of importance; outer curves typically have clearly defined edges, resulting from erosion. Inner curves, on the other hand, are much more susceptible to sedimentation, leading to floodable plains, which are, thus, less developable. Settlements that are situated on outer curves profit from their location through the development of a direct waterfront; in the past because of the accessibility to the water as an infrastructure and today because of the attractiveness of the landscape.
- Patterns emerging from an economic exploitation of the landscape. A third criteria looks into human activity that embedded itself into, and made use of, the landscape, mainly in terms of economic opportunities. As mentioned previously, direct accessibility to the river offers a distinct locational

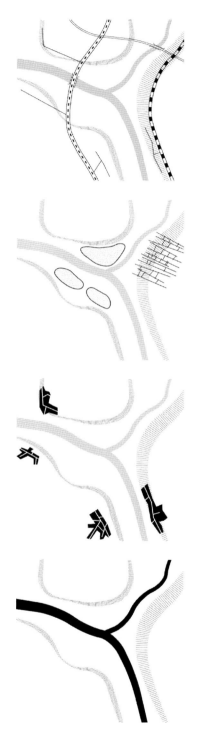

FIGURE 8.10 Schematic representation of the four design criteria.

Infrastructure Design as a Catalyst for Landscape Transformation

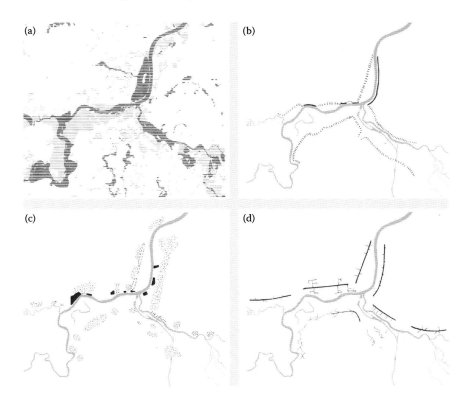

FIGURE 8.11 An overview of the four design criteria as they apply to the region. (a) The physical landscape as a structuring force. (b) The influence of morphological landscape characteristics. (c) Patterns emerging from an economic exploitation of the landscape. (d) Infrastructure projected on the physical landscape.

advantage. This explains why the waterfront south of Antwerp developed as a strip of water-bound industries. Alternatively, the landscape itself can also have an economic significance as a productive landscape. Apart from floodable lands used for forestry, a clear example of this is the clay pits near Boom. The availability of resources and the easy transport over the river made it an economic activity of importance for the entire region. As the economic importance of the river valley has declined, both in terms of its role as infrastructure or as a productive landscape, there are many brownfields emerging with a strong potential for reconversion. Another aspect of this criteria is whether urbanizations embedded in a linear topography structure are built up from a string of separate towns with open space, such as polders, in between them or whether development dynamics have caused settlements to join into one continuous strip as is the case south of Antwerp. As light rail has a lower density of stops than bus or urban tram systems, it could be used as a strategy to redefine the individual centers in a continuous strip.

- Infrastructure projected on the physical landscape. The last characteristic puts the light rail in relation to other infrastructures, raising the question of whether a bundling with landscape elements is always the right choice. For example, as the secondary road between Kruibeke and Rupelmonde lacks structuring potential in terms of spatial and civic quality and is already congested, the proposed bundling with the dyke offers a good alternative. On the other hand, a critical evaluation of the proposal south of Antwerp shows that a transformation of the existing railway is much more effective than a new line near the waterfront.

8.5 CONCLUSIONS: COMPARING THE PROPOSED DESIGN SCHEMES WITH DIFFERENT SCENARIOS AND APPLYING THE DESIGN LESSONS ON THE SCALE OF FLANDERS

As mentioned in the introduction, this research by design on landscape-based light-rail trajectories is only one of several scenarios being considered within the ORDERin'F project. The last of the formulated design criteria, which looks at the waterfront in relation to other infrastructures, is, in fact, a first step toward a comparison with the other two scenarios—the heavy- to light-rail reconversion of existing railways and the *tramification* of bus corridors. The proposed design schemes should thus not be read as a final product of which the merits are considered proven. On the contrary, they are merely testing the capacity and structuring potential of the landscape for the development of public transport–supported corridors; confronting the designs elaborated in this chapter with the other scenarios and calculating the traveler potential for some of the light-rail connections—which was done by the Flemish public transport company—allowed a critical evaluation of each corridor.

At the outcome, one of the decisive factors is whether the proposed strategy of a network based on landscape structures actually works. The effectiveness has to be proven in two ways. On the one hand, ridership potential should be high enough to justify the construction of this new infrastructure. On the other hand, its capacity as a backbone for new developments is crucial. For this, the design explored possibilities in which the landscape was considered in an inclusive way; it is not only about realizing a scenic public transport route, but also about aligning infrastructure to the physical landscape, thus mitigating its impact. Most of all, however, it's about making infrastructures an integral and integrated part of the cultural landscape.

Yet this is not the only goal of the research by design. Through the design of public transport trajectories, it has explored the morphology of settlements—whether they are broad or narrow, sharply defined or diffuse, structured by parallel or transversal infrastructures or not—resulting in possible design schemes. As such, it proposes a methodology that is radically different yet complementary to existing infrastructure planning.* On

---

* Within the existing policy framework, some of the research methodologies were tested and put to practice in the trajectory study for Brabantnet, which was commissioned to BUUR in collaboration with OSA, two of the ORDERin'F research partners, http://www.delijn.be/mobiliteitsvisie2020/pegasus _vlaamsbrabant/index.htm.

Infrastructure Design as a Catalyst for Landscape Transformation

a policy level, the challenge is to support these integrated ambitions of spatial, mobility, and infrastructure planning with a framework that capitalizes the synergies between them. As Smets argues,

> In practice, the division among the traditional disciplines and the customary forms of commission related to it, run against this perception of infrastructure as an all-inclusive landscape. ... A very large number of sectoral authorities intervene in the construction of the territory. Many of them have their own habits, their own budget, and like acting as their own principal. ... The increasing complexity is the greatest drive to alter this policy of compartments. (2001, 121)

Thus, in conclusion, the design works on a number of different levels. It is a proposal for a new mode of public transport in a region that is currently car-oriented. It is a proposal for an infrastructure that steers a concentrated urban growth dynamic that is currently diffused along an overlay of different mobility networks. And it is also a strategy to define potential landscape-supported corridors of urban development.

REFERENCES

Boussauw, K. (2011). *Ruimte, Regio en Mobiliteit: Aspecten van Ruimtelijke Nabijheid en Duurzaam Verplaatsingsgedrag in Vlaanderen*. Antwerpen—Apeldoorn: Garant, 255 pp.

Curtis, C., J. L. Renne, and L. Bertolini (Eds.). (2009). *Transit Oriented Development: Making It Happen*. Farnham: Ashgate Publishing, Ltd.

De Block, G., and J. Polasky. (2011). Light railways and the rural-urban continuum: Technology, space and society in late nineteenth-century Belgium. *Journal of Historical Geography, 37*:312–328.

De Block, G., M. Smets (sup.), B. De Meulder (sup.). (2011). *Engineering the Territory. Technology, Space and Society in 19th and 20th Century Belgium (Infrastructuur als inzet voor de organisatie van het territorium. Technologie, ruimte en maatschappij in België sinds het begin van de 19de eeuw)*, 382 pp.

De Jong, T. M., and D. J. M. Van der Voordt (Eds.). (2002). *Ways to Study and to Research Urban, Architectural and Technical Design*. Delft: Delft University Press.

Grosjean, B. (2010). *Urbanisation sans Urbanisme: Une Histoire de la "Ville Diffuse."* Wavre: Editions Mardaga, 352 pp.

Klaassen, I. T. (2004). *Knowledge-Based Design: Developing Urban & Regional Design into a Science*, Delft: Delft University Press.

Nolf, C., B. de Meulder, O. Devisch, and K. Shannon. (2011). Infusing identity into suburbia. In *Ethics/Aesthetics. ECLAS 2011 Sheffield Proceedings*, C. Dee, K. Gill, A. Jogensen, 265–266. Sheffield: Department of Landscape, University of Sheffield.

Reed, C. (2006). Public works practice. In *The Landscape Urbanism Reader*, ed. C. Waldheim, 269. New York: Princeton Architectural Press.

Ryckewaert, M., and A. Jocelyne. (2011). *Building the Economic Backbone of the Belgian Welfare State: Infrastructure, Planning and Architecture 1945–1973*. Rotterdam: Uitgeverij 010, 367 pp.

Schreurs, J., and A. Kuhk. (2011). Hybride narratieven in regionale toekomstverkenningen: Verkenning van de complementariteit van Ontwerpmatig Onderzoek en Scenario-Bouw. In *Plannen van de toekomst Gebundelde papers en bijlagen Plandag 2011*, Bouma, G., Filius, F., Vanempten, E., and Waterhout, B. (Eds.), 333–352.

Secchi, B., and P. Viganò. (2011). *La Ville Poreuse. Un Projet pour le Grand Paris et la Métropole de l'après-Kyoto*. Genève: Mētis Presses.

Shannon, K., and M. Smets. (2010). *The Landscape of Contemporary Infrastructure*. Rotterdam: NAi Publishers.

Smets, M. (2001). The contemporary landscape of Europe's infrastructures. In *Lotus International* (110), Milan, 116–125.

Strang, G. (1996). Infrastructure as landscape [infrastructure as landscape, landscape as infrastructure]. *Places, 10*(3):8–15.

Van Acker, M., M. Smets (sup.), B. De Meulder (sup.). (2011). *From Flux to Frame. The Infrastructure Project as a Vehicle of Territorial Imagination and an Instrument of Urbanization in Belgium Since the Early 19th Century (Infrastructuur als Stedenbouwkundig Concept. Het infrastructuurproject als regionale armatuur van de verstedelijking in België sinds begin 19de eeuw)*.

Verhetsel, A., I. Thomas, E. Van Hecke, and M. Beelen. (2007). *Pendel in Belgie. Deel 1: de woon-en werkverplaatsingen*. Brussel, Federale Overheidsdienst Economie, KMO, Middenstand en Energie.

Waldheim, C. (Ed.). (2006). *The Landscape Urbanism Reader*. New York: Princeton Architectural Press.

9 Beyond Infrastructure and Superstructure
Intermediating Landscapes

Sören Schöbel and Daniel Czechowski

CONTENTS

9.1 Green as Infrastructure 171
9.2 Landscape as Intermediating Level 173
9.3 Landscape Qualities 175
9.4 Landscape Strategies 175
    9.4.1 An Urban Ensemble: Revitalizing Lost Traces along the Tejo
        Estuary 176
        9.4.1.1 Context 176
        9.4.1.2 Concept 177
        9.4.1.3 Strategies 179
    9.4.2 PEATLANDscape 179
        9.4.2.1 Context 179
        9.4.2.2 Concept 179
        9.4.2.3 Strategies 183
    9.4.3 Integrating LANDSCAPES: Niches as Potential 183
        9.4.3.1 Context 183
        9.4.3.2 Concept 184
        9.4.3.3 Strategies 185
    9.4.4 Two Landscapes: One ThuRING 186
        9.4.4.1 Context 186
        9.4.4.2 Concept 186
        9.4.4.3 Strategies 189
9.5 Conclusion: Landscape beyond Infrastructure and Superstructure 190
References 190

9.1 GREEN AS INFRASTRUCTURE

When speaking of *green infrastructure* or even *nature as infrastructure*, the intention apparently is to increase, restore, or renew the societal appreciation of nature. In a society that is programmed to progress and, at the same time, dominated by the "Attention" economy, it is probably necessary to develop, from time to time, new concepts and reasons for everything that are not offered by the forces of the free market.

What does *infrastructure* actually mean? In the form commonly used in politics and the media, the term refers to the physical and institutional bases of economies. This includes the technical equipment, such as equipment for the provision of energy, transport, and communication, as well as the social institutions, such as the education system, the health system, and cultural institutions. In the provision of all that, the state has an important role, being responsible for the functioning of the national economy but also guaranteeing services of general interest. At the same time, the responsibility for the infrastructure opens up steering options to the state when, for example, as under the German constitution, equal living conditions should be guaranteed throughout the national territory.

The intention of talking about *green infrastructure* or *nature as infrastructure* now is, first, to widen a limited understanding of infrastructure, currently mainly concentrated on traffic systems and digital communication, and to point out that public and merit goods are as important for public service and for all economic activities as those goods that are directly involved in the economic exchange processes. And second, it wants to demand that infrastructure itself should become "greener" or "more natural." Against both, we would have no objection if it would not reduce nature to a material or institutional function.

Even if it seems so, we precisely do not want to rehabilitate the intangible, ideational importance of nature as Marx's economic theory would suggest it. According to him, the social relationship to nature is not solely defined in the material basis of society but only in a dialectical relationship to the superstructure,* thus, the state, the religion, the art, the science, and the prevailing social ideas that are represented in laws and paradigms but also spatially relevant in their symbols, such as significant places, monuments, etc. Indeed, it is true that we can assign many terms that we use in connection with nature to the conceptual pair *infrastructure–superstructure* (Table 9.1).

So when the *infrastructure* is stressed in comparison to the *superstructure*, this is an expression of a neo-materialistic *zeitgeist*. It is no longer about idealistic concepts for a humane living environment, according to which the social conditions should be corrected, but about technical-ecological concepts for a functioning economy. Because it is a dialectical relationship, a discourse on *green infrastructure* is finally supposed to affect the superstructure. So emphasizing the importance of nature as infrastructure means actually intending to change the superstructure. No objections can be made against this as has already been said. However, we see that highlighting *the functional* not only results in a neglect of *ideals*, but also of *something else*.

Indeed, there are also concepts of nature that we cannot clearly assign hitherto. These are terms that describe the spatial qualities, for instance, *open space* and *landscape*. Both are undoubtedly social forms of nature, which can be assigned to neither the infrastructure nor the superstructure without losing their intrinsic quality, their proper meaning. If *landscape* and *open space* are either only measured as functions or only interpreted as abstract intellectual ideals, they lose the essential property as social and spatial *forms* of nature.

---

* cf. French: *infrastructure ou base matérielle et superstructure*; Spanish: *Infraestructura y superestructura*; German: *Basis und Überbau.*

TABLE 9.1
Terms of Nature as Infrastructure and Superstructure

Superstructure
Power, polity, social awareness
Eco-policy, sustainability
Nature conservation act
Allegoric garden and park
Bourgeois public space
National parks, nature reserves
Green systems, green benchmarks

Infrastructure
Basic physical and organizational structures
Facilities and services
Agriculture
Natural resources, climate
"Ecosystem Services"

9.2 LANDSCAPE AS INTERMEDIATING LEVEL

In the 1970s and 1980s, internationally, geographers, sociologists, and cultural scientists critically assessed the economic theory of base—infrastructure—and superstructure. They realized that not only are production conditions and social awareness structuring factors of society, but this essentially applies for space. This led to the so-called "Spatial Turn" in which the role of space in the constitution of social relationships was rediscovered. However, because the spatial review of social conditions reveals not uniform classes but differences, it was necessary to leave the field of the theory of economic basis and cultural hegemonies and to deal with the practice of everyday life.

We picked out two authors who are well known among landscape researchers but seemingly fairly different, Henri Lefèbvre and John Brinckerhoff Jackson. While the French sociologist and philosopher Lefèbvre described the global urbanization at the beginning of the 1970s, the landscape theorist Jackson focused on the North American urban landscape of the early 1980s. Yet both wrote their essay studies from a perspective on the everyday worlds and criticized mainly similar phenomena: that only those who felt responsible for the social organization of space and, at the same time, for the social relationship to nature, namely urban or landscape planners, were guided by wrong spatial categories. Lefèbvre says that the industrial logic makes them blind to the processes of general urbanization. Jackson complains that the ideal of a static landscape blocks the view on the evolving vernacular-mobile landscape.

To prove this, Lefèbvre (2003) and Jackson (1984) first describe two levels—and, at the same time, historical states of social practice—as well as concepts of space and nature. The first level is that of everyday habitation and work. This spatial category is nonhierarchical, decentralized, private, and vernacular and has significantly coined

TABLE 9.2
Terms of Nature as Infrastructure, Superstructure, and Intermediating Structure

Lefèbvre		Jackson
Level G	**Superstructure**	**Landscape Two**
Centrality of wealth and power (170, cf. 78, 89, 97)	Power, polity, social awareness	Property, permanence, power (152)
Homogenizing and segregating industrial ideology (96, cf. 133)	Eco-policy, sustainability	Political ecological, social, religious static utopia (148)
Monuments, large-scale urban projects, new towns, "nature preserves" (131)	Nature conservation act	Homogenous and distinctive (152)
Industry captures nature (117)	Allegoric garden and park	Established, premeditated (148)
"town-country" distinction (79)	Bourgeois public space	Shapely, beautiful spaces: scenery, a work of art, a kind of supergarden (152)
	National parks, nature reserves	Country, nation (149)
	Green systems, green benchmarks	Boundaries, monuments, centrifugal highways (150)
Level M	**Intermediating Structure**	**Landscape Three**
Specifically "city" (80)	Specifically landscape	Balance (154, 155)
Differential and unifying space (108, cf. 125)	Balanced mobile and stable, Differential space	Both the mobility of the vernacular and a stable social order (155)
Right to (poly)centrality and the right to the city (97, cf. 150)	Natural morphologies	Environmental design: creating a new nature, a new beauty (155)
Urban ensemble of form and function (streets, squares, avenues, public buildings) (80)	Cultural landscape texture (pathways, meads and groves, water surfaces...)	
City that becomes productive by bringing together different nature (117)	Designed nature	
Spontaneous and artificial, absolute and pure facticity nature (parks, gardens) (132, cf. 26)		
Level P	**Infrastructure**	**Landscape One**
Everyday life habiting (80)	The basic physical and organizational structures, facilities, and services	Detachment from formal space, indifference to history, utilitarian, conscienceless use of environment (154)
As human nature and poesy (82)	The agriculture	Ephemeral, mobile, vernacular (148)
Agriculture settled into nature (117)	The natural resources, climate	Small, temporary, crudely measured (149)
Topoi (125)	The "Ecosystem Services"	

Source: Lefèbvre, H., *The Urban Revolution*, University of Minnesota Press, Minneapolis, MN, 2003; Jackson, J. B., Concluding with landscapes, In *Discovering the Vernacular Landscape*, 145–157, Yale University Press, New Haven, London, 1984.

Beyond Infrastructure and Superstructure

the social reality until renaissance (Jackson) resp. industrialization (Lefèbvre). The second level is the national, global, homogenizing, centralizing level that characterizes the notion of space and nature since the modern era. Then, both describe a third level that appears contemporary, everyday life–oriented and intermediating. Lefèbvre places, between the private level P of everyday life and the level G of global systems, the mixed or intermediating level M, the city in its proper meaning* as a differential space, which results from the urban tissue (Lefèbvre 1991). Jackson discovers, between a medieval, mobile Landscape One and the modern notion of a stable Landscape Two, the Landscape Three, the actual contemporary (North American) urban landscape as a mobile space that should be designed as a reasonably stable, balanced order.†

Even if the attempt seems a little foolhardy considering the authors' different times, contexts, and intentions, their levels of social production of space can be assigned to the dialectical image of infrastructure and superstructure as follows and, at the same time, picked up an intermediating level, in which now the, so far, missing concepts of nature can be meaningfully added (Table 9.2).

9.3 LANDSCAPE QUALITIES

What Lefèbvre describes as level M and Jackson describes as Landscape Three lies between infrastructure and superstructure as a separate level, precisely not opposed to the other two but mediating between them.‡ We call it *intermediating structure*. It is neither a definable space nor a category but rather an understanding, a reading of space, and, at the same time, a strategy of its development and redesign. It also represents a concept of nature, a social relationship to nature. Still following Lefèbvre and Jackson, the qualities of the intermediating level can be turned into spatial strategies that *sublate* in the triple sense—preserve, abolish, and transcend—social relationships to nature (Table 9.3).

Although these spatial strategies are all processes of interconnection, the prefix "re-" highlights the dialogical process of sublation.

9.4 LANDSCAPE STRATEGIES

In order not to get stuck, on the one hand, in discussions on performance outcome or object aesthetics of infrastructural facilities and, on the other hand, in designating landscape monuments or conservation areas, landscape architecture has to concentrate on the *intermediating* structures as described in Section 9.3. These qualities and strategies provide an overarching framework to reveal balanced and successful relationships between nature and design, between infrastructure and superstructure.

---

* "Level M (mixed, mediator, or intermediary) is the specifically urban level. It's the level of the 'city' as the term is currently used" (Lefèbvre 2003, 80).

† Jackson also refers to the term "megastructure," but, in contrast to the architectural concept, "few of us realize that there is another kind of megastructure, a megastructure in terms of a whole environment; one of the oldest creations of man. This megastructure consisting of the environment organized by man can be called the public landscape" (1966, 153).

‡ Linda Pollack (2006, 127) similarly uses "landscape as both a structuring element and a medium for rethinking urban conditions, to produce everyday urban spaces that do not exclude nature."

TABLE 9.3

From Qualities to Strategies: Intermediating Society and Nature

Superstructure

Level M	Intermediating Structure	Landscape Three
	(re)integration of Differences in Society and Nature	
	(re)cognition of Holism in Space and Nature	
	(re)construction of Morphology and Texture	
	(re)creation of Nature and Artificiality	
	(re)stabilization of Nature and Order	

Infrastructure

Throughout the years, LAREG* has researched and taught landscape as an intermediate level between local habits and generic processes, linking different—and often functionally separated—spatial requirements into "Landscapes of Holistics and Consistency." The design objectives and concepts therefore refer to and use landscape not as an object but as a spatial quality and, at the same time, as a spatial planning *principle*, for both the qualification of urbanized regions and the integration of new land uses and infrastructures.

In the following sections, we will describe selected academic projects to demonstrate how such landscape strategies are transmitted into spatial form and to discuss the role of landscape as a comprehensible interrelation as well as specific and characteristic qualities.

9.4.1 An Urban Ensemble: Revitalizing Lost Traces along the Tejo Estuary[†]

9.4.1.1 Context

The city of Lisbon is situated on the north bank of the river Tejo, a very dense and urban area whereas the south bank is characterized by diffuse horizontal suburban developments. With plans to set up a third bridge across the estuary to link the city center with its southern periphery, discussions on Lisbon's urban development have come back into focus (Figure 9.1).

Since the 1960s, Lisbon has been facing typical decentralizing and suburbanization processes, currently suffering from a decrease in population and an insidious dilapidation of the urban fabric and the historic building structure. In contrast, a suburban layer of industries and housing estates nearly wiped out the rural character of the southern periphery. "This new urban development took place without

---

* Department of Landscape Architecture and Regional Open Space at TUM.
† Master's thesis, Laurence Didier, LAREG, 2013.

Beyond Infrastructure and Superstructure

FIGURE 9.1 A third bridge across the estuary south of Lisbon will lead to new urban developments, especially in the town of Barreiro where the new bridgehead will be situated. (From Didier, L., Zwei Landschaften—Ein Urbanes Ensemble: Eine Wiederbelebung der vergessenen, verwischten Spuren an der Tejo Mündung, Master's Thesis, Technische Universität München, 2013.)

a dialogue with the existing structures of the landscape, and the long-lasting traces of the territory's formation were quickly erased by this development" (Didier 2013). Today, there are two opposite landscapes on both sides of the river that once formed an ensemble of cultural relationships (Figure 9.2).

9.4.1.2 Concept

To reestablish lost interactions between city and landscape, formative elements of the natural morphology and the cultural texture were reconstructed, and productive catalysts for local ecologies were developed. Two characteristic elements along the southern banks of the estuary were saline sites to produce salt and tidal mills to grind grains with the help of water wheels. In former times, a strip of saline sites and tidal mills could be found on specific locations along the riverbanks. Especially the saline sites were situated where little streams flow into the estuary in order to have both sweet fresh and salty water. The design proposes to reconstruct the spatial organization of the saline sites as well as the linear structure of the streams with their specific riparian vegetation so as to restore a stable framework to build upon and a relationship between the river and the landscape. New uses could be organic water treatment, algae farming, or different aquacultures (Figure 9.3).

The design concept focuses on the town of Barreiro where the planned bridgehead will be situated. Industrial development and new port facilities have transformed

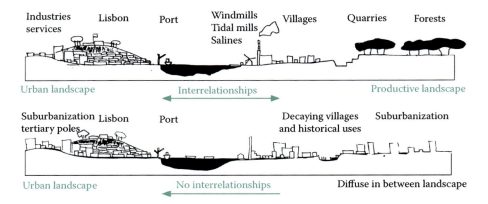

FIGURE 9.2 The interrelationships between both riverbanks are lost. Today, people commute from the southern bank to the city center. (Modified from Didier, L., Zwei Landschaften—Ein Urbanes Ensemble: Eine Wiederbelebung der vergessenen, verwischten Spuren an der Tejo Mündung, Master's Thesis, Technische Universität München, 2013.)

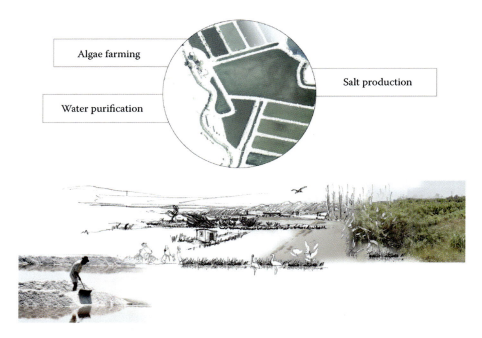

FIGURE 9.3 Lost traces of saline sites are revitalized and adapted to the 21st century, providing ecological benefits, reactivated landscape elements, and new ways of employment. (Modified from Didier, L., Zwei Landschaften—Ein Urbanes Ensemble: Eine Wiederbelebung der vergessenen, verwischten Spuren an der Tejo Mündung, Master's Thesis, Technische Universität München, 2013.)

Beyond Infrastructure and Superstructure

Barreiro's waterfront over the time and moved the historic shoreline toward the river. The original shoreline—tracing back to geomorphological processes—is exposed to mark the border of Barreiro's historic settlements (Figure 9.4).

This border leads from the "safe" land to settle on to a transitional space of a new productive saline landscape, which enables new land uses and integrates former characteristics of the place, based on a redesign and reuse of existing geomorphological, cultural, and industrial structures as an alternative to another faceless environment consisting of new generic housing and commercial estates (Figure 9.5).

9.4.1.3 Strategies

Many historic land uses are based on specific geomorphological conditions while their spatial layout originates from certain cultural techniques. In the case of the saline sites, both a *(re)construction of morphology and texture* reveals natural principles and cultural land use patterns. This reconstruction allows a *(re)stabilization of Nature and Order*. Ecological restoration and new agricultural productivity come along with setting up formative structures as a spatial framework. The new saline landscapes express a *(re)creation of Nature and Artificiality* by allowing natural processes embedded into an artificial framework.

9.4.2 PEATLANDSCAPE*

9.4.2.1 Context

The region Chernyakhovsk, part of the Russian exclave Oblast Kaliningrad, is characterized by moorland and marshes as well as by extensive fields, meadows, and forest areas. Due to the lack of social structures, Kaliningrad faces major problems: On the one hand, resettlement programs led to a strongly mixed social fabric originating from all parts of the former Soviet Union. A relatively large part of the population could not identify with the new home country and its cultural landscape. On the other hand, centralized economic planning and a lack of knowledge and appreciation of local management and craft techniques resulted in mismanagement and reinforced the neglect of large parts. The region now faces fallow arable land and run-down agricultural structures as well as intensive peat extracting in the moorlands with severe ecological implications (Figure 9.6).

9.4.2.2 Concept

The given geopolitical, historical, and social position of Kaliningrad particularly requires a sensitive approach to the landscape. PEATLANDscape reveals a cultural landscape development attached to historic structures and provides solutions to the enormous ecological, economic, and social problems of this region. The design objective was to develop a socially, environmentally, and economically adapted land use concept in which the historic qualities of the peat land cultures are connected to current social needs. The existing structures are gently renewed. This starts with a reconstruction and continuing of former path connections as well as closing thinned avenues, rows of trees, and forest residues to give access to and reconnect

---

* Bachelor's thesis, Johann-Christian Hannemann, LAREG, 2011.

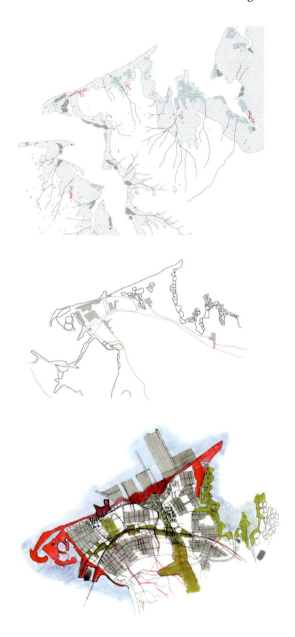

FIGURE 9.4 Geomorphological structures from the 19th century (top) and economic traces from the middle of the 20th century (center) are interpreted in a design concept (bottom). (From Didier, L., Zwei Landschaften—Ein Urbanes Ensemble: Eine Wiederbelebung der vergessenen, verwischten Spuren an der Tejo Mündung, Master's Thesis, Technische Universität München, 2013.)

Beyond Infrastructure and Superstructure

FIGURE 9.5 Design concept for the town of Barreiro. (From Didier, L., Zwei Landschaften—Ein Urbanes Ensemble: Eine Wiederbelebung der vergessenen, verwischten Spuren an der Tejo Mündung, Master's Thesis, Technische Universität München, 2013.)

FIGURE 9.6 Contemporary peat extraction in the *Angermoor*, moorlands near Chernyakhovsk. (From Hannemann, J.-C., Bachelor's Thesis, LAREG, 2011.)

the moorland to the public realm. Drainage trenches were closed to accumulate water to create new water bodies and to enable the cultivation of extensive wetlands (Figure 9.7).

An integrated approach of paludification, reed, soft wood, and pasture crops combines regional production and further processing in supply chains based on a cooperative organization with local and innovative production methods. Beyond integrated land uses, new ways of education, new infrastructure, and new sources of income arise, integrating different ages, employment groups, and education levels in order to develop a new sense of home and responsibility among the population (Figure 9.8).

FIGURE 9.7 Existing situation (top), reconstructed path structure (center), and new water bodies (bottom). (From Hannemann, J.-C., Bachelor's Thesis, LAREG, 2011.)

Beyond Infrastructure and Superstructure

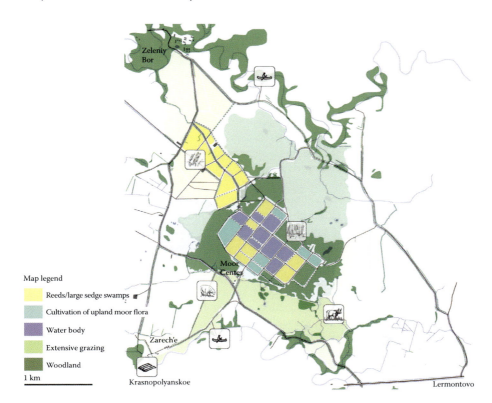

FIGURE 9.8 Design concept *Angermoor*. (From Hannemann, J.-C., Bachelor's thesis, LAREG, 2011.)

9.4.2.3 Strategies

Based on a *(re)construction of Morphology and Texture*, a new accessible and permeable spatial structure and measures for a stable natural regime of the moorlands (*[re]stabilization of Nature and Order*) prepare the ground for a *(re)integration of Differences in Society and Nature*, enabling new cooperation and allocation of responsibilities among the people and their landscape. The new moorland also fulfills the strategy of *(re)creation of Nature and Artificiality* because the spatial structure ensures different stages of natural states and succession.

9.4.3 Integrating Landscapes: Niches as Potential[*]

9.4.3.1 Context

So-called *Baraccopoli*—spontaneous and illegal settlements lacking sanitation services, supplies of clean water, and reliable electricity—are often the end point of a journey to a new country. In Rome, around 20,000 immigrants (2010) live in such informal settlements. The older shantytowns have existed for decades, but the vast

---

[*] Bachelor's thesis, Caroline Mittag, LAREG, 2010.

majority of these settlements are evacuated again and again, so there is no possibility for further development. Some of those are situated along Via Ostiense, an ancient Roman consular road. The essential function of the road is the connection of the city of Rome with the Mediterranean Sea, a transit space without a need to stop in between.

9.4.3.2 Concept

The design has taken up the current social topic of illegal settlements in the periphery of metropolises and developed concepts for the social and spatial integration of these settlements and their residents by means of landscape architecture. Integration is characterized by mutual confrontation, communication, finding similarities, and assuming a collective responsibility. However, ethnically and socioculturally segregated areas must not constitute "problem areas" but can, on the contrary, offer potential for the integration and for the productivity of an area. New cultures are weak in the context of the existing culture. Therefore, they seek niches and marginal areas but where everyday life, to a great extent, is influenced collectively. With a network of cooperative economies and adapted survival strategies, new cultures can develop relatively autonomously. This autonomy provides the opportunity for retreat, security, and advancement. The field structure between Via Ostiense and the river Tiber, as well as the forests, provides existing niche potentials combining basic needs to survive and new ways of an informal agriculture (rice, fish, forestry). New markets for economic and social exchange link the Via Ostiense to the informal settlements in the road's backyard, the fields, and the river Tiber. These markets are situated at the very few intersections along Via Ostiense where cars could turn and people cross

FIGURE 9.9 Design concept: Existing landscape structures and niche potentials along Via Ostiense between Rome and the Mediterranean Sea are used to improve and to connect *Baraccopoli* with the public realm. (From Mittag, C., Bachelor's Thesis, LAREG, 2010.)

Beyond Infrastructure and Superstructure

the road. The design concept shows how the existing niche structure can be used, designed, and incorporated into the public realm for the development of particular cultures (Figures 9.9 and 9.10).

9.4.3.3 Strategies

According to Lefèbvre (2003), the city is a differential space where differences come to light. But instead of isolating one another, these differences are active points of reference where only the confrontation gives rise to understanding. Based on improvement of accessibility and connectivity a *(re)integration of Differences in Society and Nature* establishes collective spaces allowing interaction and participation as guaranty for solidarity and tolerance. This also leads to a *(re)stabilization of Nature and Order* because people are able to take spatial responsibility using open spaces and landscapes in a cooperative and sustainable way.

FIGURE 9.10 Cooperative rice farms (top) and new market places for interaction (bottom). (From Mittag, C., Bachelor's Thesis, LAREG, 2010.)

9.4.4 Two Landscapes: One ThuRING*

9.4.4.1 Context

As one example for a European region with small-scale, decentralized structures of medium population density far away from larger metropolises, the German federal state of Thuringia is characterized by demographic change with its sociocultural and financial drawbacks. Due to climate change and "energy turn," the chance for the region's future is to increase the establishment of renewable energies, not only for a sustainable economic growth, but to enhance the landscape's potential.

9.4.4.2 Concept

Analyzing the landscape of Thuringia, one can say that the different elements often are in contrast to one another. This is indicated especially by topography and related land uses. In this perspective, Thuringia consists of two landscapes: one related to the central lowlands[†] and the other representing the mountainous areas in the south and north.[‡] The former landscape is characterized by amorphous and streamlined structures, the "Amorphous Stream Landscape," whereas the latter consists of linear and homogenous structures and therefore is called "Geometric Networks Landscape." In the interface of both landscapes (ThuRING), many opposite elements meet each other in a small space. On the one hand, the result is a high degree of landscape diversity and quality. On the other hand, this leads to vulnerability, indecision, and a diffuse superimposing of different characteristic elements. This interface area has defined borders, smooth transitions, and passages (Figures 9.11 and 9.12).

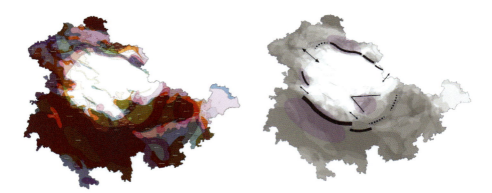

FIGURE 9.11 The overlay of various landscape characters (left) resulted in two specific landscapes and their interface (right). (From Braehler, A. et al., Design Studio IBA Thüringen, LAREG, 2014.)

---

* Anna Braehler, Hee Jin Chang, Gabrijela Tokic, design studio IBA Thüringen, LAREG, 2014.
† The Thuringian Basin (Thüringer Becken) is a depression in the central and northwest part of Thuringia.
‡ The Thuringian Forest (Thüringer Wald) and other mountain ranges like Harz, Kyffhäuser, Finne, and Windleite.

Beyond Infrastructure and Superstructure

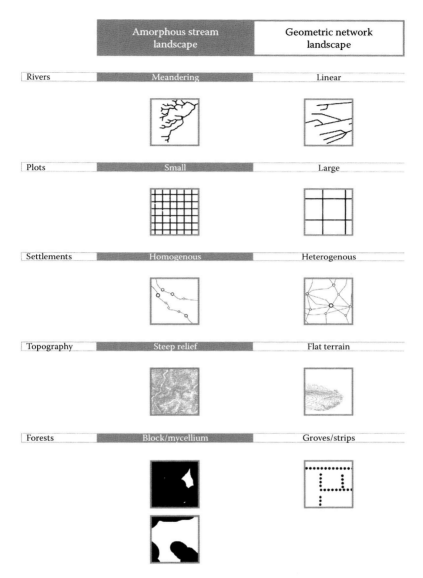

FIGURE 9.12 Matrix of differentiated basic landscape characters. (From Braehler, A. et al., Design Studio IBA Thüringen, LAREG, 2014.)

The design concept aims to accentuate the existing landscape structures and to use them as anchor points for the integration of renewable energies. The analysis results allow deriving structural rules. These rules are combined in a so-called "structural compass," which can be applied to the new energy layer as a design guide for the landscape development. Single small wind turbines and linear short-rotation plantations follow the rectangular logics of field structures and drainage systems in

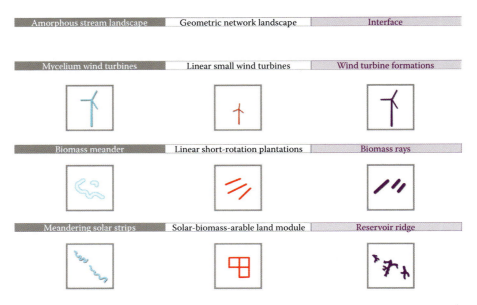

FIGURE 9.13 Matrix to allocate renewable energies to specific landscape elements. (Modified from Braehler, A. et al., Design Studio IBA Thüringen, LAREG, 2014.)

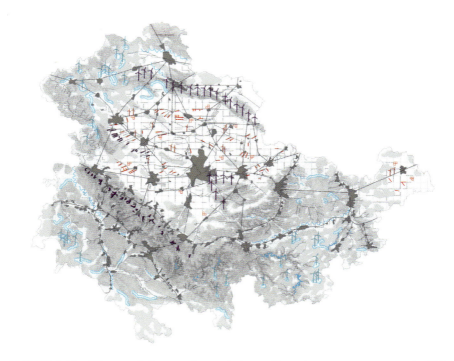

FIGURE 9.14 "Structural Compass" as a design guide to integrate new energy facilities. (From Braehler, A. et al., Design Studio IBA Thüringen, LAREG, 2014.)

Beyond Infrastructure and Superstructure

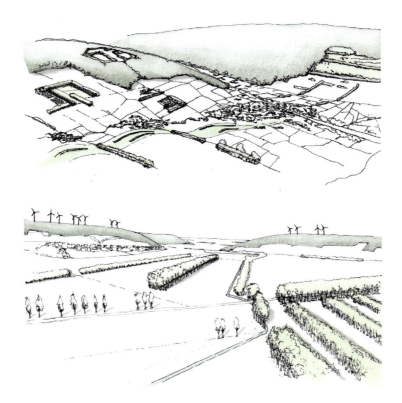

FIGURE 9.15 "Penetrations" in transition areas (top) and "divisions" along defined borders (bottom). (From Braehler, A. et al., Design Studio IBA Thüringen, LAREG, 2014.)

the "Geometric Networks Landscape." A modern triennial crop rotation enhances the wide character and enables an innovatively combined system of agriculture and energy. The dense, enclosed character and the meandering forms of the "Amorphous Stream Landscape" are the basis for solar strips, afforestation, and decentralized wind turbines in open forest clearings. "Divisions" along the clear borders and "penetration" in transition areas and passages mark the interface area. Division areas are dominated by one energy structure (biomass rays, reservoir ridge, wind turbine formations) whereas, in penetration areas, different energy structures from both landscapes meet (Figures 9.13 through 9.15).

9.4.4.3 Strategies

The energy turn will necessarily result in a transformation of the landscape. Instead of concentrating new energy facilities in regions that are supposedly already "disadvantaged," all landscapes should be open to renewable energies when these can be integrated aesthetically and even improve the quality of the landscape. In this case, a *(re)integration of Differences in Society and Nature* aims to reconcile new infrastructures with the landscape. The design concept defines specific landscapes by describing spatial characteristics, boundaries, and continuities. The character and the interrelationships of those "spatial bubbles" are enhanced by new renewable energy structures, truly a *(re)cognition of Holism in Space and Nature.*

9.5 CONCLUSION: LANDSCAPE BEYOND INFRASTRUCTURE AND SUPERSTRUCTURE

Obviously it is doubtful if the application of these strategies described actually will change the infrastructure as well as the superstructure into "green," but maybe that is not the point. What they offer is to serve as an intermediating level in the landscape—or to *be landscape* in its proper meaning as an aesthetical legible and appreciable spatial relationship. If, in contrast, new infrastructures—even if they are functionally "green"—appeal for social acceptance, the relationship between infrastructure and superstructure will be only based on power or persuasion, not on everyday life experiences or spatial qualities. In other words, even if they are "sustainable" or "green," ongoing urbanization and technical progress lead to separation when, for example, new infrastructures cause isolated and segregated spaces.

These strategies oppose a disaggregation of landscape into functionally optimized technical facilities (infrastructure) on the one side and visual impact assessments and planning concepts with the ideal of "Supergardens" (superstructure) on the other side. A conscious handle or even some kinds of *detournement* make infrastructures be a "natural" and integral component of our landscape and of people's everyday lives. This can be achieved if they follow the permanent structures of the morphology or continue knitting the texture of the cultural landscape. Thus, they are no longer seen as negative impacts that should be stored in disadvantaged regions and be compensated by restored biotopes in conservation areas somewhere else. Infrastructures do not devastate but could add something to the landscape. By adding or integrating new elements, existing relationships in the landscape could be clarified or newly created, characteristics could be strengthened, and peculiarities could become productive differences.

To intermediate, landscape must be open and collective; that means accessible to the public, enabling participation and social interaction but also available for common uses. Landscape must not be reduced to usability and quantifiable performance but also not remain only an immaterial aesthetic idea. As an intermediate level, landscape must allow and promote differences while showing diversity within a recognizable cohesive and holistic structure.

REFERENCES

Didier, L. (2013). Zwei Landschaften—Ein Urbanes Ensemble: Eine Wiederbelebung der vergessenen, verwischten Spuren an der Tejo Mündung. Master's Thesis, Technische Universität München.

Jackson, J. B. (1966). The public landscape. In *Landscapes: Selected Writings of J. B. Jackson*, E. H. Zube (ed.), 153–161. Amherst, MA: The University of Massachusetts Press.

Jackson, J. B. (1984). Concluding with landscapes. In *Discovering the Vernacular Landscape*, 145–157. New Haven, London: Yale University Press.

Lefèbvre, H. (1991). *The Production of Space*. Originally published in French under the title *Production de l'espace*, 1974. Malden, MA: Oxford Carlton Victoria.

Lefèbvre, H. (2003). *The Urban Revolution*. Originally published in French under the title *La Révolution Urbaine*, 1970. Minneapolis, MN: University of Minnesota Press.

Pollack, L. (2006). Constructed ground. Questions of scale. In *The Landscape Urbanism Reader*, C. Waldheim (Ed.), 125–139. New York: Princeton Architectural Press.

10 Landscapes of Variance
Working the Gap between Design and Nature

Ed Wall and Mike Dring

CONTENTS

10.1 Introduction 193
10.2 Variances in Fresh Kills 195
10.3 Fresh Kill Intervention 200
10.4 Fresh Kills Residency 202
10.5 Discussions 204
References 206

> By drawing a diagram, a ground plan of a house, a street plan to the location of
> a site, or a topographic map, one draws a "logical two dimensional picture." A
> "logical picture" differs from a natural or realistic picture in that it rarely looks
> like the thing it stands for.
>
> **Robert Smithson**
> *A Provisional Theory of Non-Sites, 1968*

10.1 INTRODUCTION

Between the abstractions of design and interventions in the land, there are variances, deviations, and gaps. These manifest as a physical interstice between the resistant conditions of the land and conceptualized interventions, and they reveal moments in the design process that refuse the seamless transitions from the schematic to the realized. As landscapes are repeatedly transformed through the dynamic relationships between people and the land, these gaps open up questions and spaces of creative possibility. Smithson wrote that between *site* and *non-site*, exists "a space of metaphorical significance" (1968). He proposed non-site as an abstraction of a physical geographical site that can come to represent the site but without the need to resemble it, his indoor earthworks constructing a new logic and potentially physical relationship between land and occupant.

As Smithson's non-sites were created to represent a physical site, so designed landscapes are formed that represent specific relationships between people and the land (Waterman and Wall 2012). Rather than what Corner describes as "unmediated environments" (1999, 153), landscapes exist as ideas, speculations, and representations of our relationship with the land manifesting themselves through drawings, paintings, films, diagrams, writing, and three-dimensional forms. For Smithson, the space between the *site* and the *non-site* is fundamentally metaphorical with the potential to become "physical metaphorical material" (1968); however, we argue that, in design, the variances between the schematic and the physical establish tangible spatiotemporal environments that reach beyond the symbolic.

As designers work with landscape and navigate through design processes, disjunctions can appear and contradictions can occur. This chapter examines these shifting and unstable states that open up their inherent possibility. It proposes that these variances be explored and interrogated, offering a focus to the design process rather than a peripheral concern that is ignored or concealed. It also argues that, through designing landscapes, the dynamic interactions of social and environmental processes should be taken as the principal generators of spatial forms and are not merely understood as a background that can be engaged with or enhanced aesthetically. Through focusing on the formation of space through process, the chapter questions the continued dominance of landscape by the scenographic. This visual approach, associated with the commodification of land and image for diverse global and local markets, continues despite the ambition of designers and proclamations to the contrary (Waldheim 2002). In recent decades, landscape has been recovered as a design medium and a method for engaging with the land (Corner 1999); however, there remain fundamental contradictions that require elucidation.

The site of this chapter is Fresh Kills. This is a landscape of waste on New York's Staten Island that has maintained a focus of artistic engagement for almost half a century. Through a series of studies, installations, performances, and designs, what was once the largest landfill site in the world has been worked, making explicit as well as concealing the landscape processes of waste (see Figure 10.1). This chapter

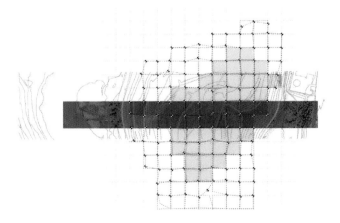

FIGURE 10.1 Between the abstract proposals for Fresh Kills and the interventions that are realized are spaces of exploration and interrogation. (Courtesy of Mike Dring.)

examines the work of these artists and designers. It reveals the questions that have opened up and the gaps left unresolved. Through examining their engagement with this vast landscape, the chapter attempts to expose the significance and potential of these moments of contradiction, ruptures in the relationship between people and land.

10.2 VARIANCES IN FRESH KILLS

The most recent reincarnation of Fresh Kills is as a public park. In 2001, *Lifescape*, a design proposal led by Field Operations, won an international design competition to remake the massive landfill site that had been closed the same year (see Figure 10.2). Conceptualized and later developed into a master plan that was published in 2006, *Lifescape* proposed to transform the 2200-acre (890-hectare) landfill site into a public park for the city (New York City Department of Planning 2006). This was the first major project win for the landscape architect James Corner, who founded Field Operations with the architect Stan Allen, leading to the *Lifescape* proposition being widely published, critiqued, and praised even before even the master plan was complete (Pollak 2002; Waldheim 2002, 2006).

Elegant layered diagrams dominate the competition and master plan documents. They present the design as a landscape of layers (see Figure 10.3), a palimpsest of proposed spaces and networks overlaid onto the engineered mounds of remediating landfill. This propositional approach appropriates and advances analytical techniques advocated by Ian McHarg (1969). As a landscape architect practicing with Wallace McHarg Roberts & Todd and teaching at the University of Pennsylvania, McHarg described a way of analyzing landscape by categorizing differing physiographic and

FIGURE 10.2 Redundant landfill machinery repurposed to display promotional landscape images within the physical site of Fresh Kills. (Rendering by James Corner Field Operations, courtesy of the City of New York.)

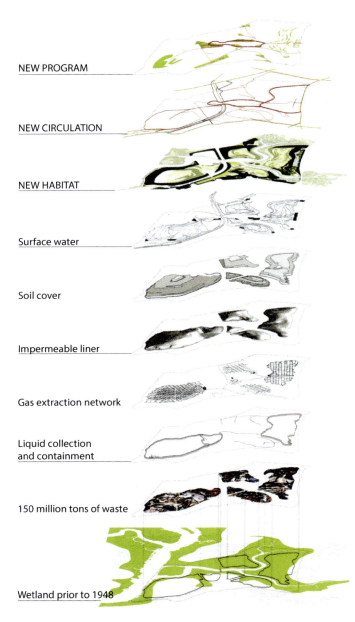

FIGURE 10.3 The layers of the *Lifescape* master plan expose separation between historic processes of waste and wetlands and the proposed new program, habitats, and circulation above. (Image used with permission of the New York City Department of Planning. All rights reserved.)

Landscapes of Variance

social values and superimposing overlay maps to determine sites for conservation and intervention. Decades before the design of the Fresh Kills Park by his student Corner, McHarg had brought together a team of specialists to interrogate the landscape of Staten Island demonstrating this method of separating out the information of the landscape into layers and then recombining them to give different readings. Described in the seminal publication *Design with Nature* (1969), McHarg was concerned with analyzing the landscape to demonstrate the "least-social-cost/maximum-social-benefit" of a proposed highway that transected Fresh Kills (1969, 35).

This technique, which has since been embraced by many designers working with landscape, was comprehensively appropriated in the design of the *Lifescape* master plan. In his mappings, McHarg presents existing conditions, such as wildlife, recreation, residential, land, soil slope, drainage, and erosion, and then overlays them to show areas for potential development or preservation (see Figure 10.4); in *Lifescape*, similar existing layers of wetlands, landfill waste, and engineered infrastructure are combined in diagrams but then strategically layered over with proposed program, circulation, and habitat (New York City Department of Planning 2006).

The *Lifescape* design for Fresh Kills Park has become an archetype in ecologically responsive designed landscape. For the design of Fresh Kills Park, Corner embraced the remaining traces of ecologically diverse wetland conditions as well as the imposition of the landfill waste that had created giant mounds of garbage and

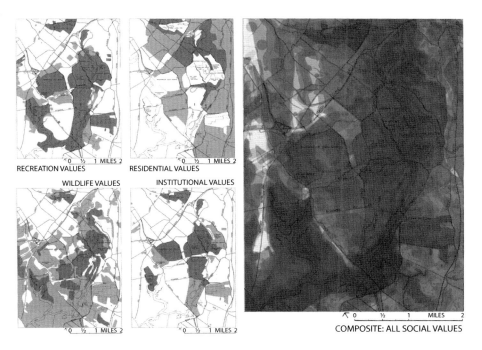

FIGURE 10.4 Layered approach to landscape analysis as advocated by Ian L. McHarg on Staten Island, New York. (Courtesy of WRT.)

damaging runoffs of leachate. Over three decades, the *Lifescape* design sets out to enhance the ecologies of the site while simultaneously capturing the productivity of waste processes to form a landscape that remediates through both plant growth and waste decomposition. Beneath this layered proposal and in place before work on the park began, an engineered clay cap and a gas extraction network draws the gas produced by this "living and breathing, potentially frightening, cyborg landscape" to provide energy for the surrounding neighborhoods (Pollak 2002, 59).

The design of Lifescape has been through layers of infrastructure, program, and habitats. "New program," "new circulation," and "new habitat" are highlighted as the *Lifescape* master plan's contribution to a landscape already transformed since the landfill operations ceased (New York City Department of Planning 2006, 12). These three propositions sit atop two other prespecified layers: surface water that flows across the four sealed giant mounds of garbage and a six-inch depth of soil that will facilitate the new habitats proposed in the master plan. Beneath this lays a two-foot-thick barrier protection material, a drainage layer, an impermeable plastic liner, a gas vent layer, and, finally, lying directly on the waste, a soil barrier layer.

This layering of the master plan accentuates the disconnection between the conditions of waste and the new park. What Linda Pollak describes as a "relatively conservative" system of impermeable layers of clay and membranes forms a divide between the history of the site and a purified vision that will emerge (2002, 59). This "reflects a desire to ignore our waste" through concealing the history of wetlands, 150 millions tons of waste, and a gas extraction network that intermix below (Pollak 2002, 59). The layering approach to the design appears an apt technique for a proposal which follows an engineered solution that ignores the potential interactions between the layers and, instead, separates them out. The conservative approach and the lack of experimentation by the designers and engineers disengage with the challenges and contradictions of this landscape: "Buried garbage does not biodegrade; it becomes inert. Slicing open the mound in 2030 might produce a still readable newspaper" (Pollak 2002, 60).

Penetrating the protective clay layer is a network of concrete nodes in a grid pattern across the new park. These geometric grids are the surface manifestation of the gas collection system that collects methane below and distributes it to the local community. Corner (2006, 31) describes grids as particularly effective for "extending a framework across a vast surface for flexible and changing development over time." The grid, he claims, provides "legibility and order" as well as "new opportunities" for a landscape of infrastructure rather than one dominated by object making. Represented in the *Lifescape* master plan, these ideas are manifest in the gas extraction and collection infrastructure that follows a notional 200-foot grid laid over the former landfill park (see Figure 10.5). However, in places, the realized grid deviates from the 200-foot diagrammatic grid, leaving many of the gas collection well heads forming a distorted arrangement. Their composition, contrasting the conceptualized grid, reveals a gap between what was physically constructed and the abstract gridded diagram.

These physical gaps raise questions. They suggest a disjuncture in the move from the schematic to the constructed, and they suggest an interruption in the design process or a resistance of the land against the abstract diagram. What caused the well

Landscapes of Variance

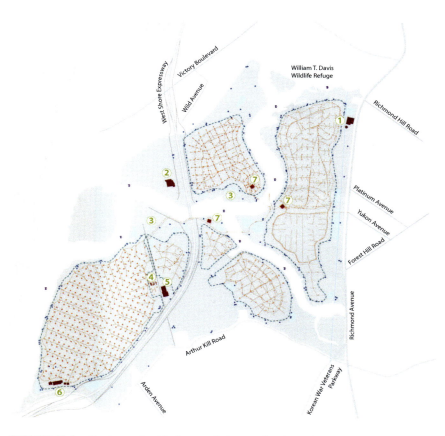

FIGURE 10.5 Engineered grid of concealed gas collection infrastructure contained beneath impermeable substrate. (Image used with permission of the New York City Department of Planning. All rights reserved.)

heads to be inconsistently placed? Did the conditions of the waste resist the imposition of the grid? What were the events of installing the grid? These gaps in the process, however, are not exclusive to the redevelopment of Fresh Kills. The difficulty of fully reconciling abstract proposals with the reality of sites is part of realizing designed landscapes. When they are not fully resolved, they are often seen as failures in the design process; these deviations, variations, and gaps are too often overlooked in reality, concealed in drawings, or smoothed over in construction.

Fresh Kills park is a landscape deployed as camouflage, creating the "magic disappearing act" described by Pollak (2002, 59–60). Fundamental to the design is an emphasis on processes of selected ecosystems and a programmatic engagement with the land. In the initial proposal, the designers describe nature as "no longer the image we look at, out there, but the field we inhabit" (Field Operations 2001, 7), reflecting an interest in image that Corner had written about two years earlier in his essay "Eidetic Operations and New Landscapes" (1999). Corner describes "landscape's dark side," where landscape is viewed as a passive object that can be controlled and possessed, and he criticizes those who overlook the "ideological, estranging and aestheticizing effects of detaching the subject from the complex realities of participating in the world" (1999, 156). In the *Lifescape* proposal diagrams of movement, programming and phasing illustrate the long-term ambitions for the park but remain offset with enticing scenographic photo collages. Despite a critical stance against the "dark side" of traditional forms of landscape representation, Corner struggles to dislodge the need to communicate through the scenographic visual image or find alternative ways to represent his designed landscapes.

Since its definition through picturesque landscape painting, landscape as a representational medium has provided a filtering and illusory effect (Dixon Hunt 1994). As a picture, these landscapes have consistently presented the world so that they can be "seen more pleasantly" and "visited more safely" (1992, 5). At Fresh Kills, the layered approach to the designed landscape offers an accessible veneer of ecological richness and purified programming, which are materially separated from perceived hazards of the waste below. This is a safe designed landscape, leaving unchallenged the contradictions that are present and either unable or unwilling to question the engineered assumptions that have already been made.

10.3 FRESH KILL INTERVENTION

Through land and environmental art in the 1960s, alternative understandings of the interrelationships between people and their environments emerged. The artistic interpretation of these relationships offers an alternative appreciation of site and formal intervention, one that moves beyond visual representations as demonstrated in Smithson's speculation of metaphorical gaps between site and non-site. Although both picturesque and land art forms can be seen as interpretive, it is the critical position that the latter occupies in framing ecological and societal concerns that gives it contemporary currency. Central to Smithson's work was his reading of entropy, a riposte to the minimal art movement of the 1960s, which he saw as embodying a systematic reduction of time (Lee 2001, 40). Originally a scientific term related to the second law of thermodynamics, entropy is described as a dissipating force within the universe, driving the physical world from a system of order to maximal disorder (Lee 2001, 39). Smithson and his younger contemporary Gordon Matta-Clark were concerned with making work that revealed its temporal, de-evolutionary nature. Entropy became a heavily worked metaphor not only in ecological terms, but in confronting issues of economics, energy, property, and consumption (Smithson 1973).

Site-specificity was also fundamental to the subjects and practices of Smithson and Matta-Clark. Smithson tended to give his work monumental significance through its indoor displacement as an act of removal that located entropy at its origin (Lee 2001, 54). Conversely, Matta-Clark's most renowned work was concerned with creating new social and artistic space through what he termed "anarchitectural" incisions (Lee 2001, 55), often resulting in complete destruction of the host, most apparent through his numerous building cuts. This "non-umental" tendency reinforced the temporal nature of his work, internalizing historical loss through entropic remove.

Landscapes of Variance

While site can be seen as a constant concern in Matta-Clark's work, the critical subject varies as does its relationship with the site.

Shot within the operational Fresh Kills landfill site, Matta-Clark's film *Fresh Kill* of 1972 offers an alternative dialectic. Contrary to his (an)architectural works in which the subject and the site are one and the same, here, the subject, represented by his truck, Herman Meydag, and the site, the then-working landfill that gives the film its title, feature as costars. The film draws on the dual meaning of the term "kill" as the watercourses that define the inner boundaries of the landfill site and the destructive acts captured within. Central to the film is the violent crash and crushing of Herman Meydag through the apparatus of the landfill operations (see Figure 10.6), the pulverized remains of Matta-Clark's beloved truck left to be buried along with the other everyday waste. Unlike many of Matta-Clark's other films that play on human relationships, *Fresh Kill* is notable for its absence of biological actors apart from Matta-Clark himself as driver and conductor along with flocks of gulls interspersed among the movie footage.

The narrative of the site is captured in the film, an endless cycle of consumption, sorting, and depositing, a violent terminal repurposed as peaceful burial ground. This inversion of the monument through the personalization and mortification of an inanimate object raises questions of value and use to both subject and site as corpse, ruin, and memorial, a place with history. The move from an alienated, universalized realm of waste processing to a place with an applied "use-value," ascribed by Matta-Clark's destructive acts and recorded through the film, articulates a paradigm for a deeper and more critical stance on the interconnections between identity and

FIGURE 10.6 Still from *Fresh Kill* video by Gordon Matta-Clark. (© 2014 Estate of Gordon Matta-Clark/Artists Rights Society [ARS], New York, DACS London.)

landscape where history and time are given greater authority in shaping future patterns and relationships with the land. Through the film and the actions within, the site and the externally perceived understanding of the site are subject to entropic processes that interrupt the previously accepted use-values and patterns. The symbiosis between urban growth and decay is revealed through the film, the site romanticized as ruinous with a rich archaeology of 20th-century consumer culture, a literal and metaphorical layered landscape.

10.4 FRESH KILLS RESIDENCY

Artist Mierle Laderman Ukeles, a contemporary of Gordon Matta-Clark, has been similarly concerned with property, consumption, and waste. Her long-term association with Fresh Kills is representative of her often epic artistic practice. Following her *Manifesto for Maintenance Art* in 1969, which was underpinned by core themes of maintenance, gender, remediation, and the environment, Ukeles began a series of art projects that ultimately brought her to interrogate the Fresh Kills landfill site itself. A proposed exhibition, *Care* (1969), drew from the manifesto and set the vision for her artistic engagement:

> Every day, containers of the following kinds of refuse will be delivered to the museum: (1) the contents of one sanitation truck; (2) a container of polluted air; (3) a container of polluted Hudson River; (4) a container of ravaged land. Once at the exhibition, each container will be serviced: purified, depolluted, rehabilitated, recycled, and conserved by various technical, and/or pseudo-technical, procedures either by myself or by scientists. These servicing procedures are repeated throughout the duration of the exhibition. (Ukeles 1969, 4)

The ambitious but unrealized exhibition aimed to intentionally break down traditional boundaries between artists, people, and land. Proposing direct interactions beyond conventional relationships and reaching beyond the constraints of formal art galleries, Ukeles' work attempts to expose awkward juxtapositions, ask difficult questions, and address societal prejudice. Following her appointment as artist-in-residence with the New York City Department of Sanitation, an unlikely meeting of utility and culture, Ukeles began a yearlong project, *Touch Sanitation* (1979). This work involved the artist personally meeting all of the 8500 sanitation workers (see Figure 10.7), documenting their stories, shaking them by the hand, and saying, "Thank you for keeping NYC alive." *Touch Sanitation* was to bring Ukeles to Staten Island's Fresh Kills, and over the following decades, she would return many times.

Although Ukeles' residency with the Department of Sanitation was unsalaried, in 1989, she received a Percent for Art commission to work on Fresh Kills through which she has developed numerous proposals and exhibitions. Her contract described her as "Artist of the Fresh Kills Landfill" (Ukeles 2002a). Ukeles initially worked on landfill sites across New York City in every borough except for Manhattan, where there are none. She developed proposals for earthworks within these sites as well as the project *Re-Raw Recovery*, which aimed to make these spaces of waste more public, opening them to "all kinds of creators" (Ukeles 2002a). Although the opposite occurred, as a series of environmental legislations made all the landfill sites except

Landscapes of Variance

FIGURE 10.7 *Touch Sanitation* performed by Mierle Laderman Ukeles across the sites of waste in New York City. (Courtesy of Ronald Feldman Fine Arts, New York, http://www.feldmangallery.com.)

for Fresh Kills inaccessible, Ukeles continued to express an interest in how these spaces are made and remade and those involved in the process.

Through her commission with the department, Ukeles was linked to the team selected to design a park for the landfill site that was announced to close in 2001. She wrote in anticipation of this engagement, "I've been waiting to get to work here for 24 years. Even though I've been absorbing the place all these years and creating various art works and texts about it, I've been waiting for a master design team to be organized to begin permanent work here" (Ukeles 2002a). The first phase was a research and reconnaissance project, *Penetration and Transparency Morphed*, a collaboration with filmmakers Kathy Brew and Roberto Guerra that extended her interest in how the site would be remade (see Figure 10.8). The work formed part

FIGURE 10.8 Mierle Laderman Ukeles documents the early remediation and containment of waste at Fresh Kills in *Penetration and Transparency Morphed*. (Courtesy of Ronald Feldman Fine Arts, New York, http://www.feldmangallery.com.)

FIGURE 10.9 *Morphing Timelines* offers a delicate addition to the gas collection well heads and to the Fresh Kills master plan, *Lifescape*. (Courtesy of Ronald Feldman Fine Arts, New York, http://www.feldmangallery.com.)

of her *Morphing Challenges* series that initially described three transformational layers of the landfill later to be joined by a fourth as the site was reopened in 2001 to accommodate the debris from the collapsed World Trade Center towers (Ukeles 2002b). Layer two defined the landfill closure and capping procedures, which Ukeles recorded through videos of the still-inaccessible post-industrial landscape set alongside interactions with those involved. What amounted to 21 individuals that Ukeles terms "pathfinders" represented the experts involved in the regeneration of Fresh Kills, including the director of landfill engineering, technicians, planners, theoreticians, and local residents (http://www.peconicgreengrowth.org). Through this work, Ukeles referenced the site's multilayered complexity not only in its reengineering but through its reconception as a public space.

As the *Lifescape* project progressed from a competition-winning entry into the draft master plan, Ukeles' project *Morphing Timelines: Energy*, as seen in Figure 10.9, was included in the *Lifescape: Fresh Kills Draft Master Plan* (New York City Department of Planning 2006). Ukeles is cited as a collaborator within the team led by Field Operations, her work framing the "Art and Culture" section, like a layer within a stacked proposal, held in place between the "Structures Plan" and the "Landscape and Habitat Plan" (New York City Department of Planning 2006, 24–29). *Morphing Timelines: Energy* reveals the gridded energy system that collects gas from the deteriorating waste below, including "tiny, delicate points of light" that were to mark the grid of methane gas well heads (New York City Department of Planning 2006, 26). Unlike her previous work, this simple project merely adds to the engineered system rather than engaging with its operations.

10.5 DISCUSSIONS

> ... there seems indifference towards penetrating beneath the mound's grassy surfaces continuing an Anglo-American tradition of deploying landscape as camouflage or a mask for abused land.
>
> **Linda Pollak**
> *Fresh Kills—Sublime Matters, 2002*

Landscapes of Variance

There are multiple narratives that define Fresh Kills. And there are differences between them and gaps within them. These interventions and interrogations of Fresh Kills on Staten Island offer narratives of ecological, landscape, architectural, engineered, and artistic process; they arise as contradictions between the envisioned and realized and represent spaces of latent potential with creative, social, and ecological gains. Extending from shortly after the Fresh Kills landfill site first opened to decades beyond its reimagination as a public park, these contrasting approaches to the site expose inconsistencies in past engagements and open up further questions of the site itself.

These approaches open up spaces in the horizontal field of the Fresh Kills park. They reveal tangible measures, as shown in Figure 10.10, between where gas collection well heads were intended and where they were ultimately constructed. They suggest subterranean conditions of waste, geology, or landform that were incompatible with what was expected through the engineered process. These gaps, which have spatial and temporal measures, can be appropriated to inform future design interventions. These successive interventions, by Matta-Clark, Ukeles, and Field Operations, are enacted in the context of what has come before and should benefit from ideas generated in the past. These gaps that open up between abstract design diagrams, the built reality, and the unpredictable outcomes could be worked to offer creative, social, and ecological gains. They infer narratives yet to be explored and potential unrealized. Future interventions could interrogate these moments as well as new conditions when the waste below becomes inert and the cap begins to break up and when social conditions inform new demands for Fresh Kills.

Instead of vertically layering up programs and processes onto the site, existing processes can be understood as redirected, adapted, and combined. The continuation of McHarg's analytical layering approach into a methodology through which

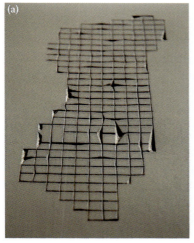 

FIGURE 10.10 Representations of spaces of variance between (a) the abstract well head grid and (b) their constructed locations across Fresh Kills. (Courtesy of Mike Dring and Ed Wall.)

new design solutions are formed, offers clarity in a process of designing landscapes. However, this conceptualization of a proposed landscape separated into layers can leave it insufficiently resolved as well as reinforcing static hierarchies of unconnected processes. Field Operations, in stage one of the master plan competition (2001), separated out the "alien" conditions of waste from the historic ecologies and photogenic social program set out in the Lifescape proposal, an approach criticized by Pollak, who sees potential in the "dark and disordered waste" (2002, 59). Understanding how these layers of process relate, connect, and influence each other could begin to overcome conventional engineered impositions. New approaches could begin to address the continued dominance of scenographic media and commodification through layering; they could challenge techniques that narrow an understanding of ecology and social programs; and inventive connections between these layers could be explored through inventive representations.

The projects of Matta-Clark and Ukeles begin, however, to meld processes—in particular, production and destruction. These artists define processes of change. This does not negate the significance of creative interventions but highlights those that are fundamentally adaptive rather than additive. These artists synthesize their conceptions with the will of the site rather than exposing the site to their creative method, a combination of design and site processes that allows the further delineation of spaces, generation of forms, and redefinition of landscape processes.

This synthesis of production and destruction and adaptive processes rather than additive could be imagined through the long-term development of the Fresh Kills Park beyond the timescale of the master plan. As Matta-Clark accelerates the deterioration of his truck through crushing and burying it in the waste so the decomposition of the waste is accelerated through the imposition of the cap. At some point in the future, as the waste becomes inert and the gas ceases to be produced, the protective clay cap and the gas collection system will become redundant. At this point, new processes could be engaged to open up the cap in a celebratory phase of growth, production, and destruction. It is this profound reorganization that Pope describes as the "promiscuous intermingling of the natural and the man-made" (1996, 211).

This chapter concludes that increased spatiotemporal site-specificity can inform a deeper response to a site through programmatic and conceptual frameworks for intervention, allowing site processes to delineate spaces, generate forms, and redefine our relationship with landscape. A greater connection between physical and conceptual layers and an extended timeframe could open up the potential of a site—interrogating moments of variance and accepting the unpredictability of unforeseen events. This technique could redirect, adapt, and combine disparate processes while multiple phases of design, implementation, maturation, and decline accept the entropic nature of the site and could be considered as opportunities for human engagement with the frequently restricted conceptions of nature and strictly concealed processes of waste.

REFERENCES

Corner, J. (1999). Eidetic operations and new landscapes. In *Recovering Landscape: Essays in Contemporary Landscape Architecture,* J. Corner (ed.), 153–169. New York: Princeton Architectural Press.

Landscapes of Variance

Corner, J. (2006). Terra Fluxus. In *The Landscape Urbanism Reader*, C. Waldheim (ed.), 21–33. New York: Princeton Architectural Press.

Dixon Hunt, J. D. (1994). *Gardens and the Picturesque: Studies in the History of Landscape Architecture*. Cambridge MA: MIT Press.

Field Operations. (2001). Lifescape: Fresh Kills Reserve. Competition proposal for design of Fresh Kills Park.

Lee, P. (2001). *Object to Be Destroyed: The Work of Gordon Matta-Clark*. Massachusetts: MIT Press.

McHarg, I. (1969). *Design With Nature*. New York: Wiley.

New York City Department of Planning. (2006). *Lifescape: Fresh Kills Park Draft Master Plan*. The City of New York. New York.

Pollak, L. (2002). Fresh Kills—Sublime Matters. *Praxis 4*: 40–47.

Pope, A. (1996). *Ladders*. New York: Princeton Architectural Press.

Smithson, R. (1968). A provisional theory of non-sites. In *Robert Smithson: The Collected Writings*, J. Flam, (ed.) 1996. Berkley, CA: University of California Press.

Smithson, R. (1973). Entropy made visible, interview with Alison Sky. In *Robert Smithson: The Collected Writings*, J. Flam, (ed.) 1996. Berkley, CA: University of California Press.

Ukeles, M. L. (1969). Manifesto for Maintenance Art. Manifesto written by artist and exhibited at the Feldman Gallery, New York.

Ukeles, M. L. (2002a). Leftovers / It's about time for Fresh Kills. *Cabinet*, Issue 6. http://www.cabinetmagazine.org/issues/6/freshkills.php (accessed October 31, 2012).

Ukeles, M. L. (2002b). Penetration and Transparency Morphed, 2001–2002. http://peconicgreengrowth.org/wp-content/uploads/2012/pdf/Mierle%20Laderman%20Ukeles.pdf (accessed October 31, 2012).

Waldheim, C. (2002). Landscape urbanism: A genealogy. *Praxis 4*: 10–17.

Waldheim, C. (2006). *Landscape Urbanism Reader*. New York: Princeton Architectural Press.

Waterman, T. and E. Wall. (2012). The landscape conversation. In *Donana: Out of the City*, G. Moretti (ed.). Milan: Maggioli.

11 Designing Integral Urban Landscapes
On the End of Nature and the Beginning of Cultures

Stefan Kurath

CONTENTS

11.1 Introduction into the Modernists' Conception of Nature and Its Contradictions..209
11.2 Adding Realism to the Modernists' Conception of Nature 210
11.3 New Practice of Planning: Interfere EVERYWHERE! 213
11.4 Designing Cultures: A Work of Cultural Extension....................................214
    11.4.1 Test Design River Delta 215
    11.4.2 Test Design Gravel Quarry 217
    11.4.3 Test Design Urban District 219
11.5 Questioning Apparent Matters of Facts Terminates the RE-Orientation in Design..221
References..222

11.1 INTRODUCTION INTO THE MODERNISTS' CONCEPTION OF NATURE AND ITS CONTRADICTIONS

In the "convention of modernism," as Bruno Latour calls it, "nature" as a category is understood as something "God-given," "self-organizing," and clearly also as a counterpart category to "society" (Latour 2000, 23). In this understanding, interventions into landscapes are generally perceived as interventions into the modernists' conception of "nature." Urban sprawl spoils the landscape, resources are exploited, and nature is destroyed. Even if contemporary landscape theory today defines "landscape" as a "dynamic system of man-made spaces," deeming the category "nature" hopelessly outdated (cf. Prominski 2004, 147), the closer definition of "landscape" continually relies on it being opposed to nature. Thus, the landscape is interpreted as "nearly entirely acquired nature" (Prominski 2004, 147), or the cultural landscape is referred to as "naturally as well as culturally acquired nature" (Körner 2005, 108). As long as this connection between landscape and the concept of nature described above cannot be overcome, the discourse regarding urbanism and landscape architecture

209

will always also find itself confronted with the call for protection of a "God-given" nature against human interventions and design measures. This "will to protect" nature as an independent entity threatened by "society" thus also drastically contradicts current requirements for urban, landscape, and environment planning within the context of the post oil and/or eco-city concepts (cf. Mostafavi and Doherty 2010), in which landscapes must increasingly be activated within and beyond the settlement areas as locations to be used for energy production, locally accessible leisure and recreation facilities, production areas for food, and compensation areas for fauna and flora in order to deal with ecological and social problems alike.

Against the backdrop of these current and future problems, a central question must be asked: How "destructive" are interventions into nature in reality? In order to answer this question, the concept of an opposition between "nature" and "society" must be reviewed critically—not in order to call reality into question, but in order to find a new access to it. Setting out from Bruno Latour's (2001) and Michael Hampe's (2011a,b) deduction that nature in the modernist conception, as an independent category, does not exist, it is the aim of this chapter to outline tactics and strategies enabling us architects, landscape architects, and urban planners, together with representatives of the natural, environmental, and engineering sciences, proactively to deal with the task assigned to us, i.e., to develop "new" urban landscapes within a cultural-disciplinary framework determined by a discourse regarding design and development.

11.2 ADDING REALISM TO THE MODERNISTS' CONCEPTION OF NATURE

When dealing with spatial reality under the aspect of impact history it proves that the categorical differentiation between "nature" and "society" is highly unfocused in everyday reality. Nature reserves, for instance, suggest that they accommodate "nature" that needs to be protected against humans. Actually, however, a nature reserve can also be reconstructed as a figuration of interplay among conservationists, flora and fauna, land owners, users, financiers, and politicians as well as technologies—applied to maintain the reserve. Similar phenomena can be observed regarding present-day cultured landscapes. The landscapes cultivated for agricultural and silvicultural exploitation are often considered partial aspects of nature or, at least, "close to nature," although they also prove to be extremely complex figurations consisting of farmer, tractor, plant seeds, agricultural theories, fertilizer, pesticides, and plots, i.e., human and nonhuman actors and factors (cf. Kurath 2011, 183). Approaching the scene from the opposite angle, the natural sciences have also recognized this reality. According to Stefan Ineichen and Max Ruckstuhl (2010, 9), the settlement landscapes were formerly considered foreign to nature and therefore hardly suitable for the development of a diversified fauna and flora. Current studies focusing on the issue of biodiversity in cities, however, show that the variety of species in many groups of relatives is actually higher in the settlement areas than, for instance, in areas of predominantly agricultural utilization approximately the same size (Ineichen and Ruckstuhl 2010, 9). Although the false concept of nature recapturing the city is often proclaimed, the assumption that the "natural," i.e., natural elements, such as stones, grasshoppers, lizards, lime, etc., always have been an essential

Designing Integral Urban Landscapes

element of our everyday surroundings is far more realistic. Currently, phenomena such as "urban farming" or "urban gardening" demonstrate diverse and innovative ways of dealing with the hybrid reality characteristic of dense urban centers, for instance, in the field of vegetable production or fish farming.*

This empirical way of dealing with networks of causal interrelations in everyday life hence reveals that reality cannot be explained by means of scientific categories, such as environment, economy, or society. The evolution of such categories thus has a different background. According to Bruno Latour, the outset is the "convention of modernism" (Latour 2007, 150). Latour describes how the project of modernism at the end of the 19th century consisted of making the world explicable by means of indisputable facts (Latour 2007, 150). In order to create states of knowledge that were as clear as possible, i.e., independent of subjective perception, a delimitation process among the fields of science became necessary. Although this differentiation process enabled the systematic production of "pure" states of knowledge, it simultaneously caused a rapid departure of scientific activity from reality as the subdivision of the world into equations and natural laws gradually cut the ties between science and the outside world (Latour 2000, 10).

A side effect of this separation from reality was an increasing trend toward ideology that transformed into a "war of the worlds"—a war among the different disciplines. In the race for ever new, seemingly indisputable facts, a struggle for control and responsibility for saving the world (e.g., global warming, urban sprawl, etc.) was sparked (cf. Latour 2004). Thus, the political negotiation process is increasingly handicapped in favor of better constructions of reality. The reason for this lies in the claim of every individual discipline, arising from the diversification, to be in a better position for explaining the world's fundamental mode of operation than the others. This includes the claim to be in a position to decide which solutions are "right" and which are "wrong." As nature, which, according to the modernists' concept, is perceived as something "God-given" and per se "good," presents itself as the greatest and thus the only source of facts in this game, it is repeatedly referred to. Yet, according to Bruno Latour, this is fatal in view of the necessity to find qualified solution strategies for urgent environmental problems. "Due to the ubiquitous threat of reference to an indisputable nature, ordinary political life is condemned to torpor" (Latour 2001, 22). In other words, nature is progressively used as an argument for defeating politics (Latour 2001, 33).

According to Michael Hampe, this is owed to the fact that noticeably ideologized sciences describe "concretely experienced interrelations with always the same terms that have become fetishes," thereby progressively neglecting precision requirements (Hampe 2011a). The ecological crisis, for instance, is related to concrete constellations of persons and objects. The analysis of such a situation requires an increased ability to differentiate. Yet this is precisely what the ideologization of nature does not allow: "Due to their abstractness ideologies are not suitable for finding solutions to such problems. To intensely recite them is an expression of the crises, and one of their causes, not a condition for overcoming them" (Hampe 2011a).

---

* See, among others, the Prinzessinnengarten in Berlin (http://www.prinzessinnengarten.net) or Frau Gerolds Garten in Zurich (http://www.fraugerold.ch).

Yet, to Michael Hampe, there exists a possible way out of the dilemma that would entail, fundamentally, to call experience into question again and to rearrange it along the lines of precision requirements. The case is urgent as the new challenges in context with post oil and eco-cities depend on overcoming this nature-related ideology in order for political solutions to be found for pressing problems.

At the outset, this rearrangement starts with the cognition gained from representing both nature reserves and cultured landscapes from the perspective of impact history. It appears that within these networks of causal relations, actors formerly considered "natural" cohabit with actors formerly labeled "artificial" or "social" (cf. Voss and Peuker 2006, 23). In this respect, everyday space must be reconsidered and described anew. Setting out from the associations of human as well as nonhuman actors described at the beginning of this chapter, everyday space must be perceived as a hybrid best described as an "urban landscape" or, even more accurately, "urban landscapes of associations" (cf. Kurath 2011, 443) within the context of the current discourse regarding urban and landscape planning.*

These "urban landscapes of associations" thus stand in a specific relation to the respective social reality, i.e., the locally existing alliances among different entities. In this respect, urban landscapes are also socio-technological urban landscapes. This combination of terms points to the fact that urban landscapes are neither socially nor technologically determined but emerge from an array of interrelations, relational networks, and transformations (Kurath 2011, 444). In this manner, such multiple causal networks certainly also override categorical differentiations always highly esteemed in the urbanistic discourse, such as "here city" and "there landscape," "here man" and "there nature," and "here technology" and "there transcendence," fundamentally challenging them. In existing urban landscapes, there exists no "opponent" but rather a collective togetherness, the figurations of which translate into spatial reality and become readable in the shape of multiple locally specific cultures. Thus they are neither "artificial" nor "natural" but "artificial-natural." This summarizes the theory of hybrid realities also reflected by the term "urban landscape" and, bringing aforementioned categories to an end, as reflected, for instance, by the term "interurb" ("Zwischenstadt") (Sieverts 2001) for the urban area between cities. Based on the same theory, the aesthetic concept of landscape proves terminally outdated. This concept originally seeks to define the "nature of an area as a beautiful landscape" and thus an aesthetic object to its own end (cf. Körner 2005, 121). Within the framework of this conception, the term "landscape" is reduced to the "nature of an area" we are looking at. Due to this idealizing and purifying categorization of "here" human being and "there" nature, object and subject are clearly divided, a division that has never existed in reality as something such as an outside position does not exist.[†]

The recognition of hybrid realities hence confirms that there is no use any more for "nature" as an analytical and scientifically purified category (Latour 2001, 41; Hampe 2011b, 298). The recognition of hybrid realities hence also confirms that

---

* Here the term "urban landscape" refers to the phenomenon "neither city nor landscape." It is not congruent with the use of the term applied in the first half of the 20th century when it referred to an urbanistic model of a "new city" as opposed to outdated urban structures (cf. Giseke 2010).

† Cf. Die seltsame Erfindung einer "Aussen"-Welt (Latour 2000, 10).

Designing Integral Urban Landscapes

there has never existed a division between nature and society. It is time to recognize the real relationships and to see reality as it is: an assemblage of human and nonhuman entities neither natural nor artificial. In his summary, Bruno Latour writes, "Thank God Nature will die. Yes, Pan the great is dead! After the death of God and Man, now finally Nature had to demise. It was just about time: otherwise we could soon not at all have used politics anymore" (Latour 2001, 41).*

11.3 NEW PRACTICE OF PLANNING: INTERFERE EVERYWHERE!

The end of nature as a category is the beginning of the cultures (plural), i.e., assemblages of human beings, things, and natural elements. Therefore, accurate tactics and strategies must, from now on, stand at the center of focus with regard to new tasks in the fields of urbanism and landscape planning, which no longer concentrate exclusively on "nature" or "society" but seek to spawn and enforce new cultures—and therefore need to interfere EVERYWHERE (cf. Latour 2001, 36)! Instead of only attending to human issues within settlement structures and the demands of flora and fauna outside the settlement areas, this new way of thinking is necessarily integral, uniting former opposites, and not seeking to separate with force and ideology. Based on an integral understanding of the urban landscapes of associations connecting the physical-material realities with its natural yet also socioeconomic, political, and cultural-mental dimensions (Eisinger and Kurath 2009, 82), interventions actually start to show productive effects as they do not destroy the "God-given" but extend causal chains in order to create the "new." Productive approaches can be sketched with the aim of improving "bad" urban landscape construction to make them "better" by adding missing ingredients, such as biodiversity, quality of residence, green space connections, and adequate density relations.†

Such a course of action, however, does not legitimate undiscerning interventions into spatial reality. On the contrary, the realism added here also reveals the delicacy and vulnerability of the system and must promote a responsible way of dealing with the issue. It is therefore imperative to identify new utilization forms, such as energy production, recreation, etc., and to analyze to what extent new, specific cultures can be designed in reality with regard to the delicacy and vulnerability of the system. In accordance with Martin Prominski, a good starting point is to perceive building as an enhancement of environments or, as outlined in this chapter, as an enhancement of cultures and thus to formulate utilization rules that also show regard for the demands of natural elements (Prominski 2004, 145).

Especially problems regarding latent momentum connected to specific interventions into existing cultures need to be attended to enable real-time reactions. Instead of resorting to the traditional assumption that planning follows the logic of cause and effect, the new creed is "a cause is a cause is a cause" (Latour 2007, 180). Therefore, the actors responsible for design and planning are asked to follow two courses of

---

* Parts of this section are taken from an essay called "Towards a Political Ruralism," which again is based on the original text for the symposium DNAI at the TU Munich.
† In reference to Bruno Latour's approach to add realism in order to aid the seemingly indisputable facts in achieving "better" constructions (Latour 2000, 24).

action, according to urban and environmental sociologist Matthias Gross (2006, 177), "(1) the revision and renegotiation of the knowledge accepted until then [Latour's established facts], and, often connected therewith (2), a new angle of interest in the constellation of actors as well as the negotiation of possibly newly emerged values and objectives." So before the first interventions into the landscape are followed by others, new constellations of interests must be mutually adapted, and thus, new knowledge must be acquired. "Here we are neither dealing with a passive adaptation, nor with a change that can be carried out randomly, but with an adjustment or a tentative mode of operation of the 'collective' ... " (Gross 2006, 177). Here Gross' essay on "collective experiments in the social laboratory" becomes the program of a new practice.

In a similar context, James Corner postulated the term "Landscape Urbanism" for a new practice of planning: "It marks a dissolution of old dualities such as nature-culture, and it dismantles classical notions regarding hierarchies, boundary and centre.... Techniques drawn from landscape—such as mapping, cataloguing, triangulating, surface modelling, implanting, managing, cultivating, phasing, layering etc.—may be combined with urbanistic techniques—such as planning, diagramming, organizing, assembling, allotting, zoning, marketing etc.—to help create a larger bag of tools than the traditional planner has had in the past" (Corner 2003, 58). This new self-image will help them "to be able to respond quickly to changing needs and demands, while themselves projecting new sets of effect and potential" (Corner 2003, 63).

In order to do justice to the hybrid realities, however, other transdisciplinary contents derived from sociology, the cultural sciences (former natural sciences), management, etc. must necessarily be added to the cross-discipline of landscape urbanism. Such transdisciplinary action theories must take into account causal inter-relationships in the production of cultures and lead to "better" constructions for the social and thus spatial figurations—i.e., constructions that show consideration for the complex and delicate dependencies of a "healthy" environmental system, also taking into account disciplinary singularities.

11.4 DESIGNING CULTURES: A WORK OF CULTURAL EXTENSION

Living and working in Vrin, Swiss architect Gion A. Caminada is dedicated to a different way of dealing with nature. In a discussion among Michael Hampe; Gion A. Caminada; and editors of *Trans* magazine Viviane Ehrensberger and Yvonne Michel, Caminada said, "In my personal experience idyll does not exist and nature can only be romanticised by people who have a distanced relation to nature. I believe that the right balance between mindfulness and appropriation is important. We should learn to perceive nature less than a commodity and resource, and in this manner seek to overcome the separation between object and subject" (Caminada et al. 2012, 39).

In his native village of Vrin in Switzerland, for instance, Caminada's works are characterized by their design that follows social needs and an intensely conducted discourse as well as profound studies of the local building culture (Eisinger and Kurath 2008, 173). His design practice hence represents a kind of translation activity, combining different, even opposing, interests and transforming the resulting contents into concepts. In this manner, Caminada creates further incentives for action with the aim of inspiring the population, the craftsmen, the farmers, etc. to change

Designing Integral Urban Landscapes

the cultured landscape along the lines of Caminada's frame of experience and reference (which encompasses the social and building-cultural content). Here designing proves an action theory for extending culturally determined conditions (work of cultural extension) (Kurath 2011, 549). Due to similar contentions, Daniel Gethmann and Susanne Hauser (2009, 11) describe the design process as a technique applied by cultures in order to advance their further development. Defined as "Mangle of Practice" (Pickering 1995, 4), the practice of designing produces syntheses among diverse fields, such as art, technology, science, or economy (Schumacher 2001, 28), which can be translated into spatial concepts via the design process—balancing culturally consolidated, disciplinary contents and levels of knowledge.

If you now perceive urban landscapes as *"neither-nor* and *both-and"* figurations of hybrid realities that can only be understood, described, and modified in relation to the specific framework conditions and associations between human and nonhuman actors, then an intervention into the urban landscapes calls precisely for such a cultural technology of designing, creating relationships among different entities, and remodeling them (Domschky and Kurath 2012, 28).

Such cross- and transdisciplinary procedures are taught at the master studio of the Zentrum Urban Landscape, architectural studies section, at the Zurich University of Applied Science (ZHAW). Here the focus lies on "relational design," which is introduced as an action theory aiming at lending urban landscapes a spatial, cultural, social, economic, and ecological and prospering qualification based on specific spatial as well as social characteristics and enriched with an agile approach to design, the will to negotiate, and diplomacy (Kurath 2011, 551). Thus, the design process not only aims at creating more beautiful urban landscapes, but also at creating prosperities within the cultural framework conditions, hence extending the chain of action as well as the value-added chain.

The work created in the design course of the master studies of architecture at the ZHAW presented in the following unfold potentials and opportunities that only become productive by designing "new" cultures.* The papers also demonstrate that, as figurations of a collective, such cultures never oppose a "nature." In contrast, human and nonhuman actors confederate to form "new" cultures. Correspondingly, these cultures cannot be discussed and judged by what they destroy but by which added values are created in the interplay between man and environment in both qualitative and quantitative respects.

11.4.1 Test Design River Delta

A respective test design was developed in the inflow area of lake Greifensee at Mönchaltorf in the Zurich Oberland in Switzerland. The riverbed of the Aabach river had, in the past, been straightened in order to gain agricultural land. The moorlands were drained, and the land was intensely used for agricultural purposes.

---

* The test designs River Delta by Adrian Berger and Gravel Quarry by Dario Papalo were written and have been published as a result of the research project "Strategien für eine nachhaltige Entwicklung von Einfamilienhaussiedlungen" SNF-NFP 54. The test design Urban District by Livia Schenk is a part of the KTI project "Planungsmethode städtebauliche Quartierentwicklungsleitbilder."

FIGURE 11.1 Concept for promoting biodiversity in formerly monocultural landscapes with a straightened river course in combination with the creation of space for work and residence. The design is based on a morphological analysis of the existing field structures. Enriched and extended, these structures yield a permanent structure as a guide for the processes of appropriation. (Courtesy of Adrian Berger, student, IUL, ZHAW, 2006.)

Designing Integral Urban Landscapes 217

FIGURE 11.2 Visualization of the coexistence and habitats of humans, fauna, and flora. (Courtesy of Adrian Berger, student, IUL, ZHAW 2006.)

Nevertheless, the river delta is a designated area for ecological compensation supported by the cantonal authorities. In an analysis map of the GIS Zurich, the delta features a number of potentials for supplementing wetlands. Thus, the inflow area of the Greifensee is defined as a conservation area.

The test design depicts a possible future for the inflow area of the Greifensee as a new type of "natural-artificial" park. It is to be financed and maintained by future detached house owners, whose recreation area it represents. With regard to regional planning, the park establishes a new public connection between Mönchaltorf and the lake (Figure 11.1).

It is the aim of such an integral understanding of urban landscape to interconnect residential areas with subnatural spaces in order to establish a coexistence, diversifying the formerly monocultural landscape in order to support biodiversity (Figure 11.2). At the focal point of this support, we find plant societies, such as moor grass meadows or sedge fields. As target species from the animal world, insects such as butterfly species (blues) or dragonflies must be mentioned; their habitats would have to be designed adequately. Reed warblers are actually not a target species with respect to conservationist requirements, yet they could be attracted by planting reed.*

11.4.2 TEST DESIGN GRAVEL QUARRY

Another test design presents a concept for sustaining the existing diversity of species in a gravel quarry in the area of the Swiss municipality of Gutenswil in the Zurich Oberland. The exploitation of gravel will cease some years from now. Constraints commit the operators to fill the then-idle quarry in order to recreate the characteristic landscape. This decision is counteracted by the knowledge that abandoned gravel quarries allow for a unique flora and fauna to develop over time due to the special conditions they provide, including many species of animals and plants worth protecting.

---

* According to representatives of the NFP 54 project: BiodiverCity: Urban biodiversity from the ecological and social science point of view, WSL Birmensdorf.

By applying the procedure described here, the biotope can largely remain undisrupted and even be developed further by the detached house owners as a complementary form of payment (Figure 11.3). By conserving and supporting biodiversity in the gravel quarry, the surrounding areas of the Acherbüel nature reserve in the side moraine landscape of Freudwil could be supported, and the locally available potentials for rough pastures could be utilized (Figure 11.4). As

FIGURE 11.3 Concept for the conservation of biodiversity in abandoned gravel quarries in combination with the creation of spaces for work and residence. Regular practice would demand filling the quarry and destroying the habitats contained in it. (Courtesy of Dario Papalo, student, IUL, ZHAW 2006.)

FIGURE 11.4 Visualization of the coexistence and habitats of man, fauna, and flora. (Courtesy of Dario Papalo, student, IUL, ZHAW 2006.)

Designing Integral Urban Landscapes 219

target species for the improvement of biodiversity, butterflies; grasshoppers; spiders and ground beetles; fire-bellied, natterjack, and midwife toads; sand lizards; ring snakes; marsh warblers; and perhaps even sand martins or nightingales could be resettled, and the free space design and vegetation could be directed toward this goal.*

11.4.3 TEST DESIGN URBAN DISTRICT

Because science hasn't dealt with the fauna and flora to be found inside cities for a long period, current research registers substantial biodiversity within settlement areas. With regard to this, the Milchbuck in the city of Zurich appears as a specially interesting area of research as two large landscapes approach each other here, the Zürichberg and the Käferberg, and previous plannings, for instance, by Karl Moser in the first half of the 20th century, featured a green corridor between the two landscapes that should cross the Milchbuck. Rudimentarily, this green corridor was applied by Hippenmeier and at least realized as a free space structure connecting the Unterstrass district with Oberstrass (cf. Kurz 2008, 342). This test design intends to create attractive habitats for fauna and flora as well as residents of the Milchbuck, consistently activating the existing free space structures by differentiation areas with extensive and intensive utilization, applying a strategy for the street spaces to construct niches for human beings, fauna, and flora, including a diverse greening of façades and roofs (Figure 11.5). The consciously chosen and designed coexistence of man, animal, and plant societies aim at improving biodiversity in the city while, at the same time, intensifying the qualities of residing close to the ground on the Milchbuck. The paper demonstrates that a diversified layering of residential and optional spaces as well as utilizations leaves space for nature, which, in return, yields added value for an attractive residential environment. Here the paper again demonstrates how existing qualities and potentials can be seized and exploited by applying a consistent, comprehensive strategy (Figure 11.6).

In an exemplary manner, these three student papers represent this "cross-border" mode of operation within the design process seeking to connect elements so far opposed. According to a traditional view of nature, it would hardly be considered feasible to promote biodiversity in existing cultured landscapes by creating new residential utilizations of all things (River Delta, Figures 11.1 and 11.2) or even to conserve existing biotopes with such measures (Gravel Quarry, Figures 11.3 and 11.4). Inversely, urbanism has, so far, hardly ever pursued a strategy of supporting fauna and flora by means of step stone concepts or by creating niches (Urban District, Figures 11.5 and 11.6). Such approaches entail a careful study of the current status, the human and nonhuman actors on location; the social and spatial singularities that need to be reconnected with new contents, such as biodiversity; habitats for flagship species; and specific residential environments, and translated into spatial reality. But the papers, which have, so far, not been implemented and only describe

---

* According to representatives of the NFP 54 project: BiodiverCity: Urban biodiversity from the ecological and social science point of view, WSL Birmensdorf.

FIGURE 11.5 Reinforcement of green space structures combined with the district structure in order to connect two isolated landscape areas within the city of Zurich. (Courtesy of Livia Schenk, student, IUL, ZHAW 2011.)

FIGURE 11.6 Visualization of the coexistence and habitats of man, fauna, and flora. (Courtesy of Livia Schenk, student, IUL, ZHAW 2011.)

ideas of new cultures, also reveal that such an integral approach may yield added values, which again commit human beings, i.e., the residents of these new cultures. During the long process of implementing such new cultures, they must take into account the described delicacy and momentums by means of maintenance, care, and observation, and they are permanently invited to revise and renegotiate values. In the scope of this new consciousness, man is no longer a consumer; he is a part of his environment.

Designing Integral Urban Landscapes

11.5 QUESTIONING APPARENT MATTERS OF FACTS TERMINATES THE RE-ORIENTATION IN DESIGN

Although the craft and project of designing, until recently considered the original skill of architecture, landscape architecture, or urbanism, is at the center of focus in the line of argument presented here, this does not mean that there is an intention to exclude other disciplines from the planning process, especially the environmental and natural sciences—on the contrary, the gathering of "natural-artificial" collectives and the translation (designing) into integral cultures demands that the "war of worlds" (cf. Latour 2004) characterized by a disciplinary manner of thinking ends. As already referred to in context with Corner's explanations, this necessitates the collaboration of different disciplines, which, according to their respective competences, without reservations, connects aspects of urban planning, urbanistics, or landscape architecture with new content, such as cultural ecology, collectively establishes contents, and critically reflects their translations into spatial reality. Here the design work must yield space to the other disciplines. Inversely, it is the task of the disciplines not dealing with design to enrich the concepts of urbanism and landscape planning with precisely such contents. But now that nature no longer exists, it will also be possible to surmount the hardly productive and actually de-cultivating RE-orientation in the field of design (re-naturalization, re-cultivation, redesign).* As the discourse here has shown, the sovereignty of the form or shape of landscapes is no longer in the hands of transcendental, comprehensively self-regulating nature, but in the hands and responsibility of such actors and disciplines that co-determine the new cultures, for instance, on the basis of geo- and settlement-morphological singularities and have internalized strategies in order to translate conceptions into spatial reality.

It is Latour's and Hampe's assumption that the idealization of nature prevents the ecological crises from being solved as it is increasingly used for preventing politics. The test designs outlined here reinforce the theory that new ways can be found to reinforce existing potentials *both-and* by engrossing the experience of relationships and accepting the real-time necessity of revising debated issues without needing to pit individual categories against each other. Thus, the cultural technique of designing is capable of uniting oppositions, such as architecture-ecology, natural science–landscape architecture, or natural science–society so far considered incompatible. In this manner, not only do new urban landscapes emerge, seeking neither to be solely natural nor artificial, but, at the same time, everyday needs of humans, fauna, and flora are connected on different scale levels. Such a connection is not conceivable without calling the convention of modernism into question as it focuses on separation and not on integration. The essay at hand thus also adheres to the idea that contemporary technologies and strategies do not lead to the abolition of the traditional but only to its overcoming and, thus, to the birth of the new, to quote Walter Benjamin freely (Benjamin 1979). Here this refers to the overcoming of the convention of modernism

---

* As a reaction to a question posed in the Call for Paper to the symposium "Designing nature as infrastructure—a question of technology?" that took place on November 29–30, 2012, at the TU Munich and yielded the motivation for this paper.

and, at the same time, the faculty to conceive of new forms of hybrid realities or forms of political ecology. The student papers produced in this context may appear naive, unrealistic, or too pragmatic. But it is not their task to deliver answers. It is their task to pose the right questions with regard to the existing ecological as well as architectural and urbanistic crises. They do so in order to question modernists' conventions and declared matters of fact with the aim to initiate a debate concerning non-modernist approaches for developing future-proof cultures (cf. Latour 2007, 150). Such a debate is absolutely essential in view of the progressive transformation of the urban landscapes and the environmental problems involved.

Will nature disappear? In spite of all, this question is justified. The answer has already been given: It disappears as an analytical category—or in Michael Hampe's words, "Nature does not exist" (Hampe 2011). But there is a need for further questioning: Does nature disappear if man and his technology intervene deeper and deeper into the entity that the natural sciences have considered their object of examination since the early days of modern science? The following answer could be up for further debate: If you stay true to the convention of modernism, the interventions of man into the environment continue to have productive effects. If realism is added to the current urban landscapes, the constructive-creative, critical-reflective, yet still future-oriented interventions of man into the environment unfold instead of passive RE-orientation productive effects—productive for EVERYTHING, i.e., the idea of cultures.

REFERENCES

Benjamin, W. (1979). *Das Kunstwerk im Zeitalter seiner technischen Reproduzierbarkeit, Drei Studien zur Kunstsoziologie*. Frankfurt: Suhrkamp.
Caminada, G. A., V. Ehrensberger, M. Hampe, and Y. Michel. (2012). Für und wider die Natur. Interview mit Gion A. Caminada und Michael Hampe. *Trans 21*: 38–41.
Corner, J. (2003). Landscape urbanism. In *Landscape Urbanism. A Manual for the Machinic Landscape*, 58–63. London: AA.
Domschky, A. and S. Kurath. (2012). Stadtlandschaft Basel, Gegenwart und Zukunft. In *StadtLandschaftNatur,* Basel: IBA.
Eisinger, A. and S. Kurath. (2008). Jetzt die Zukunft. Einschreibeprozesse soziotechnischer Stadtlandschaften. In *GAM Emerging Realities*, 154–175. Graz: GAM.
Eisinger, A. and S. Kurath. (2009). Die emergente rolle der architekten. In *GAM Urbanity not Energy*, 80–91. Graz: GAM.
Gethman, D. and S. Hauser (publishers). (2009). *Kulturtechnik Entwerfen. Praktiken, Konzepte und Medien in Architektur und Design Science*. Bielefeld: Transcript.
Giseke, U. (2010). Urbane Landschaften. In *Planen – Bauen – Umwelt*. Wiesbaden: VS Verlag.
Gross, M. (2006). Kollektive Experimente im gesellschaftlichen Labor—Bruno Latours tastende Neuordnung des Sozialen. In *Verschwindet die Natur? Die Akteur-Netzwerk-Theorie in der umweltsoziologischen Diskussion*, 165–181. Bielefeld: Transcript.
Hampe, M. (2011a). Die Natur gibt es nicht. Über Hintergründe und Folgen einer falschen Vorstellung. *Neue Zürcher Zeitung*, August 20, 2011.
Hampe, M. (2011b). *Tunguska oder Das Ende der Natur*. München: Carl Hanser Verlag.
Ineichen, S. and M. Ruckstuhl (publishers) (2010). *Stadtfauna. 600 Tierarten der Stadt Zürich*. Bern: Haupt Verlag.
Körner, S. (2005). *Natur in der urbanisierten Landschaft. Ökologie, Schutz und Gestaltung*. Wuppertal: Müller + Busmann.

Designing Integral Urban Landscapes 223

Kurath, S. (2011). *Stadtlandschaften Entwerfen? Grenzen und Chancen der Planung im Spiegel der städtebaulichen Praxis.* Bielefeld: Transcript, 2011.

Kurz, D. (2008). *Die Disziplinierung der Stadt. Moderner Städtebau in Zürich. 1900 bis 1940.* Zürich: Gta.

Latour, B. (2000). *Die Hoffnung der Pandora.* Frankfurt am Main: Suhrkamp.

Latour, B. (2001). *Das Parlament der Dinge. Für eine politische Ökologie.* Frankfurt am Main: Suhrkamp.

Latour, B. (2004). *Krieg der Welten—wie wäre es mit Frieden?* Berlin: Merve Verlag.

Latour, B. (2007). *Eine neue Soziologie für eine neue Gesellschaft, Einführung in die Aktor-Netzwerk-Theorie.* Frankfurt am Main: Suhrkamp.

Mostafavi, M. and G. Doherty. (2010). *Ecological Urbanism.* Baden: Lars Müller Publishers.

Pickering, A. (1995). *The Mangle of Practice.* Chicago and London: University of Chicago Press.

Prominski, M. (2004). *Landschaft Entwerfen. Zur Theorie aktueller Landschaftsarchitektur.* Hannover: Reimer.

Schumacher, C. (2001). Dogged by the Model of Science. *tec21* no. 13: 25–28.

Sieverts, T. (2001). *Zwischenstadt: Zwischen Ort und Welt, Raum und Zeit, Stadt und Land.* Basel, Boston, Berlin: Birkhäuser Verlag.

Voss, M. and B. Peuker. (2006). *Verschwindet die Natur? Die Akteur-Netzwerk-Theorie in der umweltsoziologischen Diskussion.* Bielefeld: Transcript.

12 Counterpoint
The Musical Analogy, Periodicity, and Rural Urban Dynamics

Matthew Skjonsberg

CONTENTS

12.1 Introduction 225
12.2 Harnessing Polemics 226
12.3 Second Nature 227
12.4 Origins of the Musical Analogy 230
12.5 How to Think Contrapuntally 231
12.6 Contrapuntal Landscapes 233
12.7 Conclusion 241
References 243

12.1 INTRODUCTION

Every thesis calls for its antithesis, and every revolution prompts a counterrevolution—this takes place within the same generation as well as across intergenerational oscillations (Ortega y Gassett 1962; Sennett 1976). Enlightenment thinkers were critical of the Humanist tradition of analogical thinking—their own encyclopedic enthusiasm was intent upon creating a *lexicon* of the world, an ambition that has been assiduously realized in contemporary Geographic Information Systems (GIS) and empirical attitudes toward industrial agriculture and managerial urbanization. However, languages are comprised of two parts—a *lexicon* and *grammar*—and analogical thinking, focused as it is on seeing relationships between parts, is particularly well suited to provide conceptual frameworks for contextual design. To harness the power of polemics, we can anticipate that at least two conceptual paradigms, polarities to one another, are needed at any given moment—and that these are best conceived of as, to paraphrase Sébastien Marot, "opposite, but not exclusive of one another" (Marot 2003). Further, as any given analogy will inevitably prompt justifiable reactions against it, I propose that we work between those two oldest and most enduring architectural analogies: the biological analogy (on growth and form) and the musical analogy (on composition and form). Of these, the biological analogy is clearly in ascendancy—see, for example, Philip Steadman's seminal *The*

Evolution of Designs: The Biological Analogy in Architecture and the Applied Arts (Steadman 2008) or Lynn Margulis' *The Basic Unit of Life* (Margulis 2010). Hence, this sustained reflection on the musical analogy, made with a view to its instrumentality for composing rural urban dynamics in relation to existing landscapes.

12.2 HARNESSING POLEMICS

A public event at the Harvard Graduate School of Design commemorating the 50th anniversary of the urban design field featured a discussion between advocates of "New Urbanism" and "Landscape Urbanism." The moderator, Michael Sorkin, concluded the polemic event with the statement, "Let's make humane, equitable, sustainable and beautiful cities ... Cities need to supply their own food, energy, water, thermal behavior, air quality, movement systems, building and cultural and economic institutions. This urban self-sufficiency is a means to political autonomy and planetary responsibility. Sustainable, equitable, and beautiful ..." (Sherman 2010). In staking out this common ground, Sorkin puts his finger on the need for urban designers to be tuned in to their responsibility to champion public interest and engage the public imagination. But it is important to consider whether cities really should be, or even could be, self-sufficient.

Johann Heinrich von Thünen's *The Isolated State* of 1826 put forth principles relating the "city of commerce" with its rural hinterlands. These principles were illustrated with a simple diagram, as seen in Figure 12.1, representing an idealized central city in an unarticulated field, surrounded by a series of concentric rings defining production zones whose distance from the city is determined by an equation relating key variables: $R = Y(p - c) - Yfm$, where R = land rent, Y = yield per unit of land, c = production expenses per unit of commodity, p = market price per unit of commodity, F = freight rate, and m = distance to market (von Thünen 1966).

The equation's simplicity supports its utility—advancing the rational principles of a self-sufficient city based on capital and rent, ostensibly conceived of as neutral and objective. But analysis of real, historic capital networks shows that the reach

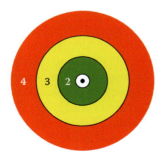

FIGURE 12.1 The Isolated State, von Thünen's model: the black dot represents a city; 1 (white) dairy and market gardening; 2 (green) forest for fuel; 3 (yellow) grains and field crops; 4 (red) ranching. The area beyond this represents wilderness where agriculture is not profitable. (Image courtesy of Matthew Skjonsberg, after von Thünen.)

Counterpoint 227

of the city's economic footprint is much broader and more uneven (Cronon 1991). This "uneven development" was identified by Marx as being a fundamental characteristic of capitalism, leading to the simultaneous emergence of concentrations of wealth and capital on the one hand and poverty and oppression on the other. In extreme cases of "uneven development," short-term economic gains often involve the permanent loss of the very ecologies and social structures on which both rural and urban rely.

In principle, of course, every artificially created thing is only other things rearranged—that is to say, displaced and recomposed. The surface extraction of materials, mining and open pit mining, vividly illustrate the "cut and fill" nature of architecture, infrastructure, and urbanization. It is important that urban designers look to the origins of the resources they work with and acknowledge that the extraction of subterranean resources, such as water, oil, and "natural" gas from distant regions causes geological fractures and leaves substantial underground voids, destabilizing the land above, depleting and poisoning groundwater, and causing sinkholes and earthquakes. Recognition of the importance of urban context must therefore broaden the conceptual scope of urban design.

12.3 SECOND NATURE

Since their historic appearance about 10,000 years ago in the Fertile Crescent, the cradle of civilization, urbanization and agriculture have been interrelated. In *De Natura Deorum* (*The Nature of the Gods* ca. 45 BC) Cicero coined the term *second nature* to describe the domain of infrastructure relating urban and rural districts within their regional contexts, writing, "… we sow cereals and plant trees; we irrigate our lands and fertilize them. We fortify riverbanks, and straighten or divert the courses of rivers. In short, by the work of our hands we strive to create a sort of *second nature* within the world of nature" (Cicero 2008). The fundamental reliance of cities on this designed *second nature* within a contextual *first nature* is illuminated by Kostas Terzidis: "In Greek, the word 'design' is *schedio*, which is derived … from the word *eschein*, which is the past tense of the word *eho*, which … means to have, hold, or possess. Translating the etymological context into English, it can be said that design is about something we once had, but have no longer" (Terzidis 2007). This recognition of design as an activity that reveals preexisting qualities and implicit relationships that are neglected, obscured, or forgotten is very different than the idea of invention or stylistic innovation generally associated with design. Terzidis points out that design and planning are terms that, while often conflated, have quite different implications. "Design is about conceptualization, imagination, and interpretation. In contrast, planning is about realization, organization, and execution … Design provides the spark of an idea and the formation of a mental image." These mental images are shaped through the dual processes of *analogy* and of *abstraction* by which concepts are derived from the classification of experiences and information. Analogy has its origin in the Greek *analogia* (ἀναλογία), meaning "proportion." To make an analogy is a cognitive process by which meaning is transferred from a relatively known subject to one less well known. For instance, in his early work, physicist Niels Bohr drew an analogy between the solar system and atomic structure. He developed a model of the atom with the nucleus at the center and electrons in orbit around it, which he compared to the planets orbiting the sun—work for which he subsequently won the Nobel Prize in 1922. Although this analogy is now known to be accurate only in a very limited sense, it provided a useful step in the right conceptual direction.

Contemporary idealization of the city as self-sufficient, on the other hand, abstracts and conceptually isolates the urban from the rural and from those very natural and cultural systems that make the city possible to begin with. In *Method in Social Science: A Realist Approach*, Andrew Sayer writes, "A bad abstraction arbitrarily divides the indivisible and/or lumps together the unrelated and the inessential, thereby 'carving up' the object of study with little or no regard for its structure and form" (Sayer 2010). The self-sufficient city—like the idealized, undifferentiated planar surface of von Thünen's *An Isolated State*—is hardly to be seen in reality. Likewise, the Cartesian grid is prized for its conceptual simplicity as an interscalar system of reference, but as a design framework, at the regional scale, it is hardly capable of responding to the complex topography of mountains and river valleys. One well-established model more capable of responding to the complex interactions between rural and urban networks and their sites can be found in computational geometry: the dual network triangulation method of *Voronoi diagrams* and *Delauney triangulation* as seen in Figure 12.2.

The Voronoi/Delaunay dual network can be read as a conceptual framework for understanding urban and rural as phases of the same civilizing force, *second nature* (see Figure 12.3).

Second nature is sited within first nature, and together, these constitute a higher-level Voronoi/Delaunay dual network. As the notion of these nested geometries can extend to three and higher dimensions, generalizations are possible in metrics other than Euclidean ones, and in principle, they resemble the periodic optimization that is seen in the "nesting" interaction of normal matter and dark matter distributed throughout the universe itself. Quantum physicists even refer to the entire set of the known laws of physics as the "landscape," echoing Bohr's interscalar analogy. Such recurring analogies highlight how the principle of periodicity can inform a conceptual

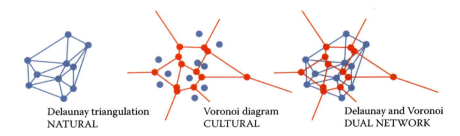

FIGURE 12.2 Dual network. Natural and cultural infrastructural networks are arranged in counterpoint to one another, together forming a dual network as *second nature*. (Image courtesy of Matthew Skjonsberg.)

Counterpoint

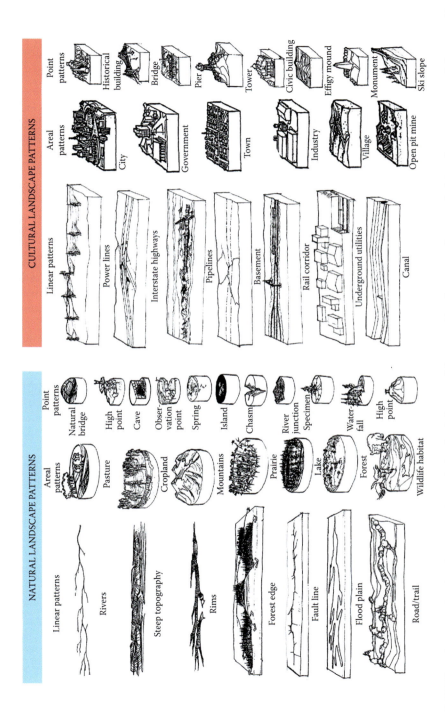

FIGURE 12.3 Natural cultural patterns. An analysis of existing conditions provides the *cantus firmus* underlying the arrangement of the natural/cultural dual network. (Image courtesy of Matthew Skjonsberg with Phil Lewis, FASLA.)

230 Revising Green Infrastructure

model of design that is both more holistic and more responsive to the underlying ecological and social realities of rural urban dynamics.

12.4 ORIGINS OF THE MUSICAL ANALOGY

Music is among the earliest and most enduring of architectural analogies. Following Cicero in the first century B.C., a period in which some of the greatest works of Roman infrastructure were established, the architecture of Vitruvius was largely based on the proportional musical investigations of Pythagoras (570–495 B.C.). In the 15th century, shortly after the rediscovery of Vitruvius, these Pythagorean principles were greeted with renewed enthusiasm. In Chapter VI of his *De re aedificatoria* (*On the Art of Building*), Leon Battista Alberti methodically develops the relationship between proportion and architectural form, writing, "We shall therefore borrow all our Rules for the Finishing of our Proportions, from the Musicians, who are the greatest Masters of this Sort of Numbers, and from those Things wherein Nature shows herself most excellent and complete" (Alberti 1965).

The term *contrapunctus*, from the Latin *punctus contra punctum* ("point against point" or "note against note"), initially appeared around 1300. In 1412, the Italian theorist Prosdocimo de Beldomandi wrote that rather than dealing with note against note individually, the composer was actually concerned with the problem of *cantus contra cantum*—one complete melody against another—thus requiring a new integration of vertical (harmonic) and horizontal (melodic) concepts (de Beldomandi 1984). With this realization, counterpoint gradually came to be recognized in its full complexity, resulting in the "golden age of polyphony" (Fux 1971, vii). It is interesting to note that it was just one year later, in 1413, that architect Filippo Brunelleschi established the geometric method of perspective drawing. Indeed, both counterpoint composition and perspective drawing illustrate the emergence of three-dimensional thinking—the composer dealing with multiple voices in time in a manner very similar to the architect's newly contextual conception of buildings within a multilayered landscape. In both cases, the breakthroughs involved new insights into proportional principles underlying physical reality. In perspective drawing, this was a system of proportions on paper that accurately related to actual sizes and distances, and counterpoint composition addressed resonant harmonic phenomena and the means of engendering it—namely the range of voice or instrument. Beyond being merely *picturesque*, these developments illustrate the capability of *design* to derive meaningful order from apparent complexity.

In 1725, the Viennese musician and theorist Johann Fux published *Gradus Ad Parnassum*, a work that was to set the standard for counterpoint composition to this day. Written in the form of a classical Greek dialogue between an elder master composer and his young apprentice, the lessons are laid out in sequence according to what had by then become the five "species" of counterpoint, from the simplest to the most complex: first species—whole note against whole note; second species—whole note against two half notes; third species—whole note against four quarter notes; fourth species—whole note against syncopated whole note; fifth species—florid counterpoint, using all the previous species. Further, based on sophisticated insights into harmonic phenomena assiduously obtained by the cumulative effort of many

Counterpoint

FIGURE 12.4 Musical analogy. The musical analogy is among the most enduring of conceptual models for architecture. Counterpoint compositions begin with the *cantus firmus*, a fragment of existing music. At the scale of the city and its regional context, this linear structure presents a compelling conceptual model for natural and cultural equity. (Image courtesy of Matthew Skjonsberg.)

generations, the set of additional rules had become highly nuanced: Consonances were now categorized as perfect or imperfect, and the octave could be subdivided harmonically or arithmetically. The possible movement of simultaneous voices in relation to one another was carefully categorized into three types of motion—direct motion (two lines moving parallel), contrary motion (two lines moving in opposite, mirrored fashion), and oblique motion (one line moving while another remains stationary)—and the use of these motions was likewise determined by a set of rules relating perfect and imperfect consonances (Fux 1971, vii). One of the most interesting rules to have developed throughout this process, insofar as it provides a direct connection back to contemporary architectural discourse, is the use of the *cantus firmus* as the basis for contrapuntal composition. A *cantus firmus* (fixed song) is an existing melody taken from elsewhere and used as the basis for the composition with each additional voice composed fundamentally in relation to it and in relation to one another. It becomes the site of interpretation, the *terra firma* of the composition, as it were, making the entire compositional procedure explicitly contextual and interpretive (see Figure 12.4).

12.5 HOW TO THINK CONTRAPUNTALLY

Discussing the contemporary relevance of contrapuntal composition, concert pianist and music scholar Rosalyn Tureck (1914–2003) asserted that its foremost feature was as a conceptual model, maintaining that there was a deep conceptual correlation in Bach's music between sound wave periodicity, ornamental figurations, melodic and rhythmic phrasing, and the overall harmonic structure of his compositions (Tureck 2000). So fascinated was she by this correlation that she made it the subject of her lifelong research and founded a series of research societies dedicated to disciplinary interaction about the subject. Among her noteworthy collaborators were evolutionary biologist Stephen Jay Gould, physicist Roger Penrose, and mathematician Benoit Mandelbrot, whose fractal geometry had by then emerged as *the* go-to analogy for understanding principles of self-similarity in the new age of complexity. In response to a paper called *The Fractal Geometry of Music* (Hsu and Hsu 1990), in which a series of fractal metaphors were suggested to exist in the music of Bach, Tureck penned a detailed response nearly as long as the article itself. "I am delighted to see the interest in the deeper levels of musical composition," she wrote, "... However the article sets down assumptions and conclusions about musical structure which are not only based in subjective interpretation but also contain sweeping blanket statements which cancel out any possibility of the validity of their claims ... Bach was clearly not concerned with fractal relations and fractal theories. The fact that the authors find a 'significant deviation' simply demonstrates the fact that their theories are ... artificially created. 'Deviations' are a very convenient way for excusing the inapplicability of a theory. This is not to say that the possibility of analysis of music in terms of Mandelbrot's theories is impossible." She finally concludes, "If one is looking for the fractal geometry of music, which I applaud wholeheartedly, then the steps taken must be based on something immeasurably more solid and accurate than what appears in this article" (Tureck 1990). For Tureck, the most promising contrapuntal analogies were fundamentally conceptual, having to do with insights obtained by what she described as *contrapuntal thinking* rather than by attempts to force analogies into a theory after the fact.

In her opening editorial essay for the first volume of *INTERACTION—Journal of the Tureck Bach Research Foundation*, she writes, "' Interdisciplinary,' 'intercommunicative,' 'interface,' and 'interpersonal' are the guiding icons in the currents of contemporary thought. These constitute varied facets of interaction. The assembly of essays on areas such as physics, mathematics and music, conventionally regarded as disparate, must speak for itself on the platform of interaction ... we have [long] known that there is a correspondence between sound waves and arithmetic. But composed music is more than wavelengths, vibrations, and number. It partakes of, indeed its very essence is dependent upon, processes of thought, form and structure. These processes are abstract; they have to do with mental concepts, not aural activities. They are not technological realizations, nor is their reception limited to physiological, aural responses and emotional, imaginative experience. The abstract processes that create music underlie the acts of musical creativity and performance, which also affect levels of both conscious and unconscious experience in reception. Abstract realms of concept, form and structure are not the sole possession of the sciences. The powerful impact of ... the arts and the sensuous blandishments emanating from the tools of the musical art tend to overshadow the abstract concepts and processes underlying them and from which their forms emerge ... On the other side of the coin, the materialistic thinking that characterizes recent centuries and which has so blatantly crescendoed in our own past century, has emphasized materials, technological tools, and specialized methodologies as prime movers over and above abstract concepts and processes" (Tureck 1997).

In her scholarship, Tureck focused on identifying the characteristics of *contrapuntal thinking* on the basis of what could be found in the music itself. In the preface to her three-volume *An Introduction to the Performance of Bach*, she writes, "... the interpretive suggestions are founded on deep principles which grow out of the music itself" (Tureck 1960a, 3). Book II features a chapter with the compelling title "How to Think Contrapuntally," in which she explains, "By contrapuntal thinking I mean the following: firstly, the ability to think simultaneously in two or more

Counterpoint

parts; secondly, the ability to envisage motives continually shifting in the parts; and thirdly, the ability to preserve the relation of the parts to each other successively as well" (Tureck 1960b, 11). Throughout the three volumes, she returns to this theme, giving special emphasis to the relationship between ornamentation and structure—often describing the nature of this relationship using terms such as "integral" and "organic." "The term 'ornamentation' gives the impression, to the modern mind particularly, of something extraneous and dispensable. In the music of Bach and his predecessors this is quite wrong. The first point to learn about ornamentation is that for this music it is indispensable, and the second is that it is as much a part of the musical structure as the printed notes" (Tureck 1960a, 7). Under the heading of "structure," she writes, "Ornaments always are and must be conceived as an intrinsic part of the structure ..." (Tureck 1960a, 8).

Familiar as she was with his work, we should not be surprised by how much Tureck's characterization has in common with Frank Lloyd Wright's conception of ornament as structural expression, which he, in turn, relates first to nature and then to the contrapuntal music of Beethoven: "Integral ornament is simply structure–pattern made visibly articulate and seen in the building as it is seen articulate in the structure of trees or a lily of the fields. It is the expression of the inner rhythm of Form ... What I am here calling integral ornament is founded upon the same organic simplicities as Beethoven's Fifth Symphony, that amazing revolution in tumult and splendor of sound built on four tones based upon a rhythm a child could play on the piano with one finger. Supreme imagination reared the four repeated tones, simple rhythms, into a great symphonic poem that is probably the noblest thought-built edifice in our world. And architecture is like music in this capacity for the symphony" (Wright 2005, 347–348). Of course, Wright was likely introduced to the idea of integral ornament through his mentor Louis Sullivan, for whom this ornament was nothing less than a bridge between nature and science, between the organic and the inorganic. Oscillating between the conceptual lens of the musical analogy and that of the biological analogy, Wright later described his ambition to develop this idea of integral ornament into more than a metaphor, making it the basis of the architectural plan itself: "But now, why not the larger application in the structure of the building itself in this sense? Why a principle working in the part if not living in the whole? If form really followed function ... why not throw away the implications of post or upright and beam or horizontal entirely? ... Now why not let walls, ceilings, floors become seen as component parts of each other, their surfaces flowing into each other. To get continuity in the whole, eliminating all constructed features just as [Sullivan] had eliminated [planar] background in his ornament in favor of an integral sense of the whole. Here the promotion of an idea from the material to the spiritual plane began to have consequences. Conceive now that an entire building might grow up out of conditions as a plant grows up out of the soil" (Wright 2005, 146–147).

12.6 CONTRAPUNTAL LANDSCAPES

The title of Reyner Banham's *Architecture of the Well-Tempered Environment* is an allusion to J. S. Bach's 48 Preludes and Fugues of *The Well-Tempered Clavier* (so called in reference to "well-tempered" tuning, an innovation enabling modulation between all major and minor keys) although it contains no references to counterpoint—or even to Bach, for that matter. But there are dozens of references to Wright's buildings, illustrating in detail the advanced integration of structural design, heating and cooling systems, lighting, passive ventilation, and solar design—now early examples of the performative synthesis of architecture and environmental context (Banham 1984). Wright's telling observations about Sullivan's ornament can also be seen as having informed his conception of architecture as the attenuation of latent qualities as potentials *already existing* in the landscape. In reference to his regional-scale project, Broadacre City (1932–1958), he described this contextual approach as follows: "… the ground itself predetermines all features; the climate modifies them; available means limit them; function shapes them" (Wright 1935). This active regard for context results in a contrapuntal interplay between architectural and topographic form—as is brilliantly illustrated by the cut and fill diagram Wright himself drew for Taliesin West (see Figure 12.5).

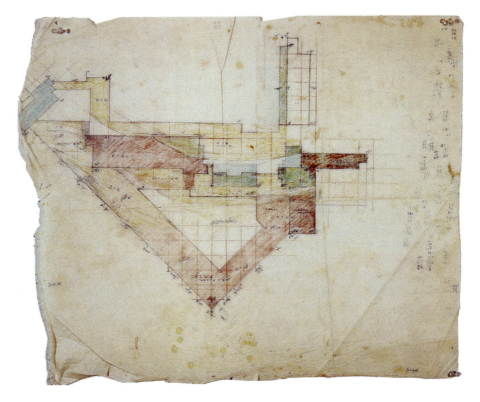

FIGURE 12.5 Taliesin West—cut and fill, 1937. Frank Lloyd Wright's cut and fill drawing for Taliesin West (gold = cut/red = fill) was established so as to incorporate as much material as was excavated—designing for neutral cut and fill, a kind of on-site periodicity. (Image courtesy of The Frank Lloyd Wright Foundation Archives, The Museum of Modern Art, Avery Architectural and Fine Arts Library, Columbia University, New York.)

Counterpoint

This exemplifies one practical architectural application of the musical analogy—both the origin and destination of displaced materials are anticipated in the design process, acknowledging, as was stated earlier, that architecture is composed by the displacement of other things. At Taliesin West, proximity of material to building was immediate: Earth cut from one part of the site was used to fill another, and the building itself was built of sand from the seasonal river "washes" and stones found on site. Similarly, Wright's designs are typically composed so as to purposefully heighten the sense of security on one hand—often earth-integrated, backed into the site's topography through such cut and fill procedures as described previously—and the sense of freedom on the other—sheltered spaces facing onto broad elevated terraces with panoramic views. Thus, the relevance of *aesthetics* to what might appear a functionalist method can again be discerned by looking to the Greek etymology of the term, *aesthesis* (αἴσθησις), which pertains cumulatively to our *senses*—the sensory organs of sight, sound, smell, touch, and taste as well as to our cognitive *sensibilities*, including those of justice, well-being, and harmony.

Therefore, a *contrapuntal* design process involves composing designs in relation to their apparent opposite and begins by identifying the *cantus firmus*—the stable, underlying ecological structure to which the design must respond and defer and the social realities that design represents performatively and symbolically. At the scale of the city, the initial steps of such design-analysis could be (a) Identify Natural Landscape Systems, (b) Identify Void Design to Be Retained, (c) Identify Cultural Landscape Systems, (d) Ascertain Natural and Cultural Boundaries, and (e) Ascertain Natural and Cultural Intersections. By emphasizing the fundamental priority of continuity in existing ecological and social systems, regional design methods such as Phil Lewis' approach to determining "where not to build" set the stage for design as an interpretation of such contextual analysis (see Figure 12.6) (Lewis 1996; note: also see Manuel Schweiger's Chapter 15 in this book).

The *contrapuntal* application of Lewis' methods can be seen in Arup's ongoing sustainability master plan for the city of Addis Ababa, which correlates rural mobility and urban river corridors and is predicated on the recognition that soils in the region have been severely depleted due largely to subsidization policies that have displaced traditional pastoralists in favor of intensive ranching methods (see Figure 12.7).

By strategically prioritizing soil and water management along with mobility throughout the city for pastoralists, the scheme demonstrates how, by turning away from notions of self-sufficiency, both rural and urban districts can benefit from the contextual composition of natural and cultural infrastructural systems.

Likewise, West 8's topographically and hydrologically proactive design for the Tulsa Riverfront addresses a site that links rural and urban districts by making the landscape itself a visual and atmospheric register of seasonal water levels (see Figure 12.8).

Restoring soils and disrupted hydrological flows, terraced retention basins yield water levels that can be read at a glance, communicating seasonal variability to

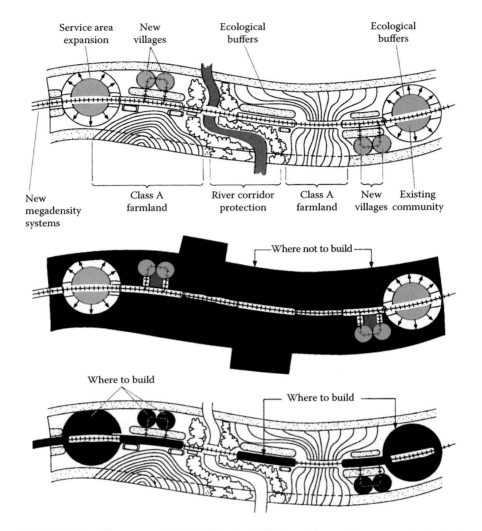

FIGURE 12.6 Where not to build. Phil Lewis, FASLA, outlines a nine-step method, prioritizing existing natural and cultural conditions on site—i.e., river corridor protection, Class A farmland, existing communities, etc.—resulting in a circumscribed footprint for development. (Image courtesy of Matthew Skjonsberg with Phil Lewis, FASLA.)

highway commuters and creating a variety of new water landscapes and habitats experienced by park visitors as visually distinct from one visit to the next. While these terraced ponds might be regarded as superficial—even sentimental—they are, in reality, integral ornament, a structural expression of the functional dynamics of the site as *second nature* (see Figure 12.9).

Counterpoint

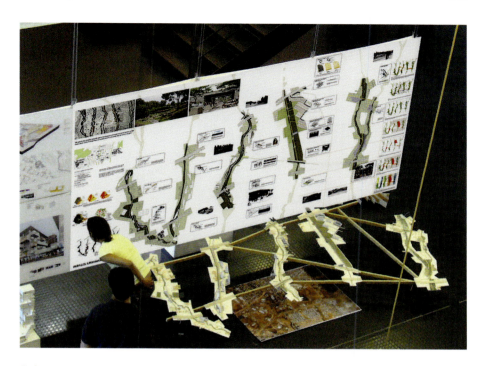

FIGURE 12.7 Parallel urbanism—Addis Ababa, Ethiopia, 2007. Addressing the strategic decentralization of cattle markets by locating them along a latent natural corridor network, the design relates both rural and urban interests and simultaneously creates safe mobility corridors, addresses erosion in the rivers and streams, improves water quality and access, and triggers public spaces with functional benefit to both adjacent and temporal parties. Initiated as a project with Prof. Marc Angélil at ETH-Zurich, it has since been taken up by Arup as the basis for the sustainability master plan they are preparing for the City of Addis Ababa. (Image courtesy of Matthew Skjonsberg with Sebastian Alfaro Fuscaldo and Noboru Kawagishi.)

Another of West 8's projects, Happy Isles, takes this *contrapuntal* logic to its extreme: the creation of a series of new, sprayed-up sand islands to mitigate storm surges off the coast of Belgium and The Netherlands. Located beyond the curve of the horizon 10–20 km from shore—and therefore not visible from the existing coast—these dune islands, measuring up to 1500 km^2, will break the increasing waves while creating new underwater habitats and respecting current sea transportation routes. As a further precaution, ingenious engineering of underwater topography causes offshore undertow to draw down the sea level at the existing coast during northwestern storms. Together with Magnus Larsson's project for artificial linear dunes to stop the accelerated expansion of the Sahara, the project indicates how infrastructural design can employ *contrapuntal thinking* in response to the polar phenomena of climate change: sea-level rise and desertification (see Figures 12.10 through 12.12).

FIGURE 12.8 Tulsa riverfront plan, West 8, 2011. West 8's design for Tulsa's Riverfront establishes strategic connections to both natural and cultural systems and can be read as a gauge of seasonal water inundation levels by those crossing the site, itself a rural urban threshold. (Image courtesy of West 8 urban design and landscape architecture b.v.)

Counterpoint

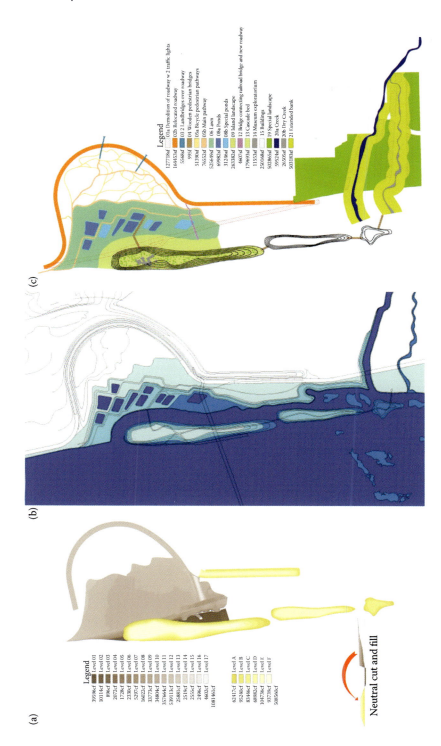

FIGURE 12.9 Tulsa Riverfront—contrapuntal planning, West 8, 2011. (a) Neutral cut and fill; (b) design for seasonal inundation and water retention; (c) programmatic integration and differentiation. (Image courtesy of West 8 urban design and landscape architecture b.v.)

FIGURE 12.10 Regional-scale periodicity—West 8, Happy Islands: Belgium, Dutch, and Flemish Coast, 2006; Magnus Larsson, Dune: Transcontinental Africa, 2008. Along with rural-urban dynamics, climate change–induced sea-level rise and desertification are also examples of periodic phenomena. These two strategies represent attempts to address sea-level rise and desertification as polar effects of climate change to retain currently inhabited regions. An alternate, lowest-risk version of this initiative would be to establish "safe zones" on the basis of those areas of the world least affected by climate change. (Image courtesy of (left) West 8 urban design and landscape architecture b.v. (right) Magnus Larsson.)

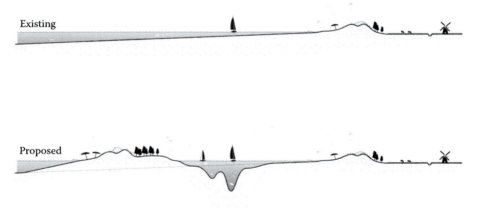

FIGURE 12.11 Happy Islands—cut and fill, West 8, 2006. The creation of the islands is made possible by a massive cut-and-fill operation, yielding protective barrier islands and its counterpart: an inland sea, whose terraced underwater contours provide varied habitats for ecological diversity in contrast to the existing, relatively monocultural condition. (Image courtesy of West 8 urban design and landscape architecture b.v.)

Counterpoint

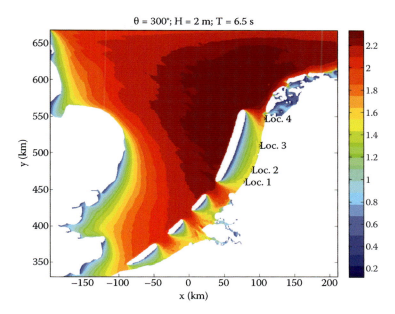

FIGURE 12.12 Happy Islands—as storm surge protection, West 8, 2006. These dune islands, measuring up to 150,000 hectares in size, will break increasing storm surge waves. Also, due to ingenious engineering of the gullies, the offshore undertow will cause the sea level to drop during northwestern storms. (Image courtesy of West 8 urban design and landscape architecture b.v.)

12.7 CONCLUSION

In his final year, Louis Sullivan (1856–1924) drew an esoteric series of graphic abstractions illustrating "organic" and "inorganic" forces, representing natural forces with curved lines and human forces with straight ones, writing, "These two elements ... are not to be considered separate conceptions to be harmonized, but as two phases" (Sullivan 1967). Again, this periodic concept can be attributed to Pythagoras, who saw the circle as related to the circuits of stars and planets and the square represented the physical world and human rationality, and to his contemporary, Heraclitus (535–475 B.C.), who conceived of the "unity of opposites," which led to the pedagogical tradition of *coincidentia oppositorum*, in which the coexistence of opposites are understood as creating productive tension. Vitruvius later developed this concept in his architecture, and Da Vinci's *Vitruvian Man* memorably illustrates the conceptual model of *dual geometries* and the aspiration to synthesize artistic and scientific knowledge. Taken together, in the tradition of *coincidentia oppositorum*, musical and biological analogies can invigorate and clarify one another—providing a *grammar* of contextual relationships to complement the discipline's dedication to the computational-empirical-managerial *lexicon*.

Borrowing the term *autopoesis* (Greek αὐτο "self" and ποίησις "creation") from the emerging field of chronobiology—which examines periodic phenomena in living organisms and their adaptation to solar and lunar rhythms—Patrick Schumacher concludes volume one of *The Autopoeisis of Architecture* with the statement, "... one might talk about an *artificial ecology* ... not only can parameters be shifted out of natural ranges but wholly new, artificial forces and their ... laws and logics might be defined ... An artificial, *second nature* can be conjured via scripted, quasi-natural laws, rich in internal resonances, as well as inter-articulations with external contexts" (Schumacher 2011). This last comment indicates how *contrapuntal thinking* can meaningfully oppose contemporary interpretations of the biological analogy, which curiously enough tend to go along with arguments for self-sufficient cities, the autonomy of architecture, and architectural *autopoiesis*—as though *second nature* could be conceived of as independent from nature itself. The opposite of *autopoiseis*, a closed process in which context might be an afterthought, is *allopoiesis*: the process whereby an organizationally open system produces something other than itself—as, for example, in an assembly line, on which each participant contributes a part to a synthetic whole. It is clear that cities are, in reality, *allopoietic* rather than *autopoietic*; they are the cumulative result of the contributions of diverse actors, none of whom are solely in control. This brings us back to Terzidis' distinction between design and planning: Until now the arguments for analogical thinking have been made from the standpoint of the designer's conceptualization of urban form and its relationship to context. Turning finally to planning, the analogy of *contrapuntal thinking*—concerning itself with multiple voices and a contextual *cantus firmus*—anticipates the *allopoietic* reality of cities and lends itself to open planning processes involving inputs from diverse actors. The projects illustrated here also represent sincere efforts in this direction, acknowledging the form of the city as fundamentally open-ended and resulting from an ongoing dialogue among its citizens—often despite resistance from the planning authorities involved. In *The Voice of Liberal Learning*, Michael Oakeshott expresses a sympathetic attitude regarding education itself: "The pursuit of learning is not a race ... it is not even an argument or symposium: it is a conversation. A conversation ... has no predetermined course ... and we do not judge its excellence by its conclusion; it has no conclusion, it is always put by for another day. Its integration is not superimposed but springs from the quality of the voices which speak, and its value lies in the impressions it leaves behind in the minds of those who participate" (Oakeshott 1989).

Whatever designs are made, what really counts is how those who inhabit the city will think of it, describe it to one another, and participate in it. To this end, the relational nature of analogical thinking goes a long way toward involving the public in the design and planning processes. Just so, as an established, historic alternative to prevalent contemporary notions of self-sufficiency, *contrapuntal thinking* provides a promising conceptual framework for the design and planning of rural urban dynamics capable of giving contradictory interests and polemical aims relational, aesthetic formal expression with enduring cultural value.

REFERENCES

Alberti, L. B. (1965). *Ten Books on Architecture*, ed. by Joseph Rykwert as a complete reprint from the 1755 edition. London: Tiranti.

Banham, R. (1984). *Architecture of the Well-Tempered Environment*. 2nd ed. Chicago: University of Chicago Press.

Cicero, M. T. (2008). *The Nature of the Gods*, New York: Oxford University Press.

Cronon, W. (1991). *Nature's Metropolis*. New York: W.W. Norton.

de Beldomandi, P. (1984). *Contrapunctus*, ed. Jan W. Herlinger. Lincoln: University of Nebraska Press.

Fux, J. (1971). ed: Mann, Alfred, *Gradus Ad Parnassum*. New York: W.W. Norton.

Hsu, K. and A. Hsu (1990). Fractal geometry of music, *Proceedings of the National Academy of Science,* Vol. 87, pp. 938–941.

Lewis, P. H. (1996). *Tomorrow By Design: A Regional Design Process for Sustainability*. New York: Wiley.

Margulis, L. (2010). eds. Brouwer, Mulder, The basic unit of life; *Politics of the Impure,* Rotterdam: NAI.

Marot, S. (2003). *Sub-Urbanism and the Art of Memory*. London: Architectural Association.

Oakeshott, M. (1989). ed. T. Fuller, *The Voice of Liberal Learning: Michael Oakeshott on Education*, New Haven: Yale University Press.

Ortega y Gassett, J. (1962). *Man and Crisis*, New York: W.W. Norton.

Sayer, A. (2010). *Method in Social Science: A Realist Approach*. Revised 2nd ed. New York: Routledge.

Schumacher, P. (2011). *The Autopoiesis of Architecture: A New Framework for Architecture*. Vol.1. Chicago: Wiley.

Sennett, R. (1976). *The Fall of Public Man*, New York: W.W. Norton.

Sherman, G. (2010). *GSD Throwdown: Battle for the Intellectual Territory of a Sustainable Urbanism*. Available at http://urbanomnibus.net/2010/11/gsd-throwdown-battle-for-the-intellectual-territory-of-a-sustainable-urbanism/ (accessed November 8, 2012).

Skjonsberg, M. (2013). eds. Farini, Nijhuis, Hidden rivers: Addis Ababa's rural urban metabolism, *Flowscapes: Exploring Landscape Infrastructures,* Madrid: Universidad Francisco De Vitoria.

Steadman, P. (2008). *The Evolution of Designs: Biological Analogy in Architecture and the Applied Arts*, London: Routledge.

Sullivan, L. (1967). *A System of Architectural Ornament According with a Philosophy of Man's Powers*. New York: Eakins Press.

Terzidis, K. (2007). *The Etymology of Design: Pre-Socratic Perspective*; Massachusetts Institute of Technology; Design Issues, Autumn 2007, Vol. 23, No. 4, p. 69; Posted Online October 5, 2007.

Tureck, R. (1960a). *An Introduction to the Performance of Bach*, Book I, London: Oxford University Press.

Tureck, R. (1960b). *An Introduction to the Performance of Bach*, Book II, London: Oxford University Press.

Tureck, R. (1990). Unpublished letter.

Tureck, R. (1997). *Interaction, Journal of the Tureck Bach Research Foundation*. Vol.1. London: Oxford University Press.

Tureck, R. (2000). In conversation with author.

von Thünen, J. H. (1966). *An Isolated State*. New York: Pergamon Press.

Wright, F. L. (1935). Broadacre City: A new community plan. *Architectural Record*, April 1935.

Wright, F. L. (2005). *An Autobiography*. 3rd ed. Petaluma: Pomegranate.

Section III

Governance and Instruments

13 A Transatlantic Lens on Green Infrastructure Planning and Ecosystem Services
Assessing Implementation in Berlin and Seattle

Rieke Hansen, Emily Lorance Rall, and Stephan Pauleit

CONTENTS

13.1 Green Infrastructure and Ecosystem Services between Theory and Practice ... 247
    13.1.1 Meaning of Green Infrastructure 248
    13.1.2 Meaning of Ecosystem Services..250
13.2 A Transatlantic Lens..251
    13.2.1 A Lens on Berlin 251
    13.2.2 A Lens on Seattle...252
13.3 Methodological Approach 253
13.4 Application of Principles 255
    13.4.1 Range of Ecosystem Services..255
    13.4.2 Coverage of Selected Green Infrastructure Principles 257
13.5 Discussing Rifts and Bridges between Planning Practice and Scientific
    Discourses..259
13.6 Conclusion 261
Acknowledgments..262
References...262
Planning and Policy Documents 264

13.1 GREEN INFRASTRUCTURE AND ECOSYSTEM SERVICES BETWEEN THEORY AND PRACTICE

Emerging in the 1990s to address environmental concerns, green infrastructure (GI) and ecosystem services (ES) are now seen as promising planning concepts for confronting current urban challenges. Although green infrastructure is usually understood as a network of natural and seminatural areas that is planned and maintained in order to provide multiple benefits for humans (Naumann et al. 2011), ecosystem services refers to the multiple benefits people can derive from nature (MEA 2005). Together, the two concepts offer the potential for holistic planning linked to the idea of sustainable development, aspirational of a balance between economic, ecological, and social demands on urban green. They are firmly present in the research of fields such as urban ecology and landscape planning with the benefits of each approach widely purported and documented.

Next to the scientific discourse, both concepts also gained attention in policy in recent years. Within Europe, the European Commission has recently advocated green infrastructure as an approach to promote ecosystem services and transform the often-defensive position of nature conservation to a more active and constructive element of spatial planning in Europe (European Commission 2013).

The broad scope represented by ecosystem services and the focus on synergies in green infrastructure planning can help to overcome sector-oriented planning and raise awareness for the multiple functions of green spaces past traditional aesthetic, recreational, health, and habitat-oriented concerns (Mazza et al. 2011). Despite the potential of these concepts for benefiting planning on all spatial scales, studies of how practitioners take up discourses around green infrastructure and ecosystem services are limited (e.g., Sandström 2002; Hauck et al. 2013; Piwowarczyk et al. 2013). Although evidence exists indicating growing city interest in green infrastructure planning and measuring ecosystem services, there are still major gaps in understanding how ecosystem services and principles of green infrastructure are being integrated into city planning.

The aim of this study is to explore the actual application of principles discussed in relation to green infrastructure and ecosystem services in urban planning through the lens of two cities with different spatial planning cultures: Berlin and Seattle. Selected plans and policy documents from both case study cities are surveyed to gauge the extent to which principles related to green infrastructure and ecosystem services are used. The intent of this examination is not to determine which city's approach is preferable but rather to understand more about the level of awareness, concern, and adoption, thus shedding light on the transmission of knowledge from theory to practice and highlighting areas that may require extra attention for bridging the divide between science and practice.

13.1.1 MEANING OF GREEN INFRASTRUCTURE

Green infrastructure remains something of an elusive concept since its introduction in the United States (Benedict and McMahon 2006; Mazza et al. 2011). Depending upon the author, it is either described as a more eco-centric or more anthropocentric approach (Mell 2008). One of the strengths of the concept is that it can be used on different scales, varying from local storm-water management projects to transnational ecological networks. Here we discuss green infrastructure as a holistic urban green planning concept on the level of cities and city-regions.

A Transatlantic Lens on Green Infrastructure Planning and Ecosystem Services 249

Although it is used in different contexts, the shared idea is that the label "green infrastructure" valorizes "green" and highlights its utility and meaning for human well-being by putting green structures and elements on par with "grey infrastructure," such as transportation networks and water treatment facilities. Green infrastructure as a planning strategy can be narrowed down to keywords such as multifunctionality and connectivity of green structures as well as multi-scale, communicative, and socially inclusive approaches (see Table 13.1).

Although these principles, separately, are not new for scientists and practitioners, their combination under the name "green infrastructure planning" can be seen as a new and innovative approach to planning—as a melting pot of state-of-the-art planning theory with a strong holistic thrust (e.g., Kambites and Owen 2006; Pauleit et al. 2011).

TABLE 13.1
Green Infrastructure Planning Principles

Approaches addressing green structure

- Integration: Green infrastructure planning considers urban green as a kind of infrastructure and seeks the integration and coordination of urban green with other urban infrastructures in terms of physical and functional relations (e.g., built-up structure, transport infrastructure, water management system).
- Multifunctionality: Green infrastructure planning considers and seeks to combine ecological, social, and economic/abiotic, biotic, and cultural functions of green spaces.
- Connectivity: Green infrastructure planning includes physical and functional connections between green spaces at different scales and from different perspectives.
- Multi-scale approach: Green infrastructure planning can be used for initiatives at different scales, from individual parcels to community, regional, and state. Green infrastructure should function at multiple scales in concert.
- Multi-object approach: Green infrastructure planning includes all kinds of (urban) green and blue space, e.g., natural and seminatural areas, water bodies, public and private green space, such as parks and gardens.

Approaches addressing governance process

- Strategic approach: Green infrastructure planning aims for long-term benefits but remains flexible for changes over time.
- Social inclusion: Green infrastructure planning stands for communicative and socially inclusive planning and management.
- Transdisciplinarity: Green infrastructure planning is based on knowledge from different disciplines, such as landscape ecology, urban and regional planning, and landscape architecture and developed in partnership with different local authorities and stakeholders.

Source: Pauleit, S. et al., Multifunctional green infrastructure planning to promote ecological services in the city. In Niemelä, J. (ed.), *Urban Ecology: Patterns, Processes, and Applications.* Oxford Univ. Press (Oxford biology), Oxford, 2011; Kambites, C., and Owen, S., Renewed prospects for green infrastructure in the UK. *Planning Practice and Research* 21, 4, 483–496, 2006; Benedict, M. A., and McMahon, E., *Green Infrastructure. Linking Landscapes and Communities.* Island Press, Washington, DC, 2006.

13.1.2 Meaning of Ecosystem Services

Although it is a slightly newer concept than green infrastructure, ecosystem services has recently gained much international attention as an approach to more actively consider and plan for the varied benefits of the environment, especially for benefits not currently valued in today's market-based systems (e.g., microclimate regulation, habitat provision, health and well-being). Particularly, the Millennium Ecosystem Assessment (MEA 2005) and The Economics of Ecosystems and Biodiversity (TEEB) reports (2010, 2011) heightened attention on the ecosystem services that nature provides. Although research on ecosystem services in urban areas is under-developed, it has been gaining momentum in recent years as has city interest in mapping and valuing some of these services, such as carbon sequestration, air pollution removal, and storm-water regulation (McPherson and Simpson 2003; Longcore et al. 2004; Soares et al. 2011).

A common critique of the concept of ecosystem services is that valuations are often monetary although not all services are marketable or easily included in value

---

TABLE 13.2

Ecosystem Services

Provisioning Services: Material outputs from ecosystems
- Food
- Raw materials
- Fresh water
- Medicinal resources

Regulating Services: Ecosystem processes that serve as regulators of ecological systems
- Local climate and air quality regulation
- Carbon sequestration and storage
- Moderation of extreme events
- Wastewater treatment
- Erosion prevention and maintenance of soil fertility
- Pollination
- Biological control

Habitat or supporting services: The provision of living spaces and maintenance of plant and animal diversity (serve as the foundation for all other services)
- Habitat for species
- Maintenance of genetic diversity

Cultural services: Nonmaterial benefits obtained from human contact with ecosystems
- Recreation and mental and physical health
- Tourism
- Aesthetic appreciation and inspiration
- Spiritual experience and sense of place

Source: Adapted from TEEB, *The Economics of Ecosystems and Biodiversity. Mainstreaming the Economics of Nature: A Synthesis of the Approach, Conclusions and Recommendations of TEEB*, 2010. Available from www.teebweb.org (accessed November 1, 2012).

A Transatlantic Lens on Green Infrastructure Planning and Ecosystem Services 251

assessments, especially cultural values (Cowling et al. 2008). Here, ecosystem services are not understood as a necessarily monetary approach (e.g., Norgaard 2010) but as a conceptual framework that can be used to more consciously consider the many different economic, ecological, and social functions that the environment provides in a similar way to how modern conceptions of green infrastructure include planning for smarter functionality, including multifunctionality of green space. To classify these multiple functions, we use the TEEB classification of provisioning, regulating, supporting/habitat, and cultural services (see Table 13.2).

13.2 A TRANSATLANTIC LENS

The case study cities Berlin and Seattle were chosen as a lens through which to examine differences in implementation of these concepts between North America and Europe. Both are rich in green space planning histories, yet they have contrasting planning cultures, urbanization dynamics, and challenges.

Berlin represents the German planning tradition with a strong emphasis on formal land use planning based on federal law. Although the term "Green Infrastructure Plan" has not yet been used in Germany for green systems on a city or regional scale to our knowledge, a lot of larger cities have recently developed or updated plans for urban green (e.g., the *GrünGürtel Frankfurt*, *Grünes Netz Hamburg*). A few examples exist of municipal ecosystem service assessments in Germany, but attention as yet has mainly been on a national scale through the TEEB initiative. Berlin makes an interesting case study as it is increasingly using informal strategic and visionary approaches due to criticism of the effectiveness of formal planning.

Seattle serves as an example from the United States, one of the few countries where numerous "Green Infrastructure Plans" have been published for cities, city regions, or larger areas (e.g., Philadelphia, Kansas City Metro, State of Maryland) and where municipal ecosystem service assessments are, although in their infancy, not uncommon among larger and more progressive cities. Seattle is known for innovative planning approaches and strong emphasis on stakeholder and community involvement (Dooling et al. 2006; Pauleit et al. 2011).

13.2.1 A Lens on Berlin

Urban and green structure planning on a city scale in Berlin started around the turn of the 20th century after a phase of rapid population growth. In the following decades, city planning and green planning were established as professions, and resulting plans and projects shaped the current green structure of Berlin (Schindler 1972). Environmental and green structure planning increased in the 1980s with wider environmental consciousness. Shortly after German reunification, city and landscape planning for a united Berlin were developed (SenStadt UM n.d.). The almost century-old idea of a citywide green space network was refreshed, and a "chain" of eight regional parks was established in 1998 to protect the suburban landscape against sprawl. Unlike other parts of former Eastern Germany, whose population declined after reunification, Berlin's metropolitan area stayed relatively stable throughout the 1990s at close to six million people (SenStadt UD and MIR 2009).

FIGURE 13.1 Berlin's recent strategic plans relating to green space development. (Reprinted with permission from the Senatsverwaltung für Stadtentwicklung und Umwelt, Berlin [Senate Department for Urban Development and the Environment for the City of Berlin], 2011, 2012; Graphic for the Strategie Stadtlandschaft Berlin [Urban Landscape Strategy of Berlin] by Projektbüro Friedrich von Borries and bgmr Landschaftsarchitekten.)

However, the relatively stable overall population figure belies the high rate of land use change and uneven growth in the area, especially in the form of suburbanization (SenStadt UD and MIR 2009). While some areas within the city and wider metropolitan area are growing, others are steadily shrinking. These disparities in development are expected to increase, on one hand, caused by suburbanization and, on the other hand, due to inner-city movements to specific districts with a high housing development potential (SenStadt UD 2012). Its population density stands at 3945 people/km^2 (Statistische Ämter des Bundes und der Länder 2012), and the city still has a share of green and open space of 41% (Amt für Statistik Berlin-Brandenburg 2011). To tackle the city's evolving challenges, including climate and demographic change and declining municipal revenues, a number of informal plans were agreed upon recently: the City Climate Development Plan, the Biodiversity Strategy, and the Urban Landscape Strategy Berlin (see Figure 13.1).

13.2.2 A Lens on Seattle

Like many major American cities, Seattle experienced a period of rapid growth toward the end of the 19th century, prompting citywide planning activities to combat mounting public health concerns. As in much of the rest of the country, planning policies in the 1950s and 1960s emphasized single-family housing and freeway development, leading to sprawl. Economic and environmental decline in the 1970s prompted regional planning efforts designed to increase the city's quality of life, such as park expansion. This period is characterized by a high degree of civic involvement, something that has come to characterize the city today. Ecological concerns strengthened in the 1990s (Dooling et al. 2006; Klingle 2008).

A Transatlantic Lens on Green Infrastructure Planning and Ecosystem Services

FIGURE 13.2 Seattle's 2005 Comprehensive Plan and the various elements included in the contents. (Reprinted from City of Seattle, 2005.)

Seattle's planning system nowadays is characterized by a high number of (informal) visions and strategies approved by politics as well as regular plan updates under broad community participation. The element of mandatory plans became more important because the Washington State Growth Management Act passed in 1990 demands comprehensive land use plans that anticipate growth and impact for a 20-year horizon and are updated regularly (King County 2013, see Figure 13.2).

Today, around 3.4 million people live in the Seattle metropolitan area (U.S. Census Bureau 2010). The population density stands at 2866 people/km^2, and the share of parks and open space is 14%. For the larger central Puget Sound region, growth to five million people by the year 2040 is forecasted (Department of Planning and Development n.d.). Further challenges the city faces are storm-water management, salmon recovery, and carbon emission reduction.

13.3 METHODOLOGICAL APPROACH

For this chapter, we chose a case study and qualitative document analysis approach. The analysis started with a search for planning and policy documents whose aim was to actively influence the urban structure, including urban green. The search was based on information taken from the cities' official websites, consultation of local experts, and planning literature in the case study cities.

From the more than 100 documents found, we selected five plans from each city to analyze in detail. The documents were chosen to represent plans relevant for the future development of green spaces in each city region. Plans from three thematic clusters were considered as most relevant:

- Comprehensive planning as constituting the general direction of spatial development
- Green space/landscape/biodiversity as sectorial planning directly addressing green space
- Environmental planning as sectorial planning focusing on a single environmental issue with indirect influences on green space (e.g., water management, climate change adaptation)

Additional criteria were that the plans are currently valid and cover a citywide or city-regional area (see Table 13.3).

Analysis of the extent to which the plans included aspects that could be seen as green infrastructure planning or referencing the ecosystem services concept first required developing a set of measurement criteria. For green infrastructure, three principles considered to be core elements of green infrastructure planning were chosen:

- *Connectivity*, referring to networks or corridors for habitats/species/recreation or ventiliation
- *Multifunctionality*, or the consideration of grouped ecosystem functions or services (i.e., provisioning, regulating, habitat/supporting, and cultural)

TABLE 13.3
Assessed Planning Documents

ID	Documents from Berlin	ID	Documents from Seattle
	Comprehensive Planning		
BER_1	Landesentwicklungsplan Berlin-Brandenburg 2009 (State Development Plan Berlin-Brandenburg)*	SEA_1	Comprehensive Plan: Toward a Sustainable Seattle (2004–2024) 2005
	Green Space/Landscape/Biodiversity Planning		
BER_2	Strategie Stadtlandschaft 2012 (Urban Landscape Strategy)	SEA_2	Parks and Recreation Strategic Action Plan 2008
BER_3	Landschaftsprogramm/Artenschutzprogramm 1994/2004 (Landscape and Species Protection Program)	SEA_3	Open Space 2100: Envisioning Seattle's Green Future 2006
BER_4	Strategie zur Biologischen Vielfalt 2012 (Biodiversity Strategy)	SEA_4	Puget Sound Salmon Recovery Plan 2007*
	Environmental Planning		
BER_5	Stadtentwicklungsplan Klima 2011 (Urban Development Plan Climate)	SEA_5	Storm Water Management Program 2012

Note: *, regional level.

A Transatlantic Lens on Green Infrastructure Planning and Ecosystem Services **255**

- *Transdisciplinarity*, which refers to the extent to which multiple departments as well as stakeholders and members of the public are included in the plan development process

Criteria for ecosystem services followed the aforementioned TEEB classification (Table 13.2) with some modifications based on ecosystem service classifications for urban areas (Niemelä et al. 2010; Piwowarczyk et al. 2013; Gómez-Baggethun and Barton 2013).

To assess the thematic scope of ecosystem service coverage, each of the 10 documents was analyzed hermeneutically and notations made for any reference to one of the 21 ecosystem services whether or not the service was directly stated. All services had to be related to a green space type or ecosystem to be considered. For example, measures in the documents for agriculture are seen as related to the service "food supply": however, if noise reduction is mentioned as a political aim but not related to the potential of vegetation to act as a buffer, it was not counted as a service. The thematic scope was aggregated for each city under the assumption that not all documents need to cover all services but that the group of selected documents should give a more or less comprehensive picture for each city in terms of green space planning and the extent to which services are addressed in planning.

In the final step, we assessed the thematic scope of the three green infrastructure principles according to the breadth and depth of their coverage. To correspond to the qualitative nature of the data, the plans were valued in broad classes from low to high. For connectivity, assessment included thematic scope as well as level of detail, such as a spatial network in the form of a map (high value) in comparison to a simple statement on the meaning of a connected network (low value). For multifunctionality, counts from the ecosystem service assessment were aggregated for each service group and the documents subsequently scored from low to high based on group values.

13.4 APPLICATION OF PRINCIPLES

The results of our study reveal to what extent the case study cities implement principles characteristic of green infrastructure planning and exhibit an understanding of the benefits provided by green space, also understood as ecosystem services. Although none of the documents analyzed appears to be heavily based on the concepts of green infrastructure or ecosystem services, references to both concepts were found.

13.4.1 RANGE OF ECOSYSTEM SERVICES

Although explicit mention of the term "ecosystem services" is found in only two documents from Berlin and none from Seattle, the hermeneutic analysis revealed that most of the documents cover the four groups of ES albeit with different thematic breadth within the classes. Figure 13.3 gives an overview per city and shows how many documents per city referred to the individual services.

The provisioning services, except medicinal resources, are frequently represented. The higher frequency in Berlin can be explained by the fact that a considerable amount of commercial forest and agricultural land lies within its boundaries.

256 Revising Green Infrastructure

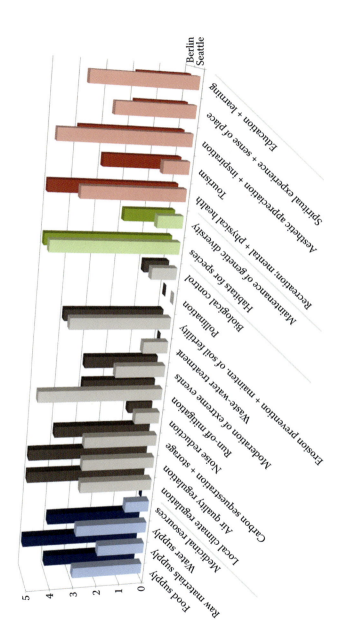

FIGURE 13.3 The frequency of ecosystem services found in Berlin and Seattle planning documents ($n = 5$ per city; blue: provisioning services, grey: regulating services, green: habitat/supporting services, red: cultural services).

A Transatlantic Lens on Green Infrastructure Planning and Ecosystem Services 257

Within the group of regulating services, the political importance of air quality and services related to climate change (climate regulation and carbon sequestration) are visible as almost all documents refer to these three services while runoff mitigation is the most frequently mentioned service in Seattle. Soil is an important topic in both areas. Services that are rarely mentioned in both areas are noise reduction, wastewater treatment, pollination, and biological control. Although noise and wastewater are mentioned as environmental issues, no connection is drawn between these impacts and the role of urban green as a potential solution. Problems with pests, invasive species, and the impacts through pest control are also mentioned in the documents from Seattle but mainly without an indication of biological alternatives.

All documents refer to the importance of urban green as habitat for species, but genetic diversity was less frequently considered.

Cultural services are frequently mentioned, most importantly for recreation and human health as well as aesthetic appreciation and inspiration. The importance of urban green for tourism is mentioned more often in Berlin, and the meaning for education and learning is highlighted more in Seattle.

13.4.2 COVERAGE OF SELECTED GREEN INFRASTRUCTURE PRINCIPLES

Three of five plans from Seattle and one plan from Berlin explicitly refer to the concept of green infrastructure. The heat maps for both cities express how extensively the three GI principles of connectivity, multifunctionality, and transdisciplinarity were considered in the analyzed documents (Figure 13.4). For connectivity,

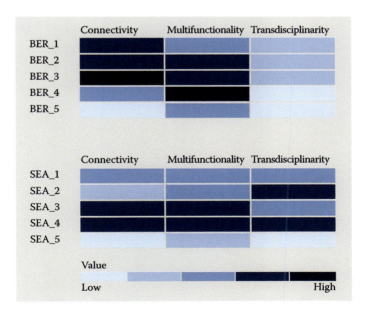

FIGURE 13.4 Values for green infrastructure principles in the different planning documents.

the Berlin documents have slightly higher values than Seattle's. Especially, the *Landscape and Species Protection Program* covers multiple aspects of connectivity related to biodiversity, recreation, and ventilation in a high level of detail. The *State Development Plan for Berlin-Brandenburg* details an open space network that shall serve multiple functions, and the *Urban Landscape Strategy* mainly focuses on a network of green and open space for recreation but also refers to the importance of habitat connectivity. Seattle's *Comprehensive Plan* also refers to an open space network as well as wildlife corridors but remains on a more general level compared to the comprehensive plan from Berlin. The *Puget Sound Salmon Recovery Plan* is ranked moderate in terms of connectivity because it only refers to ecological connectivity, which is a core aspect of the plan and discussed in detail.

The results for multifunctionality reveal that most documents have a broad thematic scope. Cultural aspects, such as recreation, are usually combined with biodiversity protection and regulative functions of ecosystems. Berlin's *Biodiversity Strategy* illustrates that biodiversity issues can be connected to a high number of ecosystem services. For Seattle, the *Open Space 2100* plan stands out as a vision for Seattle green structure based on a citizen charrette with an intentional multifunctional

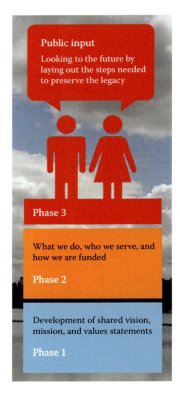

FIGURE 13.5 Public outreach and participation are important elements of planning processes in Seattle, such as in the process of developing the city's Parks Legacy Plan. (Graphic from Legacy Plan Brochure, reprinted from City of Seattle, 2013.)

A Transatlantic Lens on Green Infrastructure Planning and Ecosystem Services **259**

approach. Comparably low values can be found in both environmental plans, which deal with one specific issue, such as climate or storm water.

In regards to transdisciplinarity, the Berlin documents exhibit relatively low values. In comparison, especially the aspect of public participation is stronger in the Seattle planning context. Although formal participation in Germany is strongly based on written comments by stakeholders, in Seattle, document development often incorporated results from extensive participation activities, such as workshops and charrettes (Figure 13.5). For example, development of the *Parks and Recreation Strategic Action Plan* involved more than 70 public meetings in different neighborhoods. Although Berlin's informal strategies and plans claim that they are meant to be inclusive and are partly also designed to be implemented by the public, such as the *Urban Landscape Strategy*, the public was excluded in the process of drafting the documents. Furthermore, examples of interdepartmental project groups or plans drafted by stakeholder associations were only found in Seattle. The *Salmon Recovery Plan*, for example, was drafted by a grassroots collaborative consisting of citizens, tribes, experts, and policy makers.

13.5 DISCUSSING RIFTS AND BRIDGES BETWEEN PLANNING PRACTICE AND SCIENTIFIC DISCOURSES

Comparing the case study results with theoretical discussions on important ecosystem services in urban areas (Bolund and Hunhammar 1999; Niemelä et al. 2010; Gómez-Baggethun and Barton 2013), the discourses partially mirror one another. The regulating services local climate regulation, air quality regulation, carbon sequestration, and runoff mitigation seem broadly recognized among scientists and practitioners in these cities as are recreation and food supply.

One inconsistency found is with noise reduction and wastewater treatment. Although they are mentioned as important ecosystem services by scientists and are themes found within the planning documents, objectives and measures are addressed more at emission sources rather than the capacity of ecosystems for mitigation or restoration. To focus on emission sources is logical as less pollution requires less mitigation and restoration, but this finding might also indicate a lack of understanding or inadequate knowledge transfer of nature-based solutions for mitigation. Higher awareness of ecosystem services could expand opportunities to confront environmental pollution and find multifunctional approaches, such as Sustainable Urban Drainage Systems (SUDS). Pollination is another example of scientists seeing a higher relevance than practitioners and could represent a lack of practitioner knowledge about the benefits of supporting bees and other pollinators.

The assumption that ecosystem services that already have a market orientation such as tourism and food production have a higher focus in planning as found in a study on the presence of marine ecosystem services in planning documents of coastal cities in Poland (Piwowarczyk et al. 2013) can only partly be supported by the findings from Berlin. The frequency with which some of the regulating services are discussed also seems to follow their relationship with federal regulations, such as with storm-water management in the United States. Overall, we found that themes long discussed in nature conservation or planning for open space, such as habitats for species, recreation, or aesthetic appreciation, are often mentioned in planning documents from both cities, and awareness of less tangible issues, such as pollination or genetic diversity could be raised.

Both case study cities perform relatively well in regards to all GI principles and at least partially mirror the theoretical discourse on green infrastructure planning. Nevertheless, the cities could intensify the consideration of GI principles in all three dimensions. Transdisciplinarity—in our case, ranging from interdepartmental cooperation to collaborative planning with stakeholders and citizens—especially shows room for improvement. Although overcoming the structure and tradition of separately working municipal departments is challenging (Primmer and Furmann 2012) and extending public participation requires considerable commitment and resources, transdisciplinary planning is increasingly viewed as an instrumental component of land use management (Antrop 2006; Angelstam et al. 2013).

In relation to connectivity, the case study shows that cities like Berlin already consider the importance of an interconnected system of green and open space that serves multiple purposes. This is not surprising as green space systems, such as green rings and habitat networks, have long been discussed in Germany and many other European countries.

Unlike Sandström's (2002) conclusion for green structure planning in Sweden—that planners do not seize the multifunctional potential of urban green and still remain in traditional realms of cultural issues of urban green—the analyzed documents for green planning, such as the *Urban Landscape Strategy* exhibit a comparably broad thematic scope. As the documents analyzed by Sandström mainly originated in the 1990s, our results may be indicative of an increased emphasis on multifunctionality. The coverage of all groups of ecosystem services by both cities also supports our assessment that both cities—with potential for improvement—can be considered as familiar with holistic perspectives. However, the question of how intertwined and synergistic these different functions and services are discussed cannot be answered with our study approach.

Due to our selection of only two case study cities, the lens of implementation we have provided is limited in representativeness.* Additionally, our document analysis approach can only illuminate fragments of the planning process, namely the ones that are present within official planning documents. To get more complete insights, for example, if the level of awareness is truly higher in one city or whether the theoretical approaches might be understood by planners but not indicated in plans (see, e.g., Niemelä et al. 2010), a multi-method approach would be necessary, including interviews with stakeholders. Nevertheless, our study represents a first step toward

---

* The present study builds off of a previous study comparing implementation of the concepts of ecosystem services and green infrastructure in Berlin and New York City (Rall et al. 2012). Although that study only looked at one document per city, the Strategie Stadtlandschaft Berlin (2012) and PlaNYC (2011), respectively, and took a less hermeneutic approach to content analysis, the results supported those of this study. Specifically, the two concepts were found to be fairly widely considered in both plans although some less tangible ecosystem services, such as pollination, biological control, and, to some degree, genetic diversity were less represented. The analysis of ecosystem services presented here is part of a larger study comparing implementation of the concept across five American and European cities (Hansen et al. forthcoming).

an evaluation of gaps between theory and practice of green infrastructure planning and promotion of ecosystem services in urban areas based on empirical data that could be extended to other cities. Such an analysis can be used to deduce research gaps in terms of theory and tool development that hinder an exchange between research discourses and real-world planning.

13.6 CONCLUSION

We presented an analysis of planning documents from two cities representing different planning systems and elaborated their level of inclusion of green infrastructure and ecosystem service principles, thus providing a first step toward a discourse on current implementation of these approaches based on empirical data. We found plans in both cities address multiple aspects of urban green and can be interpreted as fitting conceptual approaches of green infrastructure planning and ecosystem services. Hence, the gaps between theory and practice seem less striking than expected, although a mixed methods approach and replication of study methods in other cities would be necessary to provide a more complete window into implementation of these concepts in planning.

Nevertheless, the gaps measured suggest that even progressive cities such as Berlin and Seattle can enrich planning processes and more consciously acknowledge the full potential of urban green planning and governance by more fully considering these conceptual approaches. Ecosystem services could be applied as a suitable concept to expand the thematic scope toward a more holistic picture in planning as they provide a framework to consider the multiple functions, services, and benefits green spaces provide. Tools for monetary valuation of ecosystem services, such as InVEST and i-Tree, have already had a strong impact on raising awareness of the manifold values of trees and green spaces. Developments in ecosystem service valuation frameworks are furthermore introducing new ways to combine multiple methods that are better able to take into account issues of social equity and value pluralism, thus addressing a long-held criticism of economically oriented ecosystem service assessments.

The concept of green infrastructure has also been helpful in considering the values of green spaces, especially regarding storm-water management. Cost-benefit assessments of storm-water management options have, in some cases, shown green infrastructure to be more cost-effective than technical gray infrastructure solutions, which strengthens arguments for green space protection, restoration, and enhancement in decision-making processes in urban planning. The concept of green infrastructure can also be used to raise awareness of issues such as interdepartmental coordination and broad public participation. Green storm-water infrastructure projects in the United States have shown the potential of green infrastructure planning to overcome "silo thinking" of different planning professions as well as to engage local communities in green planning (good practice examples from across the United States can be found in Rouse and Bunster-Ossa 2013). On a city level, green infrastructure could serve as a holistic planning approach oriented toward seeking synergies between different functions and services and which organizes their spatial relationship within a functional and physical network of green and open space.

Especially in times dominated by uncertainties related to economic crises or the effects of climate change, the concepts of ecosystem services and green infrastructure can support urban planning by offering frameworks to more consciously consider socio-ecological tradeoffs of land use decisions while seeking solutions that simultaneously provide ecological, economic, and social benefits. Thus, they offer the potential to overcome the limitations of monofunctionally oriented urban infrastructures.

ACKNOWLEDGMENTS

We thank Nadja Kabisch for contributing to the Berlin data set and Nancy Rottle for valuable insights into the planning system of Seattle. This paper was funded by the EU FP7-ERA-NET project URBES (Urban Biodiversity and Ecosystem Services, 2012–2014), which aims to fill scientific gaps on the relationship between biodiversity and ecosystem services for human well-being and to strengthen the capacity of European cities to adapt to climate change and other future challenges. A research stay in Seattle was supported by the Technische Universität München Graduate School, resp. Graduiertenzentrum Weihenstephan.

REFERENCES

Amt für Statistik Berlin-Brandenburg (2011). *Statistischer Bericht A V 3 – j/11. Flächenerhebung nach Art der tatsächlichen Nutzung in Berlin, 2011.* Available from: http://www.statistik -berlin.de (accessed November 5, 2012).

Angelstam, P.; Elbakidze, M.; Axelsson, R.; Dixelius, M.; and Törnblom, J. (2013). Knowledge production and learning for sustainable landscapes: Seven steps using social–ecological systems as laboratories. *AMBIO* 42 (2): 116–128.

Antrop, M. (2006). From holistic landscape synthesis to transdisciplinary landscape management. *From Landscape Research to Landscape Planning—Aspects of Integration, Education and Application.* Springer, 27–50.

Benedict, M. A.; and McMahon, E. (2006). *Green Infrastructure. Linking Landscapes and Communities.* Washington, DC: Island Press.

Bolund, P.; and Hunhammar, S. (1999). Ecosystem services in urban areas. *Ecological Economics* 29 (2): 293–301.

Cowling, R. M.; Egoh, B.; Knight, A. T.; O'Farrell, P. J.; Reyers, B.; Rouget, M. et al. (2008). Ecosystem services special feature: An operational model for mainstreaming ecosystem services for implementation. *Proceedings of the National Academy of Sciences* 105 (28): 9483–9488.

Department of Planning and Development (n.d.). *Land use.* Available from: http://www .seattle.gov/dPd/cityplanning/populationdemographics/aboutseattle/landuse/default .htm (accessed November 3, 2013).

Dooling, S.; Simon, G.; and Yocom, K. (2006). Place-based urban ecology: A century of park planning in Seattle. *Urban Ecosystems* 9 (4): 299–321.

The Economics of Ecosystems and Biodiversity (TEEB) (2010). *Mainstreaming the Economics of Nature: A Synthesis of the Approach, Conclusions and Recommendations of TEEB.* Available from: http://www.teebweb.org (accessed November 1, 2012).

The Economics of Ecosystems and Biodiversity (TEEB) (2011). *TEEB Manual for Cities: Ecosystem Services in Urban Management.* Available from: http://www.teebweb.org (accessed November 1, 2012).

A Transatlantic Lens on Green Infrastructure Planning and Ecosystem Services 263

European Commission (2013). Communication from the Commission to the European Parliament, the Council, the European Economic and Social Committee and the Committee of the Regions. *Green Infrastructure (GI)—Enhancing Europe's Natural Capital.* COM(2013) 249 final.

Gómez-Baggethun, E.; and Barton, D. N. (2013). Classifying and valuing ecosystem services for urban planning. *Ecological Economics* 86: 235–245.

Hansen, R.; Frantzeskaki, N.; McPhearson, T. et al. (forthcoming). The uptake of the ecosystem services concept in planning discourses of European and American cities. *Article in Review for Ecosystem Services for 2014 Special Issue.*

Hauck, J.; Görg, C.; Varjopuro, R.; Ratamäki, O.; and Jax, K. (2013). Benefits and limitations of the ecosystem services concept in environmental policy and decision making: Some stakeholder perspectives. *Environmental Science & Policy* 25: 13–21.

Kambites, C.; and Owen, S. (2006). Renewed prospects for green infrastructure planning in the UK. *Planning Practice and Research* 21 (4): 483–496.

King County (2013). *Comprehensive Plan.* Available from: http://www.kingcounty.gov/property /permits/codes/growth/CompPlan/history.aspx (accessed November 3, 2013).

Klingle, M. W. (2008). *Emerald City. An Environmental History of Seattle.* New Haven: Yale University Press (The Lamar series in western history).

Longcore, T.; Li, C.; and Wilson, J. (2004). Applicability of Citygreen Urban Ecosystem Analysis software to a densely built urban neighborhood. *Urban Geography* 25 (2): 173–186.

Mazza, L.; Bennett, G.; De Nocker, L. et al. (2011). *Green Infrastructure Implementation and Efficiency.* Final report for the European Commission, DG Environment on Contract ENV.B.2/SER/2010/0059. Institute for European Environmental Policy. Brussels and London.

McPherson, E. G.; and Simpson, J. R. (2003). Potential energy savings in buildings by an urban tree planting programme in California. *Urban Forestry & Urban Greening* 2 (2): 73–86.

Mell, I. C. (2008). Green infrastructure: Concepts and planning. *FORUM Ejournal* (8): 69–80.

Millennium Ecosystem Assessment (MEA) (2005). *Ecosystems and Well-being: Synthesis.* Washington, DC: Island Press.

Naumann, S.; Davis, M.; Kaphengst, T.; Pieterse, M.; and Rayment, M. (2011). *Design, Implementation and Cost Elements of Green Infrastructure Projects.* Final report to the European Commission, DG Environment, Contract no. 070307/2010/577182/ETU/F.1. Hg. v. Ecologic institute und GHK Consulting.

Niemelä, J.; Saarela, S. R.; Soderman, T. et al. (2010). Using the ecosystem services approach for better planning and conservation of urban green spaces. A Finland case study. *Biodivers Conserv* 19 (11): 3225–3243.

Norgaard, R. B. (2010). Ecosystem services: From eye-opening metaphor to complexity blinder. *Ecological Economics* 69 (6): 1219–1227.

Pauleit, S.; Liu, L.; Ahern, J.; and Kazmierczak, A. (2011). Multifunctional green infrastructure planning to promote ecological services in the city. In: Niemelä, J. (ed.): *Urban Ecology: Patterns, Processes, and Applications.* Oxford: Oxford Univ. Press (Oxford biology), pp. 272–285.

Piwowarczyk, J.; Kronenberg, J.; and Dereniowska, M. A. (2013). Marine ecosystem services in urban areas: Do the strategic documents of Polish coastal municipalities reflect their importance? *Landscape and Urban Planning* 109 (1): 85–93.

Primmer, E.; and Furman, E. (2012). Operationalising ecosystem service approaches for governance: Do measuring, mapping and valuing integrate sector-specific knowledge systems? *Ecosystem Services* 1 (1): S. 85–92.

Rall, E. L.; Hansen, R.; and Pauleit S. (2012). The current landscape of green infrastructure planning and ecosystem services: The cases of Berlin and New York. *Proceedings of the Symposium Designing Nature as Infrastructure*, Technical University of Munich, Nov. 28–29, 2012, pp. 160–180, Technische Universität München, München.

Rouse, D. C.; and Bunster-Ossa, I. F. (2013). Green Infrastructure. A Landscape Approach. American Planning Association. Planning Advisory Service report 571.

Sandström, U. G. (2002). Green infrastructure planning in urban Sweden. *Planning Practice and Research* 17 (4): 373–385.

Schindler, N. (1972). Gartenwesen und Grünordnung in Berlin. In: Weber, K. K.: *Berlin und seine Bauten.* Teil XI Gartenwesen. Berlin: Verlag von Wilhelm Ernst & Sohn, pp. 1–50.

SenStadt UD (Senate Department for Urban Development and Environment—Senatsverwaltung für Stadtentwicklung und Umwelt) (2012). Bevölkerungsprognose für Berlin und die Bezirke 2011–2030. Kurzfassung. Available from: http://www.stadtentwicklung .berlin.de/planen/bevoelkerungsprognose/download/bevprog_2011_2030_kurzfassung .pdf (accessed January 14, 2014).

SenStadt UD (Senate Department for Urban Development—Senatsverwaltung für Stadtentwicklung) and MIR (Ministry for Infrastructure and Regional Planning/Ministerium für Infrastruktur und Raumordnung) (eds.) (2009). *Raumordnungsbericht 2008, Hauptstadtregion Berlin-Brandenburg.* Potsdam: Brandenburgische Universitätsdruckerei und Verlagsgesellschaft Potsdam.

SenStadt UM (Senate Department for Urban Development and the Environment— Senatsverwaltung für Stadtentwicklung und Umwelt) (n.d.). *The history of open space development in Berlin.* Available from: http://www.stadtentwicklung.berlin.de/umwelt /landschaftsplanung/chronik (accessed November 5, 2012).

Soares, A. L.; Rego, F. C.; McPherson, E. G.; Simpson, J. R.; Peper, P. J.; and Xiao, Q. (2011). Benefits and costs of street trees in Lisbon, Portugal." *Urban Forestry & Urban Greening* 10 (2): 69–78.

Statistische Ämter des Bundes und der Länder (2012). GebietundBevölkerung, *Fläche und Bevölkerung, 26.9.2012.* Available from: http://www.statistik-portal.de (accessed November 6, 2012).

U.S. Census Bureau (2010). State & County QuickFacts: Seattle (city), Washington. Available from: http://quickfacts.census.gov/qfd/states/53/5363000.html (accessed November 3, 2013).

PLANNING AND POLICY DOCUMENTS

Biodiversity Strategy Berlin/Berliner Strategie zur Biologischen Vielfalt (2012). Available from: http://www.stadtentwicklung.berlin.de (accessed November 6, 2012).

Comprehensive Plan: Toward a Sustainable Seattle (2004–2024) (2005). Available from: http://www.seattle.gov/dpd (accessed November 3, 2013).

Landscape and Species Protection Program/Landschaftsprogramm/Artenschutzprogramm (1994/2004). Available from: http://www.stadtentwicklung.berlin.de (accessed November 3, 2013).

Legacy Plan Brochure (2013). Available from: http://www.seattle.gov/parks/legacy/ (accessed January 14, 2014).

Open Space 2100: Envisioning Seattle's Green Future (2006). Available from: http://www .open2100.org/ (accessed November 3, 2013).

Parks and Recreation Strategic Action Plan (2008). Available from: http://www.seattle.gov /parks/ (accessed November 3, 2013).

PlaNYC. A greener, greater New York. Update April 2011 (2011). Available from: http://www .nyc.gov/html/planyc2030 (accessed November 5, 2012).

Puget Sound Salmon Recovery Plan (2007). Available from: http://www.psp.wa.gov (accessed November 3, 2013).

State Development Plan Berlin-Brandenburg/Landesentwicklungsplan Berlin-Brandenburg (2009). Available from: http://gl.berlin-brandenburg.de (accessed November 3, 2013).

Storm Water Management Program (2012). Available from: http://www.seattle.gov/util/ (accessed November 3, 2013).

Urban Development Plan Climate/Stadtentwicklungsplan Klima (2011). Available from: http://www.stadtentwicklung.berlin.de (accessed November 6, 2012).

Urban Landscape Strategy Berlin/Strategie Stadtlandschaft Berlin (2012). Available from: http://www.stadtentwicklung.berlin.de (accessed November 6, 2012).

14 The Concept of "New Nature"

A Paradigm Shift in How to Deal with Complex Spatial Questions

Susanne Kost

CONTENTS

14.1 Introduction 267
14.2 A Mentality of Making...268
14.3 From Making Landscape to Making Nature...269
14.4 A New Paradigm in the Understanding of Nature?....................................278
14.5 Conclusion 281
References...282

14.1 INTRODUCTION

Since the end of the 1980s, a process has been taking place in the Netherlands that depicts a turning point in the development of landscape. Instead of handling complex questions and problems in a technocratic and mono-disciplinary way, a new integrative approach has come into being, which views landscape, with its multifaceted uses and functions, as a whole. This can partly be explained by the fact that the previous flood protection policy, which was designed to ensure not only the safety of open spaces, but also to provide locations for the development of housing, industry, and infrastructure, had reached its limitations, making a change in thought absolutely essential. A second process also opened up a new possibility for a change in approach to dealing with landscape: The catalyst for this process was the rediscovery of nature. In the 1980s, on the Flevoland Polder, which was emerging to the east of Amsterdam, a secondary wilderness appeared almost by chance on an area of land that was being held available for industrial purposes but not yet in use. This was the so-called Oostvaardersplassen. This spontaneous development of a nature area was celebrated in the Netherlands as "new nature" and drew the interest not only of specialists, but also of people from all walks of life. The rediscovery of nature and the increase in its importance for society can be understood as a reaction to the enormous economization of the landscape, which, for its part, creates a longing for harmony between man and nature. In the light of the rediscovery of nature, a paradigm shift in the way of planning landscape had become apparent (cf. Kost 2009).

"New nature" does not, as the term suggests, mean an undisturbed landscape. "New nature" resolved the difficult dispute about the dialectical relationship between landscape protection (preservation) and landscape development (transformation). "New nature" and the initiatives, projects, programs, and cooperation that derive from it bring both of these "adversaries" together. Landscape protection and landscape development can meet in a new practical discourse. This discourse is already reflected in varying ways in the development of landscape and in the policy for spatial development. This was new and, at the same time, inconceivable: inconceivable because nature protection had not played an active part in the development of near-natural areas in large-scale projects up to then and new because the so-called nature development projects not only established areas in harmony with nature, but also sought a connection between ecology and economy. The phenomenon of "new nature" has given rise to huge changes in the landscape and simultaneously opened up new perspectives for rural space.

The concept and the project of "making" nature provide an opportunity for theoretical discussion. Occidental culture represents the idea of subjugating nature to use it for the lives of the people and to transform it in this spirit. The "making of nature" was seen as a divine creation, which not only enabled the use of nature, but also limited it. Every kind of exploitation is limited by a "conservation mandate," in which respect for creation or, above all, religious ideas for nature is expressed. In contrast, when nature can be made, even in the case of "new nature," the perspectives and the options of planning and feasibility change. This new option enables, on the one hand, alliances of nature protection with regional development, the extraction of raw materials, flood protection, etc. and, on the other hand, it opens an awareness of what can be made, what is feasible, and thus, perhaps, of the limitation of what changes can be achieved in nature and in landscape.

14.2 A MENTALITY OF MAKING

A debate on the subject "Making of Landscape" about the future of cultural landscape and possible perspectives and instruments concerned with the development of landscape has been started and is currently running in many western industrialized nations (cf. Benson and Roe 2000; Baart, Metz, and Ruimschotel 2000; Friesen and Führ 2001; Roggema 2009; Watson and Adams 2011). This debate has a wide spectrum of subjects, ranging from the question of the future of large, disused industrial or port areas; brown coal open cast mining; and fallow agricultural plots of land to the necessity of acting on climate change and the consequences for spatial planning and landscape use. The question of how convertibility of landscape is to be understood is common to all of these subjects: Is landscape regarded as existing and unchangeable, as a cultural formative landscape that develops over the years, or generally as a project?

Whereas in other countries there are certain changes in the landscape to be seen—for example, the land reclamation in the Po Delta in Italy, the drainage of marshy landscapes in Germany in the time of Frederick the Great,* or the conversion of the

---

* Frederick the Great (Friedrich der Große) was king of Prussia in the 18th century.

The Concept of "New Nature" 269

grid system in North America at the beginning of the 19th century—the Netherlands looks back on a centuries-long continuous and systematic development of landscape. This means that, in the Netherlands, landscape has been and still is being continuously transformed. The experience and the knowledge that have been collected over centuries have produced a specific understanding of nature: a "mentality of making." Thus, the word "mentality" is not simply a colloquial labeling of a person from another nation (Italian, Austrian) or a region (Bavarian, Flemish) but refers to a development of fundamental structures, norms, and knowledge, which could be described as systematic and typical—a kind of habitus that influences thought and action. The history of a certain mentality or, rather, the "history of collective ideas" (Febvre, quoted according to Chartier 1987, 76), thus, possibly leads to another relationship to nature and, thus, to another way of treating landscape. One reason for "the mentality of making" can surely be found in the genesis of landscape in the Netherlands (land reclamation by means of drainage), based on the fundamental logic that landscape can be converted through human intervention and never, not even originally, had any intention other than that of protecting it in the course of economic utilization and productivity. This continual new creation of exploitable landscape, which has taken place for centuries, has led to landscape in the Netherlands always being viewed as formable and therefore also transformable. It is correct that the theoretical idea of unlimited control over nature is a basic principle of today's modern society, but in the Netherlands, the main reason for this lies in the genesis of its landscape. This led to the insight that, owing to the genesis of the landscape and the social relationship to nature that derived from it, each piece of land(scape) has always been more or less affected by considerable pressure for action and development. This process of permanently transforming the landscape was accompanied by an enormous variety of planning and was able to develop and establish a wide culture in planning on all levels and in all disciplines. Landscape is regarded as a communicative, multidimensional process of development wherein the different actors negotiate their interests. Landscape is a project.

14.3 FROM MAKING LANDSCAPE TO MAKING NATURE

A great deal of the Netherlands lies below sea level. For centuries, the Netherlands has been reclaiming land systematically from the water, draining it and building dykes. This is the main reason why, in the Netherlands, landscape is defined by its economic value. On the one hand, this is certainly connected with the reclaiming of land and the continual necessity of financial resources; on the other hand, it is understandable because of the existing limited area of land. The Netherlands is today the most highly urbanized country in Europe (see Figure 14.1).

As a central event, we could understand the previously described rediscovery of nature in the Oostvaardersplassen (surface area approximately 6000 hectares) near Amsterdam in the 1970s. It became one of the most important protected bird habitats in Europe. This took place mostly without human intervention (apart from drainage) and without this previously mentioned pressure. On discovering this, in a land of "man-made landscapes," a wide-ranging discussion about the meaning of nature evolved. This discussion took place in many different groups and fields and opened

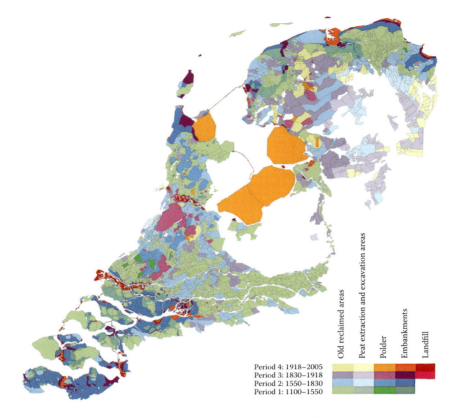

FIGURE 14.1 Since the construction of the first polder in the 11th century, the Netherlands has never ceased to reclaim land. To date, 3891 polders have been constructed. (From Geuze, A., and Feddes, F., *Polders*. NAi Uitgevers. Rotterdam: Gedicht Nederland, 2005, p. 33.)

up opportunities to strengthen nature and the development of nature. "New nature" became a mission statement of transformation, the concept of planning in policy and society.

The rediscovery of nature in the Oostvaardersplassen was accompanied by a second event, which played an important role in the planning and development of "new nature." In the 1980s, the project "Plan Ooievaar" (Stork) was invented. Plan Ooievaar symbolizes the starting point for a radical change in spatial planning. It provided an opportunity to deal with complex spatial planning tasks and regarded the landscape, with its different functions and governmental (planning) borders, as a whole. This plan gave an enormous boost to ecological planning and design in the Netherlands and changed the focus on river and flood protection.

The experience in connection with the setting up of the Oostvaardersplassen led to the establishment of the Nature Policy Plan in 1990. This plan concerned itself first of all with the status of nature preservation and nature development as a spatial planning task with the goal of creating a National Ecological Network (called EHS).

The Concept of "New Nature"

This goal was to establish 750,000 hectares of "new nature" within 20 years.* This corresponds to approximately 18% of the land area of the Netherlands. Nature and the development of nature on this scale as components of environmental planning in a man-made landscape paved the way for innovative impulses for many fields as well as in the treatment of complex problems.

The essential point is the recognition that nature development and the creation of a national ecological network had found its way into the national policy for space development, thus becoming co-equal in importance to housing, economic, environmental, and financial policies.† It was understood that intervention or, rather, changes in the named resorts would have consequences for space, too. This made it possible to reach all levels and parties involved in spatial planning and to discuss plans and activities for transforming space. Organizations for nature protection had previously seen their role as administering and conserving the natural areas; now, however, they are seeking cooperation with regional parties and an active integration of the seminatural areas with local recreation, educational, and leisure facilities as well.

At the end of the 1980s and the beginning of the 1990s, the first "new nature" project, Goudplevier, came into existence. Small, seminatural areas were recombined as larger, connected expanses by restoring agricultural areas. In addition, the "making of nature" was introduced as compensation for all kinds of building in the landscape. Not only this, but in projects such as De Geldersee Poort, which started in 1991, and Grensmaas, in 1992, long-term solutions for dealing with inland floods in the delta regions in the Netherlands have been, and are being, gradually developed to the present day. In the beginning stage of nature development projects, possibilities were sought to free, to a large extent, the ground along the rivers from buildings and other obstacles that the floods had caused and that were impeding the natural flow. Companies that had excavated raw materials out of rivers or areas close to the rivers and simply left these expanses as leftover holes are now being integrated, even during the excavation stage, into the forming and conversion of regional projects for nature development. It no longer seems possible to deal with the problems of flooding by means of technical solutions. This means (1) combining excavation for raw materials along the rivers with setting up a nature-orientated meadow, (2) combining traditional flood protection with natural pastures beside the rivers, and (3) combining nature development with local recreation and tourism. A pragmatic solution was found, thus linking the two greatly contrasting aims of nature development and the industrial excavation of gravel. The sale of the gravel is used to finance the restoration of the river ("Green for Gravel") and for implementing the concept for flood protection. Instead of

---

* In past years, the dimensions and the period of validity for the implementation of the Ecological Main Structure were adapted time and time again, owing to changing governments. This led, on the one hand, to a scaling down of the Network that was to be achieved, and, on the other hand, it meant that the implementation was delayed. In particular, those farmers on whose land the "new nature" was to be implemented fought against the Ecological Main Structure. In 2016, there is to be a provisional appraisal and, depending on the result, whether the new deadline for the conclusion of the National Network in 2021 is tenable will be decided (see also Jongman and Bogers 2009 about the state of the art).

† The 5th Nota of Spatial Development (2000–2020) contains the equal status of spatial development (previously environmental planning) with the other areas of responsibility mentioned.

a technocratic and mono-disciplinary way of handling problems, an integrative start was made (Figure 14.2).

The project Blue City (which has been in the process of implementation since 2001) is quite different compared to the first phase of nature development projects. It is the biggest nature development project with facilities for living and for leisure in which an agricultural polder that had previously been drained was then deliberately flooded. In this case, nature development can provide prospects for the future to a shrinking region (Figures 14.3 and 14.4).

The Blue City is not the only project that shows nature development can enhance the image of a region. A new layer of landscape is being laid over the top of the contaminated Volgermeerpolder (finished in 2012). This tip for toxic waste has given the location a negative image for centuries. The function of the "new nature" landscape is to encourage the development of activities for local recreation in the region, i.e., the reconstruction into a near-natural landscape endows the region and the directly affected communities with new prospects and a basis for development (Figure 14.5).

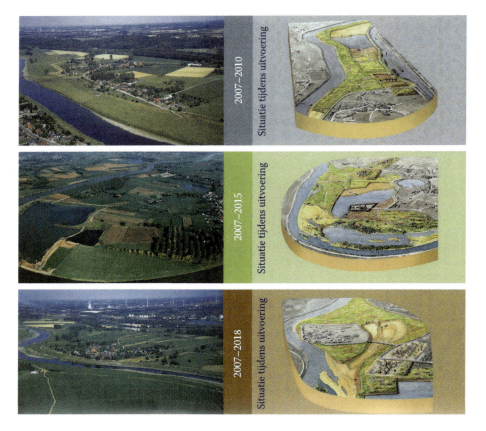

FIGURE 14.2 A selection of projects along the river Grensmaas that combine long-term flood protection, nature development, new housing projects, and tourism facilities (Grensmaasprojects: De Maaswerken).

The Concept of "New Nature"

FIGURE 14.3 The Blue City: A new image for a former agricultural landscape. (From Blue City: housing area "Het Ried," © Koos Boertjens, 2010.)

FIGURE 14.4 The Blue City: A new image for a former agricultural landscape. (From Blue City: housing area "De Wei," © Koos Boertjens, 2010.)

FIGURE 14.5 The Volgermeerpolder: a "Saba" landscape for this contaminated area as a new image. (From Volgermeerpolder: © Vista landscape architecture and urban design, Amsterdam, http://www.vista.nl.)

Both projects try to get support by promoting the positive idea and image of a "new nature." "Environment" describes everything that surrounds and interacts with us, but since its appearance in the 1960s, the concept has also revealed problems. We often connect environment with negative manifestations, such as environmental crises or environmental problems. In comparison, in "new nature," a much more positive field of activity and identity has been recognized and developed, bringing society and environmental protection closer together. This also leads to a great degree of popularity for nature protection organizations. The number of members in the largest Dutch nature protection organization Natuurmonumenten increased greatly between 1985 and 1995. Thus, in 1995, almost 5% (721,000) of the population were members (Table 14.1).

This popularity and acceptance is explained by the Dutch journalist Tracy Metz by the positive message conveyed by the concept of "new nature." For this reason, the nature protection organizations have joined in the discussions and active accompaniment of "new nature projects" and have developed themselves into powerful organizations in society.

Landscape and projects motivated by nature development have changed from an increase in nature protection areas in the beginning of this process to transdisciplinary and interdisciplinary projects nowadays. This means that more and more participants who would not be expected to be interested in nature development have become involved in this process (for example, industry). Today, nature development can be regarded as a goal for linking national and regional interests. On the national

TABLE 14.1
Natuurmonumenten Membership, 1985–1995

Year	Members
1985	235,000
1990	292,000
1995	721,000

Source: Natuurmonumenten.

level, it is important to establish an ecological network, an integrated management for the rivers, and a flood protection policy extending beyond the borders of the country. Government-stimulated programs seek complex solutions in the field of water and flood protection, actively using the dynamics of natural processes and thereby connecting the scientific approach with practical large-scale implementation. They symbolize a new step from "making new nature" in the sense of a new landscape layer (Blue City and Volgermeerpolder) to an active "building with nature." Fiselier et al. (2013, 18) summarize this process, which lasted approximately 25 years, in the following words: "Having built *against* nature, *within* nature and *for* nature, nowadays, the philosophy is to have a framework which builds with nature."

On the regional level, it provides not only the potential for solving regional problems, such as emigration and unemployment, but also for creating new regional economies, housing concepts, and images. For this reason, nature development is not only of importance for nature protection to expand and conserve natural areas, it also opens up new activities for (spatial) planners and other disciplines whose task is to structure and design a landscape, for example, flood protection and housing.

The extent to which an interdisciplinary and transdisciplinary intertwining in dealing with and solving complex questions of space in the Netherlands has already been established; thereby quite consciously choosing "building with nature" can be gauged in two government programs in particular, namely those for the development and implementation of a long-term flood protection program, both coastal and inland. The Delta Program,* anchored in a special law, the Delta Act, pursues two goals in general: to protect the Netherlands from flooding and to ensure a sufficient water supply. For this reason, the Dutch government invests one billion Euros per year. An integral part of the Delta Act is the program "Building with Nature," which came into being in 2008. This program is being implemented by the interdisciplinary and

---

* The high water and floods in the rivers in 1993 and 1995 had proved very clearly that a flood protection policy organized purely on technical engineering would no longer be able to cope with future challenges under the circumstances created by climate change and a rising sea level. The continual raising and strengthening of the dykes was no longer sufficient to cope with the pressure of increasing urbanization and land-surface soil. The growing requirement for room intensified the problem and brought about a new strategy. No longer was the principle of "pump and drain as fast as possible" an adequate solution; The new flood protection policy has developed into "retain, store, and drain."

transdisciplinary Ecoshape Consortium. The consortium is composed of representatives of the private sector (dredging contractors, equipment suppliers, engineering consultants) and of the public sector (government agencies, local authorities) as well as research institutes and universities. Their work is based on long experience in the field of hydraulic engineering integrated with the dynamics of natural and ecosystem processes. The main idea is to combine practical experience with scientific approaches and to stimulate design opportunities for "building with nature." de Vriend and van Koningsveld (2013, 6) describe the aims as follows: "New challenges associated with urbanization, economic development, sea level rise, subsidence and climate change demand an innovative approach to hydraulic engineering that aligns the interests of economic development with care for the environment" (Figure 14.6).

In one of these projects, the forces of nature have been put to use in order to repair the sandy coastline of The Hague with the help of wind and ocean currents. In 2011, a kilometer-long island in the form of a hook, consisting of 21 million cubic meters of sand, was deposited for this purpose. Wind erosion and ocean currents will gradually remove this sand, allowing the sediment to resettle along the coast in a natural way. This project, with the name "Zandmotor" (sand engine), is so successful and efficient that it could be of interest for many coastal regions throughout the world. This is more understandable when one takes into consideration that the German island of Sylt in the North Sea, for example, loses a million cubic meters of sand every year. Metropolitan coastal zones, such as Long Island or Rockaway Beach in New York, as a consequence of the horrendous damage along the coast caused by Hurricane Sandy at the end of 2012, are considering a "building with nature" strategy. The catastrophe of this storm basically pointed out the weakness of a "bulwark" strategy. This, with its breakwaters, moles, and cement walls, has been in existence for more than 100 years, standing against the sea. In the long run, and especially in very bad weather conditions, it has led to the destruction of natural (sand) stretches of coast both above and below the earth. Against the background of a rising sea level, this can have fatal consequences (Figure 14.7).

In the program "building with nature," a whole series of totally different projects have been brought into being. Their goal is to create long-term solutions for flood protection in both coastal and inland areas and to promote biodiversity, eco-systems, and the development of valuable flora and fauna. In addition, the program should encourage more development possibilities for local recreation and leisure as well as industries.

It is not a matter of finding solutions that are only developed at the office desk but, rather, solutions that are scientific and analytical and based on practical experience gained by experiments in the field. At the same time, it is a question of stimulating this new approach by actively involving stakeholders and the public and also discovering and integrating new (research) methods and models by means of an intensive and long term cooperation between science and practical experience. The program "Room for the River" (Figure 14.8) was the first complex and large-scale adaptation project in the Netherlands. It has been drawn up in a similar way, but in addition to restoring natural and ecological processes and river dynamics, it promotes the improvement of the quality of space and the relationship between living and working conditions in the Rivers Region in the center of the Netherlands (cf. Sijmons 2009, 64).

The Concept of "New Nature"

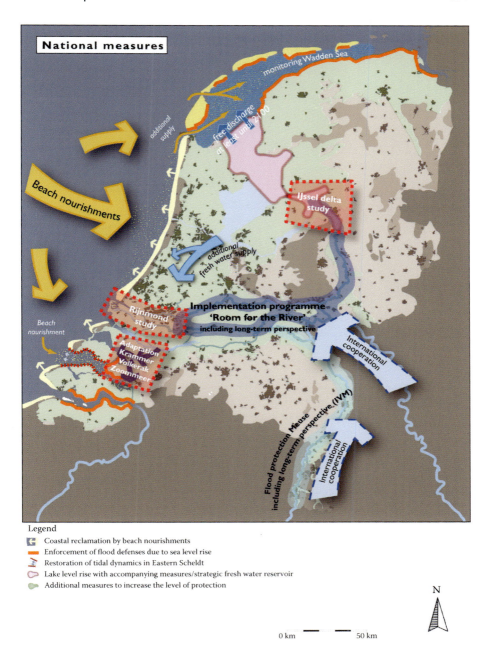

FIGURE 14.6 Flood protection—tasks and national measures. (From "Working together with water, findings of the Deltacommissie," 2008, The Hague, The Netherlands.)

FIGURE 14.7 Zandmotor (sand engine). (From Zandmotor, © Rijkswaterstaat, Joop van Houdt.)

Integrating natural processes into coastal and flood protection, whose view is reinforced by technical engineering, and into densely populated areas demands, moreover, an active involvement of many stakeholders and the public. In such more or less experimental projects, it is necessary to demonstrate how "building with nature" functions and what it means for the future in case of floods. The experience from the individual projects in so-called design, implementation, and guidelines is an important element of the program "building with nature," whence they should be transferred into other projects and fields worldwide.

14.4 A NEW PARADIGM IN THE UNDERSTANDING OF NATURE?

It is important here to return to the concept of nature in a more theoretical way and to point to the difficulty of an exact definition. As we know, there is hardly any nature in Europe, i.e., natural landscape. Nature protection maintains and develops natural landscapes, which are very important for certain species of plants and animals from the scientific point of view in relation to natural or to regional studies. These landscapes include cultivated landscapes, such as heath or pasture land. If one compares nature protected areas with nature development areas, it becomes clear that both of these forms have come into existence through human intervention and that they differ in age. Nature development areas are much closer to the concept of adaptation than to the concept of nature. But what is the case with nature protection areas, for example, a heath? Is this more natural than a nature protection area and therefore more valuable and not to be changed? The time factor seems to play an

The Concept of "New Nature"

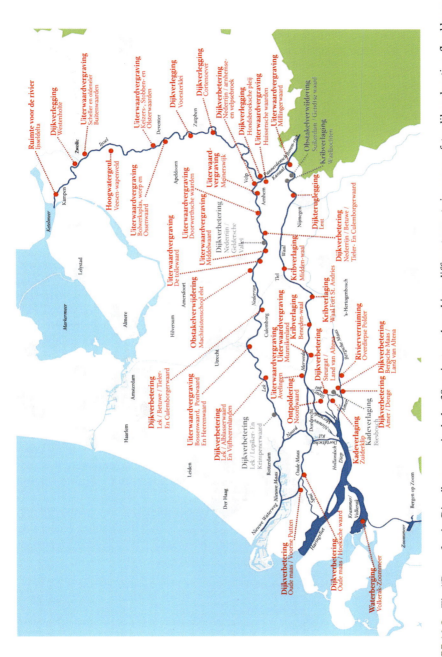

FIGURE 14.8 The "Room for the River" program includes 39 projects and combines different ways to increase safety: dike relocation, flood bypasses, lowering flood plains, green rivers, lowering of the groynes, and removing obstacles. (From De Waterstandskaart: Rijkswaterstaat.)

important role in our understanding of what nature really is; we decide when something appears valuable and worth preserving* and that it counts in our awareness as "nature" and "untouchable." Both nature protection areas and nature development areas are, however, more or less naturalized landscapes.

With the process of controlling nature and the knowledge of the feasibility of forming landscapes, a specific societal relationship to nature, characterized materialistically as well as aesthetically, has formed over the centuries. Whereas the French and English gardens led to the first extensive places of mimetic nature and aesthetic activity, nowadays the layout is in the form of "primeval forests" and "wilderness areas."

The concept of nature has always reflected the inner imagination, the emotions, and the yearnings of the individual.† It has produced an abstract image, in which emotions are individually experienced, an image that is idealized and used to compensate for reality.‡ Whereas tourism attracts people to distant places for an experience of untouched nature, the project of "new nature" creates a kind of a wilderness close to home. Nature development in the Netherlands is like the nascent picture of the landscape painter; it is, however, simultaneously painter and creator. It *makes* the landscape into a picture, according to a picture, a reference, for it translates the present understanding of nature as designed environment into a reality.

The concept of nature is also associated with what has been lost and what is past. As the natural areas were expanded, the task of the planners increased with regard to the future image of the landscape that was to be transformed. It is not surprising that the first projects took up images of a lost landscape. The desire for a more beautiful and natural landscape or, rather, surroundings arose simultaneously with these nature development projects. In this way, the landscape aesthetics became important for the regions. The creation of an aesthetic landscape is, however, only conceivable in combination with other activities (economic or security-relevant). The aesthetic value of natural areas previously held a background position and was of hardly any importance for natural protection in its role as an administrative and responsible institution. The technical value of nature protection, which represented a value that was basically only comprehensible to experts, was more important. Today, one can say that the beauty of the new landscape generally plays a substantial role for communities and regions, irrespective of the areas in which it is implemented. It is recognized that such locations have advantages when they advertise for businesses and new inhabitants.

The introduction of "new nature" as comprehensive planning with an overall discipline opened up an awareness of the entire landscape and its complex (economic and sociocultural) functions, ecosystems, and dynamics. This is a very important

---

* This can be compared with the discussion on maintenance of buildings characteristic of the 1950s. Even if it seemed senseless to maintain them in the 1980s, efforts were made to place them under protection. The time factor plays a role here, too.

† Landscape painting led to the disappearance of landscape reality and, as Wormbs ascertains, "Between the subject and object of sensory world experience, the landscape picture stretched in all dimensions of social abstraction and individual emotion" (Wormbs 1981, 16).

‡ Wormbs (1981, 32) gives the example of England supporting the capitalist method of production and industrialization for almost 300 years and, at the same time, helping landscape painting to a new peak. In a picture, the intensified contrast between town and country was perceptible in a romantic way.

The Concept of "New Nature" 281

and proper development, especially set against the background of climate change (reduction of CO_2), a rising sea level, and also the increase in flooding along the rivers. In my opinion, with the "new nature" concept, there has been a departure from the way of self-design or self-organization of ecosystems; through an active forming of multifunctional landscape, the attempt has been made to achieve the greatest possible benefit for the functions of the use of landscape that are involved. Thus, nature is simultaneously both goal and instrument.

The "new nature" concept follows the ideas of environmental aesthetics, which is more an analytical way of thinking about the necessity of combining the planned environment with aesthetics and design and the general logic of ecological design, learning from and working with nature.

Linking ecology, economy, and planning has led to ecology being rediscovered and defined as a component of an urban culture. Gross (2010), with the help of a multitude of projects, shows how recursive learning processes in association with nature and landscape can be implemented, especially in urban contexts, when careful planning is brought together by involving natural, dynamic processes and an open attitude toward surprises. This is made clearer, for example, in the discussion on "Eco-System-Services," how important natural and seminatural landscapes are, especially against the background of the preservation of biodiversity in the highly developed industrial nations, and how greatly the concept of sustainability has penetrated into society and established itself.

Today, "building with nature" means having an understanding of the complex ways of how landscape spaces and ecosystems function, being careful with resources, using the processes and dynamics of nature sensibly, and lastly, developing long-term solutions in an interdisciplinary and transdisciplinary way.

14.5 CONCLUSION

The whole process from the rediscovery of nature to "building with nature" shows, in a more analytical way, that "nature" or, rather, landscape is becoming an object to be bartered with. This idea of exchangeability shows a flexible but also a random way of dealing with (natural) expanses. In principle, it makes every kind of utilization in every location conceivable; this is because "new nature" can be remade as compensation. Ecologists in the Netherlands look on this development rather as "selling out" the existing value of nature. Nature or, rather, landscape is made suitable for the present needs and so becomes a fashionable plaything. It is, of course, possible to discover a potential for danger in how the making of nature is understood, thereby calling all nature areas into question. But, in the end, it is a matter of understanding the development of nature as an alternative to kinds of utilization that are now obsolete or to a traditional flood protection policy as well as seeing a chance for new links between nature, ecology, society, and economy. It is the attempt to depart from the path of complete regulation, controlling groundwater levels, and the development of vegetation and to combine the dynamic processes of nature synergistically by means of complex and interdisciplinary planning. This process marks a paradigm shift in dealing with nature and landscape areas and appears to be a fundamental and important learning process for the modern age.

At the present time, ecological knowledge, scientific discoveries, and data on inshore waters, biotopes, river dynamics, etc. flow, particularly in an analytical way, into the concept of restoring nature and the preservation of species as well as into the development of different kinds of natural spaces and, in this way, support, above all, the process of decision making in politics and planning. Active nature development can achieve more than ensuring the prerequisites for the preservation of species: It can also be understood as a form of land utilization that takes up spatial, politically driven climate and related problems as well as having a direct function in matters of recreation and leisure. Some projects in the program "building with nature" show that consciously using and favoring natural processes in flood protection can also lead to the development of more economical measures in the long run. This demands, however, close cooperation, preparation, and accompaniment of such projects in order to investigate these effects long term and to be able to make a prospective statement. In this case, the concept of "new nature" opens up the possibility of combining scientific ecological data, methods, and tools in the early stages of concept development with the capabilities and skills of the designers and vice versa (cf. Klijn and Vos 2000, 153). This would enable a better understanding of how ecological processes in the field of constructed ecosystems function and how forming them can be combined to make a profit (new research fields and spheres of activities). Perhaps the answers to the question of how to form and manage space in the future can be found in the context of sustainable development. An experimental approach that is oriented toward processes is in the interests of Urban Ecology and is particularly valuable in towns. It also promotes interest for it could encourage inhabitants to integrate and participate. It can also make clear and understandable, in a particular way, what has been developed by ecologists, engineers, and designers. "It offers a platform for a new model of collaboration between designers, scientists, developers and community members, engaging multiple stakeholders in a shared exercise of creating new knowledge" (Urban Omnibus 2013).

REFERENCES

Baart, T., T. Metz, and T. Ruimschotel (2000). *Atlas of Change. Rearranging the Netherlands*. Rotterdam: NAi Publishers.

Benson, J. F., and M. H. Roe, eds. (2000). *Landscape and Sustainability*. London: Spon Press.

Chartier, R. (1987). Intellektuelle Geschichte und Geschichte der Mentalitäten. In: Ulrich Raulff, ed., *Mentalitäten-Geschichte. Zur historischen Rekonstruktion geistiger Prozesse*. Berlin: Verlag Klaus Wagenbach. 69–98.

de Vriend, H., and M. van Koningsveld (2013). *Building with Nature: Thinking, Acting and Interacting Differently*. Dordrecht: Ecoshape, Building with Nature. http://www.ecoshape.nl/files/paginas/ECOSHAPE_BwN_WEB.pdf (accessed December 12, 2013).

Fiselier, J., S. Dekker, and H. Thorborg (2013). *Duurzaam Bouwen met Natuur. Nieuwe ontwikkeling is sprong in de waterbouw*, Barneveld: Land+Water Nr. October 10, 2013: 18–20.

Friesen, H., and E. Führ, eds. (2001). *Neue Kulturlandschaften*, Cottbus: Brandenburgische Technische Universität.

Gross, M. (2010). *Ignorance and Surprise. Science, Society, and Ecological Design*. Cambridge, London: MIT Press.

Jongman, R., and M. Bogers (2009). *Current Status of the Practical Implementation of Ecological Networks in the Netherlands*. Wageningen UR: Alterra, (http://www.ecologicalnetworks.eu/documents/publications/ken/NetherlandsKENWP2.pdf (accessed December 10, 2013).

The Concept of "New Nature"

Klijn, J., and W. Vos (2000). A new identity for landscape ecology in Europe: A research strategy for the next decade. In *From Landscape Ecology to Landscape Design*. Dordrecht, Boston, London: Kluwer Academics Publisher, 149–161.

Kost, S. (2009). *The Making of Nature. Eine Untersuchung zur Mentalität der Machbarkeit, ihre Auswirkung auf die Planungskultur und die Zukunft europäischer Kulturlandschaften. Am Beispiel Niederlande.* Marburg: Metropolis.

Roggema, R. (2009). *Adaptation to Climate Change: A Spatial Challenge.* Dordrecht, Heidelberg, London, New York: Springer.

Sijmons, D. (2009). Room for the river. *Topos 68*: 61–67.

Urban Omnibus (2013). *Experimental Landscapes: Alexander Felson on Ecology and Design.* http://urbanomnibus.net/2013/03/experimental-landscapes-alexander-felson-on-ecology-and-design (accessed October 18, 2013).

Watson, D., and M. Adams (2011). *Design for Flooding. Architecture, Landscape, and Urban Design for Resilience to Climate Change.* Hoboken, New Jersey: John Wiley & Sons.

Wormbs, B. (1981). *Über den Umgang mit Natur. Landschaft zwischen Illusion und Ideal*, 3rd ed. Basel, Frankfurt/M.: Stroemfeld, Roter Stern.

15 Ecological Network Planning
Exemplary Habitat Connectivity Projects in Germany

Manuel Schweiger

CONTENTS

15.1 History 285
    15.1.1 Background...285
    15.1.2 From Planting Hedgerows to Planning a Comprehensive
           Ecological Network 286
15.2 Definitions..287
15.3 National Biotope Network 289
15.4 Specific Examples of Network Plans on Regional and Local Levels..........293
    15.4.1 Examples of Baden-Württemberg and Karlsruhe............................293
    15.4.2 Example of Bavaria 295
15.5 Conclusions..297
Acknowledgments..299
References..299

15.1 HISTORY

15.1.1 BACKGROUND

Both land consumption and intensive agricultural land use displace plants and animals and lead to the loss of habitats. In addition, the formerly intimately knotted network of natural, seminatural, and cultural ecosystems is becoming fragmented. As one reaction to these processes, demands to reconnect habitats have become prominent in Germany.

A so-called Ecological Network has been anchored in the Federal Nature Conservation Act since 2002 and aims at enabling the reconnection of habitats. The current amendment of the Act stipulates the development of an Ecological Network on 10% of land area, which, in addition, should enhance the connection of the European Natura 2000 network of protected nature sites.

The effects of climate change on biological diversity makes the concept of Ecological Networks particularly topical. Still left to clarify is the role habitat connectivity plays considering climate change. The main questions are which species potentially could profit by connecting habitats and which specific requirements must be considered when designing a transnational Ecological Network (Reich et al. 2012).

Today, the prevailing opinion is that, in the face of increasingly extreme weather conditions, the connection of habitats is a prerequisite for maintaining biodiversity (Opdam and Wascher 2004; Dias et al. 2003; Ssymank et al. 2006).

The positive effect of preservation, regeneration, restoration, or redevelopment of a workable connection in the landscape is undoubted. Conceptual thinking in this area started with the highly regarded scientific publications by MacArthur & Wilson (1963, 1967). The theory of Island Biogeography proposes that the number of species found on islands is determined by the size of the islands and their distribution in space as is the probability of extinction. This concept serves as a basis for developing ecological models as a prognosis for the development of populations (Levins 1969). At least theoretically, this offers the possibility of predicting in which spatial attribution habitats must be designed and what management must take place to enhance the connectivity between local populations. Thus, an important theoretical basis for planning Ecological Networks is given.

However, relations between both habitats of the same type and also those of different types must be regarded. In this context, the relationships between natural, near-natural, and culture-based areas are expressly included (Ullrich 2008). Generally, flora and fauna have sustainable viable populations only if

1. There is the possibility of exchange, dispersal, and migration
2. There is enough space for finding food, shelter, and regeneration (Burkhardt et al. 2002)

15.1.2 FROM PLANTING HEDGEROWS TO PLANNING A COMPREHENSIVE ECOLOGICAL NETWORK

Initial demands for building structures of habitat networks in Germany go back to the time of the development of environmental thinking at the turn of the 20th century. At first, these demands aimed for bird protection only, which was one main field of attention in the early nature conservation movement. Against the background of the low bird population density in isolated trees and shrubs, von Berlepsch had concluded already in 1899 that they should be connected in some way (Drobnik et al. 2013). In 1936, the demand of 10% of land area being reserved for enhanced habitat connectivity was formulated for the first time by the landscape architect Alwin Seifert, the later *Reichslandschaftsanwalt* in the Third Reich (Drobnik et al. 2013). This 10% demand was echoed 66 years later in the Federal Nature Conservation Act of 2002 in order to build up an Ecological Network in Germany, following a longstanding request by, among others, the German Advisory Council on the Environment (SRU 1994).

Ecological Network Planning

In view of the ongoing loss of species since the 1970s, scientists demand the establishment of an Ecological Network in Germany (e.g., Blab 2004; Heydemann 1986; Jedicke 1997; Mader 1979; Mader et al. 1986; Mühlenberg et al., 1987). In practice, the idea of habitat connectivity was first implemented on a small scale. Connecting residual forest patches by planting hedges became especially popular. It became clear only later that, in many of these projects, aspects were overlooked that partially counteracted the goal to stop the loss of species diversity (Geissler-Strobel et al. 2000; Ullrich 2008).

For example, a hedge can be mainly used by comparatively common species of forests or forest edges that are in no danger of extinction. And even for these species, it is by no means certain that a newly planted hedge offers better dispersal opportunities. But worse is to come: Wooded structures may have a fragmenting effect on species of open habitats as they can function as an impenetrable barrier. Consequently, laying out habitat-connecting elements always must consider all correlations between the neighboring habitats.

The actual consensus is the penetrability of the landscape often also and sometimes even mainly depends on the quality of the surrounding landscape matrix. This quality determines if the newly created corridor will be used by species migrations. Consequently, today's concepts of Ecological Network Planning do not focus on connecting single biotopes by creating linear or punctual elements (one exception is wildlife crossings for the Autobahn). Instead, the landscape must be designed in a way that offers enough resources at least for subpopulations of certain species. It is therefore essential to provide suitable habitats on sufficiently large patches.

For the time being, the approach of Ecological Networks lost popularity because of failures in the implementation of habitat networking on a small scale by creating corridors and stepping stones. Meanwhile, it is common in practical planning to provide several partial habitats simultaneously.

Landscape must meet some minimum requirements to be amenable to colonization by flora and fauna. On the one hand, these requirements refer to the quality of the habitat, which is mainly determined by the size, form, and composition of the single biotopes and, finally, the integrity of the biotope complexes. On the other hand, they refer to its position in a space, on which it depends if an area can be part of connectivity axes or, due to its spatial proximity, can support the Ecological Network as a linking element (Burkhardt et al. 2002).

Besides intraspecies relationships—as possibilities for migration to support exchange and therefore connections between biotopes of the same type must be available—there are also relationships between different species. In addition, a flow of energy and substances takes place between different types of biotopes. In order to safeguard or to restore this complex of interactions and interrelationships in landscape, characteristic complexes of biotopes and landscapes must be available in sufficient capacity (Ullrich 2008).

15.2 DEFINITIONS

According to § 21 (1) of the Federal Nature Conservation Act, an Ecological Network serves to sustainably protect species of fauna and flora and their habitats, biotopes, and communities and also to maintain, restore, and develop ecological interactions. Apart from legally protected areas and biotopes, all sites belong to an Ecological Network that contributes to its function. By one popular definition, an Ecological Network consists of the following components:

- **Network core (composed by core areas)** should offer stable habitats for indigenous species and is an important building block of an Ecological Network. The core contains remaining natural and seminatural areas and is often surrounded by **buffer zones**, which should prevent negative influences by the neighboring intensively used landscape on the core. **Development areas**, which are equipped with the necessary potential for being developed to a core area, may be situated nearby.
- **Connecting elements** are areas that should ensure and facilitate the genetical exchange between populations of the core zone and also the process of migration, expansion, and reintroduction. Depending on the specific requirements of the species, they can be built as **stepping stones** or **corridors**.
- The **surrounding landscape matrix** is the pattern of elements in the landscape within a specific area under consideration. If there are certain elements in focus, the expression refers to the pattern of all the other elements in the surrounding landscape. The matrix should be less hostile and therefore more passable for organisms. Through an environmentally favorable extension of land use minimum quality standards could often be fulfilled (Burkhardt et al. 2002; Ullrich 2008).

Basically there are two different types of Ecological Networks: Species-oriented habitat networks focus on the specific requirements of target species, and multifunctional habitat networks try to address many landscape functions when reconnecting the remnant habitats (Haaren and Reich 2006). The following will be focused on the second type, which is also addressed in the Federal Nature Conservation Act.

It is necessary to develop a complete system of habitat networks on different spatial levels:

- (Inter)national: large-scale connectivity axes, taking into account species with very high spatial demands and migratory species
- Regional: connectivity axes on a regional scale, passable within landscapes and natural areas
- Local: complexes of biotopes, connection of single biotopes

These different scales are important for two main reasons. First, the spatial demands vary extremely depending on the species that are analyzed. Regarding a ground beetle species, local complexes of biotopes will be decisive, but for the preservation of migratory bird populations, it is necessary to bear in mind a network of available natural habitats that goes beyond international borders.

Second, the different scales are important for the planning purposes: (Inter-) national and regional planning of Ecological Networks often are instruments to

Ecological Network Planning

maintain biodiversity by attracting attention in the fields of policy and regional planning. In contrast, local planning takes place on a very concrete scale with clearly defined areas. Here the target is immediate implementation.

In conclusion, the main target of planning Ecological Networks on a large scale is to preserve spaces from counteracting planning, and on a small scale, the patches actually should be designed or preserved in a way that they can fulfill their function in the network.

In the following chapter an overview will be provided on how systems of habitat networks are conceptually developed in Germany. Because of the heterogeneity of the different approaches, for instance, in the single *Bundesländer* or in the different planning regions, it will focus on two types of planning strategies: bottom-up and top-down. Both approaches have their advantages and disadvantages, which also depend on the spatial level of planning.

15.3 NATIONAL BIOTOPE NETWORK

Germany is a federal state, and for both the policy area of nature conservation and in regional planning, there is a competing legislative competence between the federal state and its *Bundesländer*. This circumstance makes it even more complicated to establish a generally accepted Ecological Network (see Haaren and Reich 2006; Leibenath 2011). In fact, there are very heterogeneous plans in the single *Bundesländer* for implementing an Ecological Network that differ, for example, in scale, method, and size and differentiation of the network components, and the data basis used is not comparable either (Fuchs et al. 2010). Consequently, the plans of the *Bundesländer* can't be combined into one Ecological Network for Germany.

To react to this deficit, the German Federal Agency for Nature Conservation (BfN) launched a research project on designing a Germany-wide Ecological Network on a plot-by-plot basis (Fuchs et al. 2010). Because the implementation still is the responsibility of the *Bundesländer*, the concept can't be a binding plan. On the contrary, it should be a service given to the *Bundesländer* as a strategic framework to assist them in their activities for regional connectivity concepts.

In addition, a Germany-wide Ecological Network on a plot-by-plot basis can counter, with its expert advice, other plans in Germany that may lead to impairment of the quality of landscape and its habitats for indigenous species. An example for a counteracting plan is the Federal Transport Infrastructure Plan, which shows the planned *Bundesstraßen* and *Autobahnen* for Germany at the moment without regarding demands for species migration. On this Germany-wide scale, an Ecological Network visualized on the map (see Figure 15.1), the content of which can be understood easily, could help in that, even in counteracting plans, the need for habitat networks will be taken into account. As well, it can be a basis for specific decisions as in creating technical connecting elements (for example, wildlife crossings for the *Autobahn*). These are important to counteract the ongoing fragmentation of habitats, especially of species with high space demands (Ullrich 2008).

In the course of two federal R&D projects, PAN GmbH developed (in cooperation with the universities of Kassel, Hannover, and Kiel) a general concept for a national biotope network. Nationwide important areas for the network and axes of national

FIGURE 15.1 Connectivity axes and core areas of the Germany-wide Ecological Network for humid and dry complexes of habitats of open country. Core areas (brown colors), special protected areas for birds, which are of high importance for target species of the Ecological Network (red hatched) and connectivity axes for humid (blue line) and dry (red line) habitats and the Green Belt (green line) of the former Inner German Border. (From Federal Agency for Nature Conservation 2013; Fuchs, D. et al., *Länderübergreifender Biotopverbund in Deutschland - Grundlagen und Fachkonzept*. Naturschutz und Biologische Vielfalt 96. Bonn: Federal Agency for Nature Conservation, 2010; Spatial Base Data © GeoBasis-DE/BKG.)

Ecological Network Planning

and international importance were identified and shown on a map. As already mentioned, the existing plans of the *Bundesländer* (as far as implemented) are mainly based on different approaches and are hardly comparable. Consequently, their data basis couldn't be used for a selection of areas or axes/corridors of a Germany-wide Ecological Network. Unfortunately, there are hardly any Germany-wide consistent data available because of the federal competences. First and foremost, the data of the mapping of legally protected biotopes of the *Bundesländer* were used to determine existing areas belonging to the network core although there are some restrictions concerning, especially, the actuality, quality, and comparability of the data. But at least the data could be brought together in a uniform database. The resulting pool of areas was analyzed according to the recommendations of Burkhardt et al. (2002) with the parameters size, quality, degree of fragmentation, and occurrence of target species (see Table 15.1):

- For the Ecological Network in **woodlands**, the priority data set was "CORINE land cover" of the European Environment Agency. CORINE means "coordination of information on the environment" and offers an inventory of land cover in 44 classes and is presented as a cartographic product at a scale of 1:100,000 operationally available for most areas in Europe. For analyzing the fragmentation of woodlands surveyed, traffic census data was consulted.
- For the selection of areas in **open country**, the preprocessed data set of protected biotopes of the *Bundesländer* was used as basic geometries. The size of habitats was evaluated by building complexes, which combine all biotopes of one type with a maximum distance of 200 m.
- The differentiation of suitable **river sections** followed the criteria of characteristic, completeness of the complexes of biotopes, and defragmentation. Finally, all suitable river sections were rated concerning their size. A database for the analysis was structural and biological quality surveying and mapping, which is available for all rivers in Germany of a certain size, and additionally, the differentiations of the sites of community interest according to the Habitats Directive of the European Union was integrated.

In addition to core areas, search areas also were determined, which should be the sites of first choice for establishing reconnecting elements as development areas or stepping stones and should, in any event, be prevented from defragmentation. For defining these search areas, an algorithm was applied that was integrated in a Geographical Information System (GIS). The method called HABITAT-NET (Hänel 2007) combines requirement types (representing different ecological groups of species with a terrestrial distribution) with distance classes (considering the individual mobility of the requirement type).

The habitat networks have been developed separately for species of dry, wet, and forest biotopes, differentiating between different distance classes (narrower and wider functional areas), which also defines particularly suitable development areas (see Figure 15.2).

TABLE 15.1

Valuation Parameters Used to Measure the Importance of Sites for the Germany-Wide Ecological Network—Implementation of the Recommendations of Burkhardt et al. (2004: 28)

Open Country		Woodland		River Sections
All sites included in the data set of selective biotope mapping of the *Bundesländer*		All woodland areas that have less than 10% of unnaturally composed forest rated by Glaser and Hauke (2004)		All river sections that are longer than 5 km
and		and		
include biotopes of open country				
and		of which only one third of total area is split by fragmentation elements		
of which only one third of total area is split by fragmentation elements				
and that either are		and that either are		
situated in complexes (max. distance of 200 m) with a total area of biotopes of more than 1000 ha	**or** situated in complexes (max. distance of 200 m) with a total area of biotopes of more than 200 ha **and** are situated in a site of the EU Habitats Directive with typical habitats of open country (conservation status at least B)	bigger than 5000 ha	**or** bigger than 1000 ha **and** have a proportion of more than 50% of (potentially) near-natural individual areas rated by Glaser and Hauke (2004)	situated in a site of the EU Habitats Directive with typical habitats of rivers (LRT32) whose proportion of total area of the site must be bigger than 5 ha
				and that either are
				rated in the structural quality surveying and mapping with grade 1, 2, 3, or 4 while those sections rated with 4 mustn't be longer than 2 km **or** rated in the biological quality surveying and mapping with grade I, I–II, or II **or** in immediate vicinity of sites for the Germany-wide Ecological Network (open country or woodland)

Source: Fuchs, D. et al., *Natur und Landschaft* 82 (8): 345–351, 2007.

Ecological Network Planning

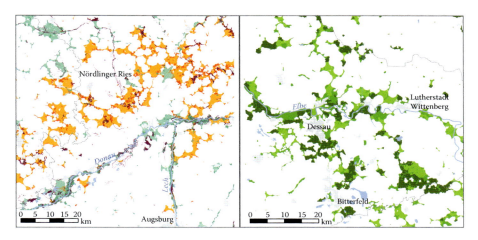

FIGURE 15.2 Search areas for networks (partial): left for dry (orange) and humid (turquoise); right for woodland habitats (light shades of green). The sites of national importance for habitat networks are shown in violet (left) and dark green (right).

This example of designing an Ecological Network followed a top-down approach, which means that an overarching main concept with axes and search areas was defined. Based on this plan, further planning should more precisely define the areas on smaller scales to obtain clearly defined areas for implementation. In most of the *Bundesländer*, which is responsible for implementing Ecological Networks, this approach is followed too as we can see in the following example of Baden-Württemberg.

15.4 SPECIFIC EXAMPLES OF NETWORK PLANS ON REGIONAL AND LOCAL LEVELS

15.4.1 EXAMPLES OF BADEN-WÜRTTEMBERG AND KARLSRUHE

In Baden-Württemberg, initially a habitat-network concept was developed in 2007. Therefore, a statewide coherent method was applied (PAN GmbH 2007, unpublished). In 2009, the method was taken up for the first time on a smaller scale, and a habitat network was planned for the city of Karlsruhe. The data basis (partly only in analog form available) consisted of 34 expert reports, surveys, and mappings, inter alia:

- Statewide mapping of legally protected biotopes
- Mapping of grasslands in the administrative district of Karlsruhe
- Woodlands obtained from the Official Topographic-Cartographic Information System (ATKIS) of Germany were calibrated with the Potential Natural Vegetation

- Delimitation of Woodlands and rivers within sites of community interest of the EU Habitats Directive and special protected areas for birds by ATKIS data
- Specific records of target species

All 34 data sets were, if necessary, digitized and then overlaid. The consequence was a very sophisticated working process of validation and rectification of the data set. The following analysis is then based on main types of biotopes that were determined before. The resulting biotopes were rated in three categories (very good, good, and average) by the criteria "characteristics" and "size." The total evaluation results in the aggregation of both categories. Based on this rating, the network situation was analyzed separately for the network types humid grasslands, oligotrophic sites, and extensively cultivated landscapes. The woodlands were regarded separately along with an analysis of connection potentials. Basing on the total evaluation, the sites were split for each network type into core areas and development areas (see Figure 15.3).

Core areas were defined when the total evaluation of sites was very good, good, or average.

Development areas were defined when the result of the total evaluation of sites was too poor for functioning as a core area in the Ecological Network, but still there is a potential for developing the site (for example, through expanding the size of the site or through optimizing the current land use).

FIGURE 15.3 Map excerpt habitat network of Karlsruhe containing core and development areas (colored representation) of the different network types (blue = humid grasslands, turquoise = humid woodland, orange = oligotrophic sites, olive green = dry woodlands) and also targets: creating network corridors (arrows) and renaturation of barriers (dotted yellow zigzag line).

Ecological Network Planning

In the planning process further elements of the Ecological Network were identified:

As a **connection element**, core areas of a different network type were defined that provide similar local characteristics with a specific functional relationship. Connection elements for the network type "humid grassland," for example, are sites of the biotope type "extensively cultivated mesotrophic grasslands," "humid shrubs," or "fen wood."

Priority areas were defined, containing a high density of high-quality biotopes of one network type. These areas are of special importance for the Ecological Network because of their size, quality, biotope density, or the distribution of a certain species. Consequently, priority areas can function as a center for migration and as a hideaway for the corresponding flora and fauna.

Areas with special importance for the Ecological Network also have a high density of biotopes available. Usually, they are somewhat smaller than priority areas, or the proportion of high-quality core areas is definitely lower. Nevertheless, these areas support the habitat network and function as a center for migration and as a refuge for species with low space demand or serve as a habitat for species that need a complex of different biotopes.

Search areas have a high potential for the prevailing development target available because of their natural conditions. These were identified on basis of landscape units, topography, and also based on soil data.

Finally, **deficiency areas** were defined as areas with only few core areas, but which have

- A high location potential for a certain network type (position within the search area).
- A certain developing potential for one of the network types.
- At the same time, there is a lack of this type of biotopes.

15.4.2 EXAMPLE OF BAVARIA

We have seen two examples of a top-down approach on national and on state/regional levels. If the database is available, it is also possible and maybe expedient to follow the other way round. This bottom-up approach is partly implemented in Bavaria.

The planning process starts on a small scale with specific sites being identified as suitable core areas. Based on this selection, targets for regional and state levels will be formulated. Only the total space requirement for the Ecological Network will be defined a priori for each district of natural areas.

The starting point is the Bavarian *Arten- und Biotopschutzprogramm* (ABSP, species and biotope protection program), which is issued for each rural district (*Landkreis*) of Bavaria and is more or less regularly updated. Not only the quality of existing biotopes and species reports are evaluated, but also so-called target areas are defined, which can be regarded as development or search areas of the Ecological Network. Based on these target areas, formulated targets should be fixed on a spatial level for all parts of the district with a special importance for nature conservation. On the one hand, target areas should cover all high-quality biotopes in the district; on

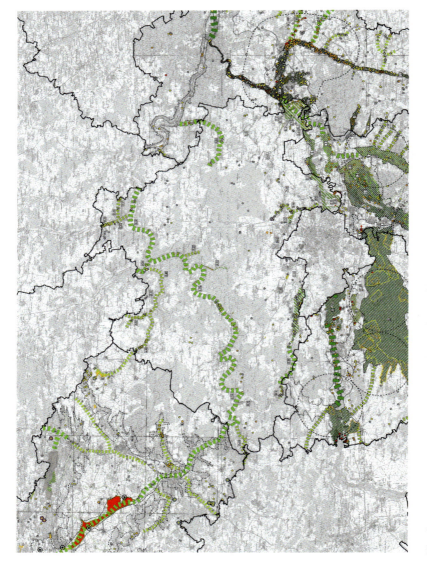

FIGURE 15.4 The bottom-up concept: target areas and axes in neighboring districts (borders in black) combine to a network on a larger scale (Bavarian species and biotope protection program ABSP for the districts of, clockwise from lower right, Kelheim, Pfaffenhofen a. d. Ilm, Neuburg-Schrobenhausen, Eichstätt, and Weißenburg-Gunzenhausen).

Ecological Network Planning

the other hand, they should consider the given natural conditions. In most cases, both aspects are taken into account. Consequently, a target area can contain biotopes for which similar targets are developed and/or which are habitats of the same species: for example, oligotrophic grassland with an adjacent thermophilic broadleaf forest. At the same time, target areas can comprise habitats with different maintenance demands, which are connected by their natural conditions, for example, biotopes on fen soils. In all cases, the target areas are defined in a way that, apart from the existing biotopes (core areas), sites also are included on which biotope development is possible and should be taken into account first and foremost. For each target area, the amount of new biotope area that should be restored or developed for strengthening the Ecological Network is determined. The sum of all the biotopes of the core area and those which should be developed in the target area should correspond to the total space requirement for the Ecological Network, which was defined a priori for each district of natural areas.

This procedure makes sure that, in the process of defining development areas, already detailed information about the condition and the availability can be included. Not least because of this the lower and higher nature conservation authorities are included in the process for all target areas. When the updates with this method for all ABSP are completed, they can be combined to a more or less comprehensive plan for an Ecological Network in Bavaria (see Figure 15.4).

15.5 CONCLUSIONS

The Germany-wide Ecological Network can communicate appropriately complex ecological relationships in a way that can be considered easily by other plans. Especially the infrastructural language of the map allows even planners outside of the ecology area to integrate the results in their work. Consequently, the Ecological Network can prevent, with its expert advice, planning results that have counteracting targets and may lead to impairment of the quality of landscape and its habitats for indigenous species. This also applies for network plans on local or regional scales; nevertheless, the implementation of specific plots here is of primary importance.

The main problem of network planning in Germany is to relate the different plans in their different scales. The main reason for this is due to the federally organized nature conservation policy and the missing coordination. Consequently, in several *Bundesländer*, the functional concept of a Germany-wide Ecological Network was received as a rival plan to the existing plans on a local or regional level rather than assistance for their activities for connectivity concepts. The Germany-wide concept provides the critics with a target anyway as long as it is based on a partly limited and outdated database, which was necessary to use for having a Germany-wide consistent data set.

In the same way, alternative approaches, such as the Federal Plan for Wildlife Paths from the Nature and Biodiversity Conservation Union Germany (NABU) (Herrmann et al. 2007), follow (even in the concept name) the infrastructural approach to communicate the planning results. But this species-oriented habitat network deals only with a sectoral aspect of habitat networks, which follow expertise on the large mammals wild cat, red deer, wolf, lynx, and otter.

The following basic requirements should be met for strengthening the concept on the national level and to stop the ongoing isolation of valuable habitats:

- The *Bundesländer* must be ready to adapt their plans on state, regional, and local levels to a common concept approach or at least search for a close coordination with the neighboring plans especially at the borders of the *Bundesländer*.
- The functional concept of a Germany-wide Ecological Network needs to be strengthened. Even if there are divergences to the plans in the *Bundesländer* in certain details, which are due to the different databases, still this tool is of a very high importance for representing the interest of nature conservation on national and partly on international levels. For this aspiration, the backing of the *Bundesländer* and of the German nature conservation associations is needed.
- A stronger link between the scientific and conceptual basis of habitat networks is needed. Prioritizing landscape functions and the selection of target species require better coordination (also between the different planning scales and the *Bundesländer*) (Haaren and Reich 2006).
- In the future, it will be necessary to coordinate the methods of generations of relevant databases (especially the mapping of protected biotopes of the *Bundesländer*) and to update them regularly. This would have a positive effect not only for habitat network planning, but also for numerous other plans, concepts, and researches on nature conservation on a national scale. In this context, also Haaren and Reich (2006) suggest that more authority in nature conservation should be transferred to the national level.

On the regional and especially on the local level, the database often is significantly better than on larger scales. If specific sites are regarded, numerous sources and expertises can be integrated in the planning, which, however, includes an extensive workload. At the same time, exceptions can be taken into account in the planning process. Both aren't possible in a large-scale plan as it would exceed the scope of work. It can be expected that the planning on local and regional scales more or less mirrors the factual situation and consequently can show deficiency areas of habitat networks effectively. But here again, as we've already seen in context with the Germany-wide Ecological Network, the quality of the planning depends on the quality of the database.

If there were a consistent high-quality database, which could be used for different scales and which could be complemented by additional available sources on a local scale (top-down approach), the transferability of results but also the comprehensibility of the concept would increase. Not least for the specific implementation of local habitat networks, in which the cooperation of the local landowner often is a limiting factor (Haaren and Reich 2006), more transferability and, therefore, more transferable experiences together with conclusiveness and transparency for nonspecialists, would facilitate the preservation and development of areas for an Ecological Network.

Ecological Network Planning

Because of the database, at present, the bottom-up approach delivers results that are closer to reality than those of top-down plans. Nevertheless, the workload is very high, and it is not realistic that this effort could be done comprehensively for the whole of Germany in the foreseeable future. As long as there is neither a consistent high-quality database for the whole of Germany nor all local habitat network plans completed with a consistent method, the above shown functional concept of a Germany-wide Ecological Network works at least on a metaphoric level of nature designed as infrastructure and, in this function, can be used as a powerful tool in planning coordination and policy advice.

ACKNOWLEDGMENTS

Many thanks to all involved colleagues at PAN GmbH for providing their work results. First of all, thanks to Daniel Fuchs whose support exceeded the expectable level by far.

REFERENCES

Blab, J. (2004). Bundesweiter Biotopverbund—Konzeptansatz und Strategien der Umsetzung. *Natur und Landschaft* 79: 534–543.

Burkhardt, R., H. Baier, U. Benzko, E. Bierhals et al. (2002). *Empfehlungen zur Umsetzung des 3 BNatSchG "Biotopverbund."* Naturschutz und biologische Vielfalt 2. Bonn: Federal Agency for Nature Conservation: 84 pp.

Dias, B., S. Diaz, and M. McGlone. (2003). Interlinkages between biological diversity and climate change: Advice on the integration of biodiversity considerations into the implementation of the United Nations Framework Convention on Climate Change and its Kyoto Protocol. *CBD Technical Series* (10): 154 p.

Drobnik, J., P. Finck, and U. Riecken. (2013). Die Bedeutung von Korridoren im Hinblick auf die Umsetzung des länderübergreifenden Biotopverbunds in Deutschland. 346. BfN-Skripten. Bonn—Bad Godesberg: Federal Agency for Nature Conservation. http://www.bfn.de/fileadmin/MDB/documents/service/Skript_346.pdf.

Fuchs, D., K. Hänel, J. Jeßberger et al. (2007). National bedeutsame Flächen für den Biotopverbund. *Natur und Landschaft* 82 (8): 345–351.

Fuchs, D., K. Hänel, A. Lipski et al. (2010). *Länderübergreifender Biotopverbund in Deutschland - Grundlagen und Fachkonzept.* Naturschutz und Biologische Vielfalt 96. Bonn: Federal Agency for Nature Conservation: 191 pp.

Geissler-Strobel, S., G. Kaule, and J. Settele. (2000). Gefährdet Biotopverbund Tierarten?—Langzeitstudie zu einer Metapopulation des Dunklen Wiesenknopf-Ameisenbläulings und Diskussion genereller Aspekte. *Naturschutz und Landschaftsplanung* 32 (10): 293–299.

Glaser, F. F., and U. Hauke. (2004). Historisch alte Waldstandorte und Hudewälder in Deutschland. *Angewandte Landschaftsökologie* 61: 52–53.

Haaren, C. von, and M. Reich. (2006). The German way to greenways and habitat networks. *Landscape and Urban Planning* 76 (1): 7–22.

Hänel, K. (2007). Methodische Grundlagen zur Bewahrung und Wiederherstellung großräumig funktionsfähiger ökologischer Beziehungen in der räumlichen Umweltplanung. PhD thesis, University of Kassel. https://kobra.bibliothek.uni-kassel.de/handle/urn:nbn:de:hebis:34-2007121319883.

Herrmann, M., J. Enssle, M. Suesser, and J. Krueger. (2007). Der NABU-Bundeswildwegeplan. http://trid.trb.org/view.aspx?id=842813.

Heydemann, B. (1986). Grundlagen eines Verbund-und Vernetzungskonzeptes für den Arten- und Biotopschutz. *Akademie für Naturschutz und Landschaftspflege (ANL), Laufener Seminarbeiträge* 10/86: 9–18.

Jedicke, E. (1997). Biotopverbund im Wald: Läßt sich dieses Naturschutz-Konzept von der freien Landschaft übertragen? *Gerken & Meyer*: 90–96.

Leibenath, M. (2011). Exploring substantive interfaces between spatial planning and ecological networks in Germany. *Planning Practice and Research* 26 (3): 257–270.

Levins, R. (1969). Some demographic and genetic consequences of environmental heterogeneity for biological control. *Bulletin of the ESA* 15 (3): 237–240.

MacArthur, R. H. and E. O. Wilson. (1963). An equilibrium theory of insular zoogeography. *Evolution*: 373–387.

MacArthur, R. H. and E. O. Wilson. (1967). *The Theory of Island Biogeography*. New Jersey: Princeton University Press.

Mader, H.-J. 1979. Biotopisolierung durch Straßenbau am Beispiel ausgewählter Arten— Folgerungen für die Trassenwahl. *Berichte der Akademie für Naturschutz und Landschaftspflege (ANL)* 3/79: 56–63.

Mader, H.-J., R. Klüppel, and H. Overmeyer. (1986). Experimente zum Biotopverbundsystem— tierökologische Untersuchungen an einer Anpflanzung. *SR Landschaftspflege und Naturschutz H. 27, Bundesforschungsanstalt für Naturschutz und Landschaftsökolgie (Hrsg.), Bonn-Bad Godesberg,* 136 p.

Mühlenberg, M., D. Müller, R. Graulich, J. Stein, and B. Heydemann. (1987). Konzeptentwicklung und Möglichkeiten praktischer Umsetzung von Biotopverbundsystemen. *SR angewandter Naturschutz* 5: 14–31.

Opdam, P., and D. Wascher. (2004). Climate change meets habitat fragmentation: Linking landscape and biogeographical scale levels in research and conservation. *Biological Conservation* 117 (3): 285–297.

PAN GmbH, Planungsbüro für angewandten Naturschutz GmbH, 2007. Arbeitshilfe zur Biotopverbundplanung Baden-Württemberg. Munich, unpublished.

Reich, M., S. Rüter, R. Prasse, S. Matthies, N. Wix, and K. Ullrich. (2012). *Biotopverbund als Anpassungsstrategie für den Klimawandel?* Naturschutz und Biologische Vielfalt 122. Bonn-Bad Godesberg: Landwirtschaftsverlag: 230 pp.

SRU. (1994). Umweltgutachten 1994 des Rates von Sachverständigen für Umweltfragen. Für eine dauerhaft-umweltgerechte Entwicklung. Unterrichtung durch die Bundesregierung 12/6995. Umweltgutachten. Wiesbaden: Rat von Sachverständigen für Umweltfragen, SRU.

Ssymank, A., S. Balzer, and K. Ullrich. (2006). Biotopverbund und Kohärenz nach Artikel 10 der Fauna-Flora-Habitat-Richtlinie. *Naturschutz und Landschaftsplanung* 38 (2): 45–49.

Ullrich, K. (2008). Biotopverbundsysteme. *AID-Heft* 1459: 1–54.

16 Planting the Desert
Cultivating Green Wall Infrastructure

Rosetta Sarah Elkin

CONTENTS

16.1 Introduction 301
16.2 Conditions 304
16.3 New Deal Shelterbelt 307
16.4 3 North Shelterbelt Project, China 311
16.5 *Populus simonii*, Simon's Poplar, Chinese Poplar 315
16.6 Great Green Wall (GGW), Africa 316
16.7 *Accacia* sp. 318
16.8 Conclusion 319
Common Abbreviations 320
References 320

16.1 INTRODUCTION

The global challenge of rapidly declining vegetative cover is being addressed by massive replanting projects that cross territorial, political, and cultural boundaries. By considering two contemporary examples in different stages of cultivation—"The Great Green Wall" in the Sub-Sahara Africa and the "3 North Shelterbelt Program" in China—a perspective is offered that highlights the tension between engineering infrastructure and cultivating healthy ecosystems. Considered together, these projects aim to replant more than 170 million hectares of land that is classified as semiarid, arid, or hyperarid. The tradition of planting deserted land is an ancient practice, most often initiated as a response to local climatic variation. The contemporary tradition tends to attribute the need to plant as a consequence of human activities. Although the global concerns surrounding the threat of desertification act as the impetus for both megaprojects, desertification is not offered as the framework for this discussion as it is often misused and confused with drought (Thomas 1993). Instead, both initiatives are presented as a form of planted infrastructure, most often specified within the prevalent framework of "green infrastructure," a dominant theme within current policy making. These projects represent the largest horticultural projects the world has ever considered, which categorically of green removes them from the discourse of green infrastructure and opens a discussion of territorial geopolitics. In both cases, a principal species acts as the foundation for planting an entire region and structures

301

the associated conditions of each site (Figures 16.1 and 16.2). Therefore, the varying frameworks that allow new plant cover to be introduced will be studied to form a perspective of each project, exploring the role of plants from innovative seed mechanics to regional bionetworks. Rather than desertification, the biological and ecological processes themselves are offered as the basis for understanding the goal of each project. In other words, the political frameworks that enable processes (such as desertification) to become permanent are revealed when addressing projects at this scale. Subsequently, this category of infrastructure is exposed as a totalizing system rather than framed as a series of projects. The speculative argument offered here is that planting deployed at a local scale considers micro-conditions and species suitability more prudently than the agencies that describe and articulate the complications to a wider audience across territorial scales.

Afforestation refers to the deliberate conversion of non-forest land to forestland, which does not involve re-establishment (replanting)—including the deliberate transformation of agricultural land back to forest—but rather, afforestation mobilizes tree planting through environmental authority and value statistics; planting one tree is fine, but planting millions is better. This classification confirms that the undertaking is not rooted in renewal or local conditions, nor in intensifying existing

FIGURE 16.1 *Populus Simonii var. fastigiata* planted as a traditional shelterbelt. (Image by Frank Nicholas Meyer; reproduced with permission from Harvard Arnold Arboretum Library.)

Planting the Desert

FIGURE 16.2 The agency of species. (a) Poplar in China. (b) Acacia in Africa. Each species displays adaptive virtues that are exploited in order to accomplish the coverage across an entire region. (Image courtesy of author.)

processes, but as a specifically human-induced incentive—an instrument of urbanism and a prime example of green infrastructure. Infrastructure offers a framework when referring to public works projects that are created in support of an industrial economy (Bélanger 2010). It also has historic significance as a military procedure and tends to be used in contemporary design discourse to describe any substructure that supports development. Finally, soil erosion is offered as the opponent, and infrastructure is offered as a means to arrest decline and amend the land in preparation for food production or to shelter the surrounding communities from airborne particulate. In the case of planted infrastructure, scale can be measured at the level of the farmstead, the province, and as a result of increasing water scarcity, the continent. Although there is seductiveness to suggesting a unifying theory, the process of desertification is often misunderstood when it is reduced to such a totalizing order. Additionally, projects that are introduced through large-scale planning measures do not encourage the required sensitivity between various arid and semiarid conditions and the associated vegetal dynamics. The supplementary protection strategies are often misaligned with local needs, plant-soil requirements, and their mutual dependence. If design professions aim to have impact on the scale of the territory by taking into consideration the principles of an entire system as a series of long-term processes, then how can we contribute as designers in the face of global environmental policy on planting trees in grasslands?

16.2 CONDITIONS

The United Nations Convention to Combat Desertification retains authority over the definition of desertification: "land degraded in arid, semi-arid and dry sub-humid areas resulting from various factors, including climate variation and human activities" (UNCCD 1995). This characterization of the land as something to combat or contest is intrinsic to the objectives outlined by the definition. In fact, most academic discussions of the topic position desertification as a threat to the "natural" environment and tend to characterize the issue using similar military terminology (Figure 16.3). The main battle presented is the tension between unproductive land encroaching upon productive land. The notion that the land is actually non-static is recognized, but it is not described through the lens of biological evolution or ecological succession. Instead, the definition offered by the UNCCD outlines a specifically human-induced condition that insinuates

FIGURE 16.3 Aridity index adapted from United Nations "Map of Desertification" (1977). The first official map that positions desertification as a risk. (Image courtesy of author.)

Planting the Desert

a disobedience of nature and highlights a potential conflict. The link between population growth and environmental degradation has theoretically existed since the origins of technology, and so it is safe to assume that erosion and development are symbiotic. Further, this is supported by the global migration from agrarian to industrial production (Odum 2007, 54). Is desertification the result of a bond between attrition and progress, or is it an urban hypothesis offered in support of expanding economies? The term desertification was first popularized in 1949 by the French ecologist André Aubreville in his book *Climats, Forêts et Désertification de L'Afrique Tropicale* (1949), in which he specifically linked the impact of man to the undesirable loss of cover:

> It is the process of deterioration in these ecosystems that can be measured by reduced productivity of desirable plants, undesirable alterations in the biomass and the diversity of the micro and macro fauna and flora, accelerated soil deterioration and increased hazards for human occupancy. (33)

According to Helmut Geist in *The Causes and Progression of Desertification*, there are over a hundred existing definitions of the term (2005). This observation offers us an immediate perspective regarding the historical complexity of the issue and the complications of a single authority when discussing dynamic ecosystems. Therefore, the term desertification is not always appropriately applied as few definitions explicitly refer to whether it is actually a permanent state or if it is artificially exaggerated as a metropolitan construct as opposed to an ecological condition characterized by drought. Although desertification may be the consequence of exploitative human practices, it is also a complex process that proceeds to varying degrees across climatic regions and diverse locations. Aubreville's contribution is critical as it goes on to refer not only to process, but also to event and cyclical patterns. This is based on his perspective as a botanist and his expertise in tropical rainforests. Aubreville's research was in Subtropical forests, where colonial silviculture was denuding the land, exposing rich topsoil to the elements. The capacity for destructive forestry practices to generate soil scarcity is at the root of his descriptions and the impetus for crafting such an authoritative term. It was also effectively linked to improvement strategies and logistics that supported the spread of colonization. However, "green wall" infrastructure (afforestation) is not replenishing cover in degraded land but is being introduced into grasslands and prairies that experience drought as an environmental and biological imperative. The term desertification seems to now have contemporary associations, which have falsely elevated it to the status of other exhausted semantics such as "sustainability" or "biodiversity." It would appear that as soon as the United Nations (UNEP, UNCCD) declares an issue within the scope of conservation, the subject itself becomes obscured by the agencies that were created to represent it (Figure 16.4). In other words, it is the regional geomorphology rather than the agent that expresses the loss that is critical to addressing the subject and, further, to proposing explanations.

The more arid the environment, the more likely it is that vegetation under distress will not recover through natural succession (Gurevitch and Scheiner 2002). Therefore the bond between vegetated and non-vegetated surfaces frames the topic in arid regions; and the relationship between species and establishment

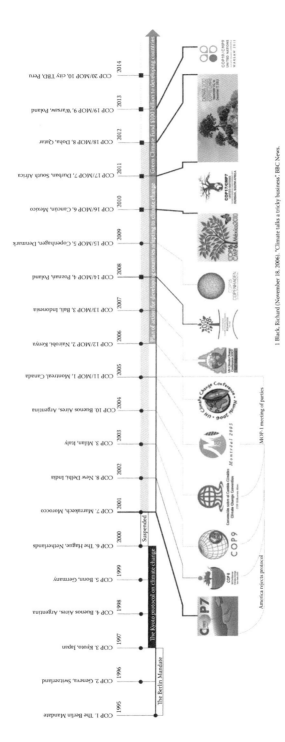

FIGURE 16.4 The evolution of an ideology from the planetary to the terrestrial. The United Nations Climate Change Conferences offer a formal setting for the Conference of Parties (COP). The COP meets to review and assess the most recent scientific, technical, and socioeconomic information produced worldwide in order to understand the advancements made internationally on climate change. The COP does not conduct any research, nor does it monitor climate-related data or parameters. Since 2001, these meetings have been fashioned with a logo in order to generate a comprehensive image of the major issues. This chart illustrates the evolution of the brand from the framework of an interconnected global issue represented by the planet to the idea of potential local resolution, represented by the symbolic tree. This shift in ideology occurred following the COP-12, the first conference held in Africa. (Image courtesy of author.)

1 Black, Richard (November 18, 2006). "Climate talks a tricky business." BBC News.

Planting the Desert 307

technique is of critical importance. From a planning perspective, the realization of large-scale replanting is hinged on the capacity for the technique to be implemented without skill while attaining a high rate of survival (Guo et al. 1989). Therefore, the methodology rarely considers the species itself, its development over time and space, and, further, its botanic associations. The tree species are merely selected based on a compromise between their adaptability and resistance to drought, and it is widely acknowledged that restoring vegetative cover is the most effective technique for slowing erosive states and securing productive soil. The species being deployed grow rapidly and adapt to difficult conditions easily, these characteristics do not contribute to a slower and more systemic ecology that would increase soil stability over time. Instead, trees are established as tools, an infrastructure that is directly associated with industrial economies albeit offered under the rubric of environmental protection.

Plants are the most successful agents when weighing the challenges of economy and restoration of degraded soils, since they are inexpensive and offer an alternative to vacancy. However, the specific manner in which dryland vegetation can recover from drought is in marked contrast to how it responds to degradation (Thomas 1993). Drought is actually a meteorological term, which is attendant to rainfall, and degradation is a specifically soil-based disturbance, defined through measures in organic composition. Degradation often occurs through human (grazing) or natural (wind) disturbances that have clear denuding effects on the land. Therefore, the articulation of local conditions is crucial to applying the appropriate replanting technique, and a definition of micro-condition will define species suitability. Although this appears instinctive, the sheer scale of regional planting projects does not support the prospect of acting in a precise and targeted manner. These projects are not being framed through the experiences of industrial infrastructure simply because live matter (trees) are the ingredient of defense—packaged with the added benefits of carbon offset, environmental amendment, tax incentive, and a sense of common good. Quantities of trees are not analogous to quantities of concrete within the current cultural domain. The positivist qualities of a tree distract criticism and reproach from the true authorities that are practicing afforestation. An infrastructure that it is sold as an absolute system—be it biotic or abiotic—cannot consider a gradation of influences or a sequence of ranges.

16.3 NEW DEAL SHELTERBELT

The practice of shelterbelt planting has roots in the American Midwest as part of Franklin D. Roosevelt's soil conservation initiative to conquer the effect of the dust bowl in the American Great Plains. The stated goals of the project were clear, defining a territory that was imagined as a defensive "green wall" to suppress airborne particles, prohibiting agriculture and framed as a local measure that could simultaneously protect the stable urban regions from erratic rural erosion. In the first U.S. Forest Services report, the design was proposed as "... shelterbelts one hundred feet wide and not more than one mile apart, in a 100-mile wide belt from the Canadian border to the Gulf of Mexico" (LSF 1935). FDR's shelterbelt project remains the most comprehensive planting project ever attempted and was implemented to varying degrees of success. It also represents half the trees ever planted in America, a fact that does not account for the number of trees that remain nor the quality of that stock (Maher 2008). Its accomplishment was clearly not botanic; rather, it was social as it effectively unified thousands of unemployed urbanites in collaboration with thousands of rural farmers, mobilizing an industrial infrastructure of tree planting in grasslands—promoted through the lens of conservation. The Civilian Conservation Corps (CCC) was initially proposed as a temporary New Deal Agency, which would support local farmers in the struggle to conserve precious topsoil. Tree planting and land management represents the nexus of interactions between society and the environment as thousands of men enlisted and were sent across the country. The work of the CCC represents the largest federally funded transformation of the land in history, and although much of the work spanned the coasts, the critical tree-planting efforts were deployed in the Prairie states—the American grasslands (Figure 16.5).

Grasslands are the largest terrestrial ecosystem on the planet and provide a tremendously high carbon storage capacity. This is due to their fibrous and deep root system (often between 4 and 5 m), which is inherent to most grass species, storing carbon deep in the ground. The more carbon stored, the higher the capacity of the soil to actually retain water. If the soil can successfully retain water, it contributes positively to the overall water table and, as a result, the dry seasons can contract as opposed to lengthen. It is generally understood that global grassland systems perform as enormous carbon sinks and are essential to overall climate stability. Grasslands occur where rainfall is too low to support a forest but higher than that which results in desert life forms. Generally, this means between 10 and 30 inches of precipitation, depending on climate and seasonal distribution. When deserts are artificially irrigated and water is no longer a limiting factor, the type of soil becomes the concern. However, if the soil nutrients are favorable, then deserts can be extremely productive due to the ongoing availability of sunlight (Gurevitch and Scheiner 2002). The plants that are adapted to deserts may be annuals, which only grow when water becomes available, flowering and multiplying in low-lying mats, or succulents, which have high storage capacity and thicken to hoard supplies, or, finally, shrubs, which branch copiously and shed their leaves to avoid wilting. All these plants have an imbedded capacity for dormancy in order to withstand fluctuations in water availability or drought. Aridity is demarcated through moisture availability and is historically associated with insufficient rainfall to support trees or woody plant life. Afforestation or "green wall" projects are pouring trees into former grasslands that are selected for their singular purpose of establishing cover as quickly as possible, endorsing trees as a superior ecology.

A shelterbelt planting is a technique that consists of planted rows of woody species. The design typically aims to vary trees for height and leaf structure. Generally, a shelterbelt has three main components: a dense layer of conifers to reduce wind velocity, tall broadleaf trees to extend the area of protection, and low shrubs to slow particulate matter (Figure 16.6). The basis for recommending species alters considerably depending on the particular needs of the site and the region. Regardless of use, the most critical requirement is that the species be adapted to the demands of limited moisture availability, followed by a strong resistance to local climatic stresses (Ritchie 2002).

Planting the Desert

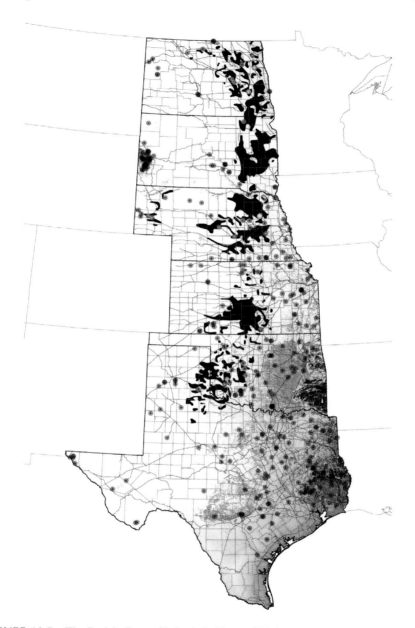

FIGURE 16.5 The Prairie States Shelterbelt. Texas, Oklahoma, Kansas, Nebraska, South Dakota, and North Dakota represent the Prairie States and the site of FDR's Shelterbelt program. Points represent Civilian Conservation Corps Camps, and the planting regions are indicated in black. (Image courtesy of author.)

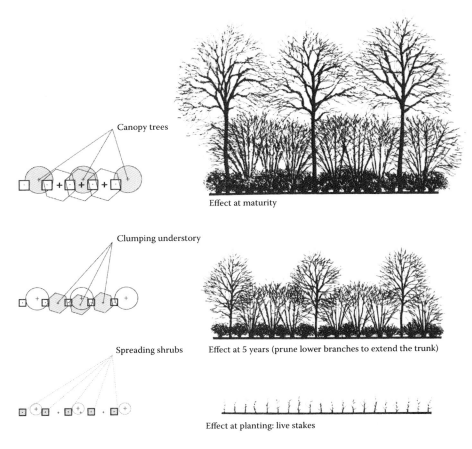

FIGURE 16.6 Typical shelterbelt considerations. (Image courtesy of author. Adapted from Lake States Forest, Experiment Station, *Possibilities of Shelterbelt Planting in the Plains Region*, 1935.)

In the past, the long-term survival and effectiveness of shelterbelts in semiarid areas has been dependent on the combined factors of proper establishment, sound design, and careful species selection—variables that are highly dependent on stewardship. These influences are well documented as founding principles, driven solely on positive results tied to agricultural productivity of adjacent land. The fundamental planting design for FDR's "Shelterbelt Project" was charted by the principles of ecology: resistance, micro-conditions, and diversity but, most critically, it was calculated with the input and ongoing support of a custodian—in this case, a local farmer or land owner with the support of the CCC. Despite even the vast ambitions of species diversity, local specificity, and enlisted labor, the project is reputed for its environmental failures, in particular its cultivation of opportunistic and highly disturbance-adapted species, which continue to prosper across the continent. As a territorial strategy, it was highly effective, positioning "green" as the principal bridge between conflicting

Planting the Desert   311

issues, promoting the planting of trees as beneficial to both society and the environment—the tree stands as the symbol of the individual and the nation.

16.4   3 NORTH SHELTERBELT PROJECT, CHINA

The dust storms in Beijing are notorious: not only do they cripple urban infrastructure, damaging railways, highways, and aqueducts, but the airborne particles often coalesce into hazes that utterly obscure visibility. Locals call these storms "The Yellow Dragon." The situation in China is the most illustrative circumstance of rural dynamics affecting urban populations because arid or semiarid land occupies the most significant percentage of land in China. It is currently estimated that 262 million ha of land in China is disturbed by overgrazing, soil erosion, and increased salinity (Zhang 1996). This accounts for 27% of the total land area (or 79% of total arid lands). To give an idea of scale, the issue affects more than 400 million regional inhabitants and impacts more than 100 million people in urban populations of the Bohai Economic Rim, including Beijing, Tianjin, Tangshan, Shenyang, and Qingdao. The shelterbelt forests currently wind their way from the north in Heilongjiang province through more than 500 counties in 13 provinces to Xinjiang. The project claims to extend 23.5 million ha in grids that can be up to 3 km in width (FAO 2002; Xu et al. 2006). This artificial ecosystem is often described as a conservation project in Chinese media (much like FDR's program), which presents an interesting discussion in terms of an appropriate lexicon for both the discussion of designing at regional scales and of planting as a strategy. Because anthropogenic disturbance is considered to be the direct and major cause of soil erosion and vegetation loss in China, it becomes possible to forecast that the afforestation projects will grow along with expanding urbanism (Zhang 1996). The twin forces of economic prosperity and massive migrations are critical measures to this growing condition, in which time, scale, and circumstances are coalescing to create a new environmental paradigm.

The Chinese government has known about the escalating issue of airborne sand since the early 1970s, and it was this awareness that prompted the "3 North Shelterbelt Project" (SFA, FAO 2002). Deng Xiaoping launched the 3 North program in 1978, aspiring to create an extensive network of plantation forests across Northern China, an expanse also called the 3 North region (Figure 16.7). The 3 North project is one of six major developments supported and overseen by the Chinese Forestry Administration (SFA): The Program for Shelterbelt Development along the Middle and Upper Yangtze River, The Coastal Shelterbelt Development Program, The Farmland Shelterbelt Network of The Plains, The Natural Forest Conservation Program, The National Program against Desertification, and the Taihang Mountain Afforestation Program. The proposal describes a "restoration" of more than 42% of the total area of the country (Figure 16.8). This ambition marks the largest unified design project the world has ever seen and the most extensive act of horticulture embossed on terrestrial ecology. The stated objectives also propose productive forest cover, aiming to increase production of wood supplies and fuel wood to meet current and increasing timber demands. Despite this claim, the project relies on a network of locally governed provinces, each with particular cultural, economic, and

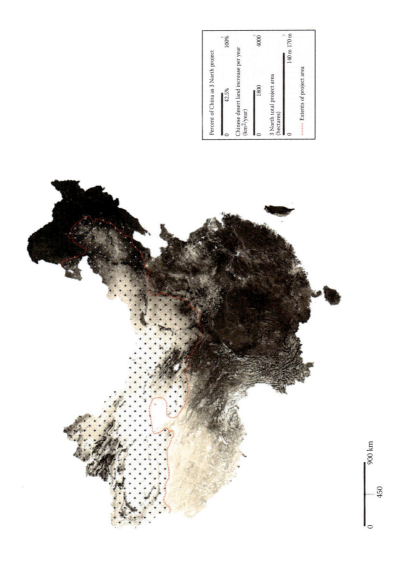

FIGURE 16.7 China 3 North Shelterbelt. The Chinese government has known about the escalating issue of airborne sand since the early 1970s, and it was this awareness that prompted the "3 North Shelterbelt Project." Deng Xiaoping launched the program in 1978, aspiring to create an extensive network of plantation forests across Northern China, an expanse also called the 3 North region. (Image courtesy of author.)

Planting the Desert 313

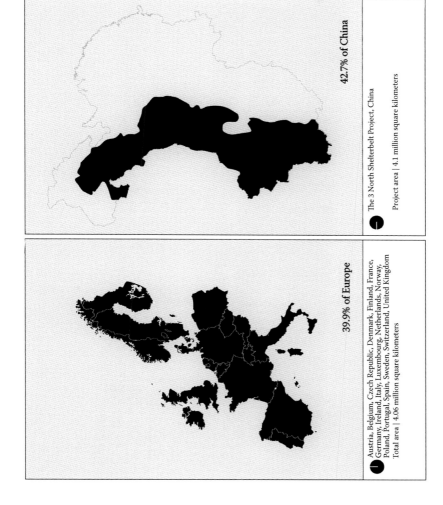

FIGURE 16.8 Scale comparison. The comparison between the scale of the project as outlined by the SFA (State Forestry Administration) in China, contrasted with the same scale as a percentage of Europe. (Image courtesy of author.)

geomorphological pressures. The demands are simply too multifarious to allow for absolute statements of intent; further, the project does not anticipate any form of owner agreement or instruction. The SFA has published results of 10- and 20-year studies of the 3 North project (Figure 16.9). There are varying claims about the success rates, but it is generally accepted that the survival rate is 15% (Wang et al. 2010).

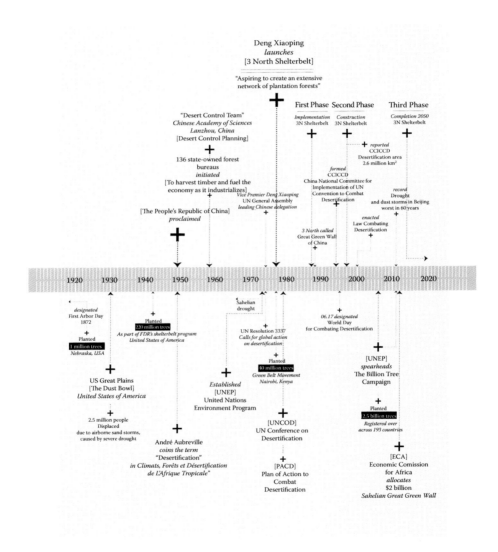

FIGURE 16.9 Timeline of desertification. The combined forces of intergovernmental agency, media frenzy, and population anxiety congregate around the threat of desertification. Common ground is achieved and monitored by posturing tree planting numbers as a means to an end, sponsoring the commodification of nature and the abstraction of territory. (Image courtesy of author.)

Planting the Desert

The shelterbelts themselves comprise an outer belt 500 m wide with a fence along the perimeter to restrain sand. The plantings are often inserted in chessboard patterns and enclose an area that is presumed to be for future agricultural use. The subsurface is prepared by laying out a 2-m-wide gravel platform between the rows of trees that are planted as bare rootstock. This planting procedure is heavily dependent on labor and community involvement, a limiting factor that was recognized early in the project. Citizens and military alike are recruited to contribute significantly through enlistment, persuasion, and by instilling a sense of partnership in a common environmental enterprise.

16.5 *POPULUS SIMONII*, SIMON'S POPLAR, CHINESE POPLAR

The description of a species offers an alternative reading of the land on which it originated almost as if its history acts as a narrative for describing local geography, terrain, average precipitation, and soil composition. Poplar trees are well known for their versatility to site, adapting to difficult, wet, dry, windy, exposed, or all-around challenging conditions. Not surprisingly, they are found throughout the world and remain the genus of choice in most afforestation projects. *Populus* is the most widely used forest tree genus in genetic modification studies and is regarded as a model tree in forest genetics for its range of varieties and ease of reproduction by cloning (NPCC 1996). The Chinese poplar, *Populus simonii*, is particularly well adjusted to both hot and cool environments, including deserts, revealing its fortitude as a native species. Correspondingly, *P. simonii* has been historically recorded through an association with human activities in the drylands of China (Wang et al. 2010, 19). It grows anywhere and as a result of its resilience, remains the major woody species in the 3 North Shelterbelt program. The species' early success is hinged on fundamental poplar characteristics, including easy propagation, rapid initial growth, and available seed. The scale and spread of poplar varieties is a new chapter in the story of the Chinese poplar, coinciding with the cultivation of arid land. Currently, China accounts for 73% of the world poplar plantation area, which consists exclusively of three main species: *P. simonii*, *P. deltoides*, and *P. nigra* (FAO 2006). According to the State Forestry Administration, poplar accounts for more than 60% of the shelterbelt stock used from 1991 to the present. If one accepts that the official number of forest plantation is 24 million ha, then more than 14 million ha of poplar have been planted to date, illustrating a powerful and disturbing dependence on a single species.

The State Forestry Administration claims that poor survival rates are lowering expectations and forcing a reinterpretation of species selection and diversity (SFA, FAO 2002). The 3 North project is entering its second phase of development and concluding the first phase of assessment. Adverse factors have been recorded that highlight the limited genetic diversity but also describe poor nursery conditions, insufficient site preparation, and low maintenance in large-scale plantings. A report by the Chinese Agriculture University states that most afforestation programs have actually contributed to environmental degradation in arid and semiarid regions (Cao et al. 2008). The most recent developments in Chinese programs do not question the obligation or scale of the tree-planting program itself but tend to emphasize the need for improved technology or techniques to increase the overall rate of success. These advancements include aerial seedling bombs, dune fixation through particle injection, and plastic-lined tree pits to capture water. It would seem that as the desert advances, the techniques for suppression invite considerable bioengineering and infrastructural promises that exclude human agency.

16.6 GREAT GREEN WALL (GGW), AFRICA

The Sahel is both a geographic region and a climatic range, which spans the African continent between the Red Sea and the Atlantic Ocean. Often referred to as a "transition zone" between the Saharan desert and semi-tropical savannah, it is actually a semiarid region with its own particular vegetative qualities, which transpire in episodes linked to seasonal rainfall. These sporadic conditions historically fostered nomadic pastoral systems, which were closely aligned with the shifting terrain of drylands and the possibilities of irrigation. Therefore, current trends toward permanence do not necessarily align themselves with the cultural or ecological history of the region, which are defined through patterns of movement. Geist recognizes the differentiation in dryland conditions between Central Asia and Africa, indicating that the African drylands do not suffer drought in the same way as the vegetation is highly dynamic and resilient. He also elaborates that it has evolved along with the dynamics of human influences for millennia (Geist 2005, 168), The vegetation tends to respond to human impact in sporadic and unpredictable ways. Accordingly, this is the suggestive rationale for their extreme resilience. Evolving highly responsive root systems and extended dormancies is particular to desert species. Additionally, meager spacing is essential to ecosystem function. Trees are found in isolation or at great distances from one another in order to eliminate subsurface water competition. Trees may appear deprived from a ground-level perspective, but below the surface, they may be developing, simply conserving their energy by limiting exposure. As a system, trees are most often found in association with shrubby grasslands and certainly not as tightly planted grids. Human occupation of the Sub-Sahara and Sahelian desert goes back more than 100,000 years and was held not only through vigilant nomadic routes, but through careful soil and water management (Mortimore 1998). In the current context, the community no longer structures the impact of human habitation, nor is it organized within an environment that relies increasingly on aid or imports. The discussion of productive plant resources can only be useful if it acknowledges a gradient of environments that are interconnected and interdependent. This is especially problematic within a complicated border condition that divides eleven countries.

There are a remarkable number of plans being researched and tested or deliberated and discussed in terms of how to counter the trend of drought in the African Sub-Sahara. However, the current discussions of greening the Sahara fail to acknowledge this history of pulsing occupation. In its place, a future of embedded permanence is projected. Climate modification schemes have been proposed since the 1930s, instigated by the American "Dust Bowl" when the United States first promoted its shelterbelt defense program (Glantz 1977). These technological or environmental cure-alls include green wall projects but are supplemented by

Planting the Desert   317

more ambitious plans, including precipitation augmentation through cloud seeding and flooding large inland depressions to create Lake Sahara. In 2005, President Olusegun Obasanjo, President of the Republic of Nigeria, officially proposed a green wall for the Sahara and promoted it as the only solution to desertification (UNEP 2005). It was anticipated as a further implementation strategy to satisfy the request of the United Nations Convention to Combat Desertification (UNCCD) from the UN earth summit in Rio (1992). The regional site was selected through high-definition vegetation mapping, which identified the area between the Sudano-Sahelian savannas and the Sahelian shrublands. The official goal released by the New Partnership for Africa's Development states, "The intent is to erect a physical barrier made of trees over a 15 km-long area linking Dakar to Djibouti—7000 kilometers—in order to stop desert encroachment and protect human and natural systems south and north of the Sahara against the adverse effects of desertification on their economic and social development" (Asante 2006) (Figure 16.10). The intricate authority of each agency is a complex and multifaceted array of acronyms

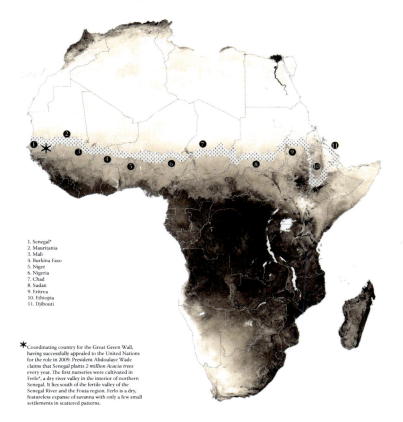

FIGURE 16.10 Africa's Great Green Wall. From Senegal to Djibouti, the Great Green Wall is coordinated by the New Partnership for Africa's Development states, and the first pilot projects are being developed in Senegal, where they have already started publishing tree counts that extend into millions. (Image courtesy of author.)

and agreements. The GGW, still in its formative years, is a proposal being continually reworked according to the growing global concerns and the economic realities of the region. The Community of the Sahel-Saharan States (CEN-SAD) and African Union (AU) claim to be leading the project and in 2011, created the Pan African Agency of the Great Green Wall (PAGGW). This newest agency confirms the project as the most contemporary re-greening initiative on the planet. For policy purposes, each country is actually responsible for adopting and implementing the plantings on their respective soils. At the same time, funding is being secured in sizeable amounts, hundreds of millions from the FAO (Food and Agriculture Organization United Nations) and ECA (Economic Commission for Africa) among others. The number of stages, countries, and agencies that are involved obscures the capital and its associated distribution. It is simply not clear how the money is allocated or being applied to the ground. It seems reasonable to conclude that the funding is caught in the planning phase, which is defended as a necessary first step to implementation. Agency aside, the GGW project is a proposal to consolidate the battle against desertification across eleven countries from West to East Africa. Therefore, the initiative is also an experiment in geopolitical collaboration, not to mention the implicit cultural and religious discrepancies of defining the need for green infrastructure. Is it possible to protect human systems from climate change through a unified dryland replanting initiative?

16.7 *ACCACIA* SP.

The GGW should be considered a theoretical project because the ideas and proposals were only endorsed by the African Union in 2007. In August 2012, the participating countries met in Burkina Faso to strategize and define the framework and deployment approaches. Yet another acronym appears to be publishing additional reports of those proceedings, GGWSSI (The Great Green Wall Initiative of the Sahel), representing the countries rivalry for the funds associated with tree planting projects. Meanwhile, a vegetation policy has been released that identifies and catalogues 37 woody species that are found locally and display drought tolerance. The genus *Acacia* accounts for eight species in this diminutive list, ranging from shrub to tree varieties. However, there are 1250 species in the genus *Acacia*, which are found across the African continent, all of which have been introduced from South Africa or Australia. The extensive history of the *Acacia* and its complicated taxonomy is not pertinent to this discussion; what is relevant is that trees present an especially complex and relatively steady component to arid zones of transition and movement. The first and most common type in the Sahel is deciduous *Acacia* sp., producing the greatest foliage during the rainy season and slowly declining afterwards. A second type is the *Faidherbia albida* (Thorny Acacia), which has an inverted leafing cycle, shedding all of its foliage in the rainy season, and a third type retains its leaves throughout the year (Mortimore 1998). The Thorny Acacia is not only useful for its anomalous growing cycle, but it is also rich in nitrogen, phosphorus, and other nutrients. Plants that grow beneath these trees benefit from their annual leaf fall, which fertilizes the soil and counteracts soil acidity (Ritchie 2002). Locals recognize the benefits, which is why any visitor can observe the trees

Planting the Desert

interspersed with cropland in many parts of East Africa. But the tree is absent from the vegetation list identified by the GGWSSI. There is no documentation or justification to provide further clarity on the issue at this time except it can be assumed, based on other initiatives, that it is the slow rate of growth working against its inclusion. The assumption that the entire Sahel can be renewed with a dependence on such a small diversity of species is not only unrealistic, but also potentially dangerous and certainly cannot be considered under environmental frameworks. In particular, it ignores the tremendous gains of planting three or more tiers of vegetation and, in particular, grasses that promote microbiotic relationships in the soil. Therefore, both the micro-conditions and the artificial spacing are critical to planning at this stage of the process. Because soil erosion is heavily dependent on spatial arrangements (Forman 1995, 458), it is critical to couple the type of erosion with the species, especially across a continental scale. Additionally, reflecting processes through patterns can ensure that the habitat represents the customs of a region and has an authentic role in sustaining local life and resources.

16.8 CONCLUSION

Green infrastructure is a term that has lost relevance due to a lack of precision and specificity, revealing the deficiencies of color-coded planning. Although it presumes to be analogous to constructing soft resolutions to otherwise hard solutions, the term nevertheless lacks a clear definition sponsoring the commodification of nature and the abstraction of territory. The design profession and, more particularly, Landscape Architecture is positioned to articulate the parameters of this so-called green expression through frameworks that are scalar and strategic and exhibit spatial significance. When applied to territorial and political spheres, the perils of this lack of characterization become more acute. "Green wall" infrastructure endorses the use of billions of seedlings, which lends itself to a quick fix within the broader environmental context. In order to challenge this discourse, it is important to recognize that trees themselves are being positioned as a distraction from the associated power structures that are promoting their planting. With climate issues becoming more and more globally visible, this is an ideal time to draw attention to the practice of landscape amendments as they relate to infrastructural scales. The framework of greening is a specifically urban classification, and its application to remote sites is remarkably problematic because contextual specificity, recognition of local conditions, and gradients of concentration are lacking resolution. In the case of afforestation, human agency is fundamental, instigating a conversation about long-term maintenance and support for artificial affiliations to take root, rejuvenate, and thrive. Afforestation offers a technological or industrial tactic as opposed to a long-term sequence of biological associations or ecological scenarios. This puts tremendous energy—economic and biological—in cultivating growth rather than sustaining relationships. Is there any question that trees are embedded in our cultural consciousness and deeply linked to progress and social responsibility? The issue of control becomes even more apparent when tree planting is presented as a survival strategy and considered the only available solution to save affected people and land. We suffer further when we link greening projects to a do-good mechanism used to offset our urban guilt. Control and defense measures must be replaced with social, economic, and biological measures that are driven by design principles.

COMMON ABBREVIATIONS

CCICCD: Chinese Committee for Implementing UN Convention to Combat Desertification
CEN-SAD: The Community of Sahelo-Saharan States
ECA: Economic Commission for Africa
FAO: Food and Agriculture Organization United Nations
GGWSSI: Great Green Wall of Sahara Sahel Initiative
NEPAD: New Partnership for Africa's Development
NPCC: National Poplar Commission of China
PAGGW: Pan African Agency of the Great Green Wall. Member states: Burkina Faso, The Republic of Djibouti, The State of Eritrea, The Federal Democratic Republic of Ethiopia, the Republic of Mali, the Islamic Republic of Mauritania, the Republic of Niger, the Federal Republic of Nigeria, the Republic of Senegal, the Republic of Sudan, the Republic of Chad.
SFA: State Forestry Administration of China
UA: African Union
UNCCD: United Nations Convention to Combat Desertification
UNCOD: United Nations Conference on Desertification
UNEP: United Nations Environment Program

REFERENCES

Asante, S. K. B. (2006). *Implementing the New Partnership for Africa's Development (NEPAD): Challenges and the Path to Progress*. Accra: Ghana Academy of Arts and Sciences.
Aubréville, A. (1949). *Climats, Forêts et Désertification de l'Afrique Tropicale*. Paris: Société d'Editions Géographiques, Maritimes et Coloniales.
Bélanger, P. (2010). Redefining infrastructure. In *Ecological Urbanism*, M. Mostafavi with G. Doherty, eds., 332–350. Baden: Lars Müller.
Cao, S., L. Chen, Z. Liu, and G. Wang. (2008). A new tree-planting technique to improve tree survival and growth on steep and arid land in the Loess Plateau of China. *Journal of Arid Environments* 72, 7: 1374–1382.
Food and Agriculture Organization of the United Nations (2006). *Global Forest Resources Assessment 2005—Progress Towards Sustainable Forest Management*. FAO Forestry Paper No. 147. Rome.
Forman, R. T. T. (1995). *Land Mosaics: The Ecology of Landscapes and Regions*. Cambridge; New York: Cambridge University Press.
Geist, H. (2005). *The Causes and Progression of Desertification*. Aldershot, England; Burlington, VT: Ashgate.
Glantz, M. H. (1977). *Desertification: Environmental Degradation in and Around Arid Lands*. Boulder, CO: Westview Press.
Guo, H. C., D. G. Wu, and H. X. Zhu. (1989). Land restoration in China. *Journal of Applied Ecology* 26: 787–792.
Gurevitch, J. and S. M. Scheiner. (2002). *The Ecology of Plants*. Sunderland, MA: Sinauer Associates.

Lake States Forest, Experiment Station (1935). *Possibilities of Shelterbelt Planting in the Plains Region. A Study of Tree Planting for Protective and Ameliorative Purposes as Recently Begun in the Shelterbelt Zone of North and South Dakota, Nebraska, Kansas, Oklahoma, and Texas by the Forest Service; Together with Information as to Climate, Soils, and Other Conditions Affecting Land use and Tree Growth in the Region.* Washington, DC: U.S. Government Printing Office.

Maher, N. M. (2008). *Nature's New Deal: The Civilian Conservation Corps and the Roots of the American Environmental Movement.* Oxford; New York: Oxford University Press.

Mortimore, M. (1998). *Roots in the African Dust: Sustaining the Sub-Saharan Drylands.* Cambridge; New York: Cambridge University Press.

National Poplar Commission of China (NPCC) (1996). *National Report on Activities Related to Poplar and Willow Cultivation (1991–1996).* Document prepared for the XXth Session of the International Poplar Commission, Budapest, Hungary.

Odum, H. T. (2007). *Environment, Power, and Society for the Twenty-First Century: The Hierarchy of Energy.* New York: Columbia University Press.

Ritchie, K. A. (2002). *Shelterbelt Plantings in Semi-Arid Areas.* Amsterdam; New York: Elsevier, 1983.

State Forestry Administration of China (SFA), Belgium Development Cooperation (BEL), and FAO. *Afforestation, Forestry Research, Planning and Development in the Three-North Region of China.* Technical Project Review Document (1990). Tongliao, China.

Thomas, D. G. (1993). Sandstorm in a Teacup? Understanding Desertification. *The Geographic Journal* 159, 3: 318–331.

United Nations Convention to Combat Desertification (1995). *Down to Earth: A Simplified Guide to the Convention to Combat Desertification.* The Secretariat for the United Nations Convention to Combat Desertification (UNCCD). Bonn, Germany.

United Nations Environmental Program (2005). *Status of Desertification and Implementation of the United Nations Plan of Action to Combat Desertification: Report of the Executive Director.* United Nations Environmental Program (UNEP), Nairobi, Kenya.

United States President (1993–2001, Clinton). United States Congress Senate Committee on Foreign Relations, and Interim Secretariat for the Convention to Combat Desertification. (1997). *United Nations Convention to Combat Desertification in Those Countries Experiencing Serious Drought and/or Desertification, Particularly in Africa.* Geneva: Interim Secretariat for the Convention to Combat Desertification.

Wang, X. M., C. M. Zhang, E. Hasi, and Z. B. Dong. (2010). Has the Three North Forest Shelterbelt Program solved the desertification and dust storm problems in arid and semi-arid China? *Journal of Arid Environments* 74, 1: 13–22.

Xu, J., R. Yin, Z. Li, and C. Liu. (2006). *China's Ecological Rehabilitation: Unprecedented Efforts, Dramatic Impacts, and Requisite Policies.* in Ecological Economics Vol. 57.4: 595–607.

Zhang, J. (1996). Achievements and problems in the construction of the "Three North Forest Shelterbelt Systems." *Forestry Economics* 2: 11–20.

17 Designing for Uncertainty

The Case of Canaan, Haiti

Johann-Christian Hannemann,
Christian Werthmann, and Thomas Hauck

CONTENTS

17.1 Introduction 323
17.2 General Circumstances..324
    17.2.1 Backgrounds of the Humanitarian Disaster 324
    17.2.2 Aftermath of the Quake..325
    17.2.3 Rise of a New City..325
    17.2.4 Urban Crisis..327
17.3 Background of the Research Project 330
17.4 Site Characteristics of Canaan...330
17.5 What Should Be Done in Canaan? 333
    17.5.1 Development Framework for Canaan 333
    17.5.2 Toward an Inclusive City 335
    17.5.3 Showcases from Haiti 337
17.6 Residents and Academics 339
17.7 From Thinking to Acting and Back..339
    17.7.1 General Assembly and First Community Workshops in Onaville ...340
    17.7.2 Second Round of Community-Based Activity 345
    17.7.3 Outlook of Academic Action..346
17.8 Preliminary Conclusion 347
References..348

17.1 INTRODUCTION

When the devastating earthquake hit Haiti in the late afternoon on January 12, 2010, its rising, fragile democracy was, once again, profoundly reminded of its own weakness. The magnitude 7.0 quake—its epicenter closely adjacent to the capital Port-au-Prince—hit the most densely populated region of the country causing the death of more than 300,000 people, injuring 300,000 more, leaving more than 1.3 million people homeless, and destroying large parts of the country's habitat and infrastructure. Three days after the catastrophe, Port-au-Prince's public spaces had been covered with tarps and tents by displaced people fearing further building collapse caused by aftershocks (Farmer 2012; USGS 2013). Relief and reconstruction in the aftermath of the catastrophe revealed many lessons to learn for future post-disaster management in Haiti or around the globe. Many mistakes had been committed by hasty decisions, lack of communication, and collaboration deficits among both international and national actors. One of the most striking was opening a camp to relocate displaced families in the remote outskirts of Port-au-Prince—causing the rise of Canaan, a vast urbanization that was declared the "biggest land grab of Latin-America" (Valencia 2013). A manifestation of institutional weakness, the case of Canaan will show if Haitian and international decision makers (in joint efforts with the Haitian civil society) will take the opportunity to develop livable and resilient neighborhoods or if the country will be stuck proceeding down the path toward further uncontrolled sprawl. This chapter tries, in Sections 17.2 to 17.5, to uncover preconditions, perspectives, and threats to reflect on; in Sections 17.6 and 17.7, academia's possible role is discussed in the context of community-based development in Onaville, one of the sectors of Canaan. On the basis of exemplary samples, the authors examine how volunteers and students can contribute to an improvement of the neighborhood's living environment by supporting participatory activity, knowledge-exchange, and self-aid.

17.2 GENERAL CIRCUMSTANCES

17.2.1 BACKGROUNDS OF THE HUMANITARIAN DISASTER

Comparing the incredibly high loss of lives and level of destruction with disasters in other geographical regions—causing much less damage—the importance of local political and historical context becomes evident.* Both will help to better understand the high vulnerability and low disaster preparedness of the population and the Haitian government.

After several bloody revolts, the former French colony of Haiti became, in 1804, the world's first free slave republic. During 20th century, the young republic endured a foreign occupation, followed by unstable, nondemocratic regimes and dictatorship. In the late 1980s, Haiti entered an unstable political decade of persistent unrest, left-oriented popular movements, and profound political destabilization leading to an international "aid-embargo" on the Haitian government.† Large waves of migration hit Haiti's professional sector: In public sectors in particular, skilled personnel left for nongovernmental organizations (NGOs) or searched for their fortune abroad.‡

---

* Brief but detailed historic background information can be found in the chapter "A History of the Present Illness" in Paul Farmer's great book *Haiti after the Earthquake* (New York: Public Affairs, 2012).

† The origins of violent riots have been basically class conflicts between old elites and a rising popular movement—claiming equal political, civil, social, and economic rights for Haiti's poor majority. For further information on the "aid embargo" and its consequences on public services, refer to Farmer (2012) and Ramachandran and Walz (2012).

‡ Due to political persecution of dissidents during the Duvalier dictatorship, big parts of Haiti's professional and intellectual elites had already left the country.

Designing for Uncertainty

By the year 2004, the United Nations Peacekeeping Mission MINUSTAH was set up to restore "a secure and stable environment" (UN Peacekeeping 2013). On the eve of the earthquake on January 12, 2010, the government—lacking budget and skilled staff to provide most necessary services to the public—hardly could have been able to anticipate this environmental and humanitarian disaster. Lack of public health care, food, and safe drinking water had, to that point, been causing daily loss of lives. Spatial planning regulations and building codes had not been adopted for decades (Farmer 2012; Forsman 2010; Ramachandran and Walz 2012).

17.2.2 AFTERMATH OF THE QUAKE

During the weeks following the earthquake, Haiti gained huge international support and aid. The public sector pledged, during an international conference in New York, donations of $10 billion for humanitarian relief and reconstruction until the end of the year 2012* (UN Office of the Special Envoy for Haiti 2011). One year after the earthquake, Oxfam's Martin Hartberg et al. stated that the "international community [UN agencies, the Interim Haiti Recovery Commission IHRC,[†] and countless NGOs] has too often ... undermined good governance and effective leadership" (2011, 7) by bypassing the Haitian people, local government, and ministries during post-earthquake reconstruction (or more accurately since the 1980s) (Id.). By the end of 2012, only slightly more than 50% of the pledged donations had been disbursed[‡] (UN Office of the Special Envoy for Haiti 2012). *Fatal Assistance* (directed by Raoul Peck, 2012)—an observant documentary of the international relief and recovery campaign—further aggravates accusations against the "international community." Raoul Peck reveals countless major issues in communication, management, and in the coordination of remits.[§]

17.2.3 RISE OF A NEW CITY

By April 2010, the decision was made to relocate 6000 people from a camp in the capital—facing flooding and mudslide risk during the starting rainy season. People were moved far from basic services, social ties, or employment facilities to the Internally Displaced Persons (IDP) Camps "Corail-Cesseless," a place potentially

---

* Huge private transfer payments, which had been donated after the earthquake, are not included.
† The Interim Haiti Recovery Commission (IHRC)—cochaired by former U.S. President Bill Clinton— was established by April 2010 to bundle and manage worldwide relief efforts.
‡ Although NGOs and private contractors appear to be primary intermediate recipients of funding, the Haitian government received less than 10%. Less than 1% went directly to Haitian NGOs and companies (Ramachandran and Walz 2012; UN Office of the Secretary-General's Special Adviser on Community-based Medicine & Lessons from Haiti 2013).
§ To give two striking examples: Few international organizations focused on the prestige-lacking debris clearing in Port-au-Prince in the aftermath of the earthquake while other works were carried out twice. By November 2010, still more than 10 million cubic meters of debris from damaged buildings remained in Port-au-Prince, hampering relief work and reconstruction efforts (Jean-Louis and Hilaire 2011; *Assistance Mortelle* 2012). The other example, the opening of camp Corail in the northern outskirts of Port-au-Prince and its resulting consequences for the urban future of Haiti will be one focus of the chapter.

endangered by flooding, set up by the "international community" in a desert environment 15 km north of Port-au-Prince (IOM 2010; Richener 2012). After opening the camps and, therefore, announcing expropriation for public utility of 5000 ha of land,* the "international community," in conjunction with the Haitian government, "opened a Pandora's box" (Jean-Christophe Adrian, UN-HABITAT, cited in Haiti Grassroots Watch, HGW, 2013a). Informal development started immediately following diverse motivations from the side of new arrivals: A first wave of earthquake-affected, displaced families spontaneously occupied land in the desertic zone of Canaan[†] to escape deficits in camps and *bidonvilles* (underprivileged neighborhoods, slums) of the capital. Others took advantage of the situation to obtain land informally, to participate in unhampered speculation on real estate in the arising city, or to escape rural regions, heading for the city[‡] (HGW 2013a,b; IOM 2010; Richener 2012). During a site visit in June 2013, Prime Minister Laurent Lamothe estimated there to be 300,000 residents in the area[§] with only 10,000 living in the formal sectors of Camp Corail (Figure 17.1) (IOM 2013b).

Every day, new families arrive in the zone, increasing the pressure on people, land, and the environment.[¶] But housing is nothing new to this area: Before the 1940s, the plantation Corail-Cesseless cultivated sugar and sisal. In the late 1990s, the Haitian firm NABATEC S.A. started projection of an Integrated Economical Zone of more than 1000 ha. The main landholder at that time prepared a master plan for so called "Habitat Haïti 2020"—containing industrial, commercial, administrative, and residential clusters[**] (International Finance Corporation 2011). The invasion of the vast terrain by April 2010 set an abrupt end to NABATEC's plans, leaving the government and humanitarian agencies in the worst imaginable situation: Confronted with the tremendous task to stop "slumification" in Canaan, the latter also needed to face compensation for expropriation over an estimated US$64 million

---

* The declaration of the Decree for Public Utility was published in the Official Journal of the Republic of Haiti, *Le Moniteur*. Different interviews with plot holders in Onaville and Delmas in 2013 gave us the impression that the declaration had been widely seen as a free land offer to the people.

† In the Decree for Public Utility, the area is designated Corail-Cesseless. As the term Canaan is used widely by the population and authorities, it will be used to denote the entire zone, including the quarters Canaan, Village Moderne, Jérusalem, Mosaïque, the camps Corail Sector III and IV, and Onaville.

‡ Personal interviews in Onaville revealed that many residents but especially young men recently migrated from rural parts of the country to Canaan in search of Port-au-Prince's higher educational and economical opportunities.

§ The press conference from June 17, 2013, can be accessed at https://soundcloud.com/laurentlamothe /point-de-presse-du-premier-7 (accessed October 27, 2013). The Displacement Tracking Matrix (DTM) of the International Office for Migration (IOM) reports—with slightly different delimitations—a number of 14,101 households with 64,378 individuals (IOM 2013b).

¶ More detailed information on driving factors and the occupation history of Canaan can be looked up in Noël Richener's *Reconstruction et environnement dans la région métropolitaine de Port-au-Prince: Cas de Canaan ou la naissance d'un quartier ex-nihilo* (Port-au-Prince: Groupe U.R.D., 2012).

** Ranked best out of 21 sites in Haiti by the International Finance Corporation, the implementation was planned by a public-private partnership between NABATEC, external investors, and the government of Haiti—which was sighted to provide basic social and technical infrastructure (International Finance Corporation 2011). During that same time, ONA (the National Office of Old-Age Insurance) implemented the gated housing project "ONA-Ville" for governmental employees in the most eastern part of the territory.

Designing for Uncertainty

FIGURE 17.1 Hundreds of thousands of residents have occupied the barren hills of Canaan since the earthquake. (Courtesy of Johann-Christian Hannemann, 2013.)

(HGW 2013a,b). Different governmental institutions and primarily the Unit for Housing Construction and Public Buildings (UCLBP)* have been confessing the indispensable necessity to provide basic infrastructure to the zone as far-reaching relocation will be infeasible. Nevertheless, different grave issues, such as the legal status[†] and the lack of budget for initial intervention, have prevented interventions so far.

17.2.4 Urban Crisis

The emergence of Canaan is surely one of the most striking reflections of Haiti's housing reality: It stands exemplarily for the state of emergency in urban planning, housing, and environmental policy. Increasing discrepancy between rural and urban areas led to unbowed migration toward the capital—fostered by economic crisis and mono-central supply of basic services[‡] (Forsman 2010). Missing

---

* UCLBP was formed after the earthquake: first, to reconstruct destroyed public buildings; second, to coordinate housing interventions and set up the new National Housing Policy; and third, to help displaced people to return to their neighborhoods and to reconstruct them (UNDP 2013).
† Declaration of Public Utility on expropriated land prohibits permanent construction or plot selling by Haitian law to undermine potential nepotism (personal communication with Odnell David, UCLBP, May 31, 2013).
‡ For the last 20 years, on average, 75,000 migrants have been moving annually to the metropolitan zone, arriving mostly in not formally planned, peri-urban settlements (World Bank 2007).

answers on affordable housing for the urban poor and the middle class have been paving the way toward a de facto normality of informal housing in Haiti's urban development—leaving a majority of urban residents for years with a lack of basic infrastructure and public services. Needless to say, informal peri-urban sprawl had been fostered by unstable political decades (Id.) and even more so after the earthquake, which left more than a 10th of the Haitian population homeless for months and years (IHRC 2011). Long before Canaan, vast occupations in the steep slopes of Morne l'Hôpital at the southern edge of Port-au-Prince had been showing similar tendencies: Extremely dense *bidonvilles* have been developing since the 1980s as a result of a political and judicial vacuum, contradictory legal dispositions, and lack of operational capacity. Today, the settlement stretches across three municipalities hosting about 500,000 habitants (Richener 2013).

The "informal" occupation is also omnipresent in the lowlands of Cul-de-Sac, which had once been a main area of agricultural cultivation. Dramatic urban expansion of the capital along major streets had marked the beginning of the occupation of quarters like Croix-des-Bouquets, Tabarre, and Bon Repos in the north of the metropolitan zone (Saffache 2002). Today, these neighborhoods cause uncontrollable problems regarding traffic, public health, security, and natural hazards by substituting forever one of the country's only lowlands with the potential for large-scale agriculture.

Today the emergent "city" of Canaan stands at the crossroads between the chance to relieve Port-au-Prince's urbanization pressure, on one hand, and the threat to become a large underserviced city on the other (Figure 17.2)—destabilizing Port-au-Prince's urban and environmental development. Although the earthquake has enhanced the process, Canaan is just one more manifestation of the capacity in Haitian society to build its own habitat by "auto-construction."* This rather enforced self-aid is not only apparent in the housing sector or in the emergence of micro-economic activity, but also in the greening of the environment. Even in the harsh climatic and environmental conditions of Canaan, residents have begun notable efforts to improve their livelihood by planting home gardens with fruit and timber trees (Figure 17.3). Can these self-aid capacities be better supported in the future to develop a healthy living environment in Canaan?

---

* By buying lots through informal land markets, purchasing construction material, and paying labor, Haitians have already invested more than US$90 million in the urbanization process in Canaan (personal communication with Maggie Stephenson, UNO-HABITAT, May 30, 2013).

Designing for Uncertainty

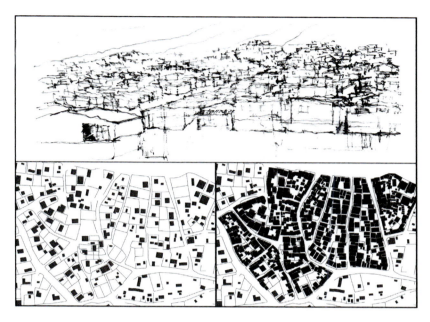

FIGURE 17.2 Density scenarios of upper Onaville: structural plan of existent foundations and houses (left image); potential future consolidation along streets and pathways (upper and right images). (Courtesy of Johann-Christian Hannemann, 2013.)

FIGURE 17.3 Onaville has been greened by its residents; the foothills of Chaîne des Matheux are heavily marked by landslides. (Courtesy of Mariana Aramayo Donoso, 2013.)

17.3 BACKGROUND OF THE RESEARCH PROJECT

Urban Strategies for Onaville is the interdisciplinary project platform of a small research group, consisting of five master's students and one PhD student at the Technical University of Munich (TUM) working on strategies for a more disaster-proof urban development of Canaan's subsection Onaville. The project arises from collaboration between the chairs of Landscape Architecture and Public Space, Sustainable Urbanism, and Urban Water Systems Engineering.* The Onaville neighborhood had been selected as priority area of research in Canaan as it shows the condition of having numerous environmental risks—flooding, erosion, and landslides—along with potentially favorable zones for urbanization. The research group has further been able to build on established social relationships and confidence in big parts of the neighborhood as Haitian volunteers of the nonprofit organization TECHO Haiti there have been contributing to community-building since the year 2011.[†]

TECHO, as well as small grants from the German Academic Exchange Service DAAD and the TUM Task Force for Development Cooperation, enable the integration of internships and fieldwork in Haiti. As the researchers work entirely on their own initiative, they stand, to a certain degree, outside of economic and political constraints and remain flexible to react on changing conditions due to community-led activities. Out of their personal interest in the Canaan case, students contribute as volunteers by providing their personal resources to project organization and fieldwork in Onaville.[‡]

17.4 SITE CHARACTERISTICS OF CANAAN

Located at the interface between the fertile lowlands of Cul-de-Sac plain and the foothills of the deforested mountain range Chaîne des Matheux, Canaan ranges over about 20 km² of the municipalities Croix-des-Bouquets and Thomazeau. To the west, the urbanization is bordered by the bay of Port-au-Prince; to the north, the moderate to steep slopes of Morne-à-Cabris shape a physical barrier. Due to land speculation, this border is already getting eroded—leading to occupation of the

---

* The project was initiated by Christian Werthmann in his role as a Senior Hans Fischer Fellow under the auspices of the Institute of Advanced Studies. Werthmann provides external supervision and project coordination. Project coordinators are Prof. Regine Keller and Dr. Thomas Hauck (Chair of Landscape Architecture and Public Space), Prof. Mark Michaeli (Chair of Sustainable Urbanism) and PD. Dr. Brigitte Helmreich (Chair of Urban Water Systems Engineering) at the Technical University of Munich. Further technical assistance is offered by Dr. Wolfgang Rieger (Chair of Hydrology and River Basin Management, TUM), and Prof. Dr. Boris Schröder (Department of Landscape Ecology, TUM; since August 2013 professor at TU Braunschweig).

† The youth-led nonprofit organization TECHO Haiti opened its local office in Haiti in January 2010 to provide transitory shelters after the earthquake as well as to overcome poverty and social exclusion by joint work between residents and young Haitian volunteers. Until 2013, TECHO Haiti was able to mobilize more than 6500 volunteers and to construct more than 2200 transitory shelters in six Haitian neighborhoods. Following a socioeconomic survey, TECHO provided aid to vulnerable families in Canaan and constructed more than 1480 homes. Since February 2011, the organization fosters community-building, self-aid, and local governance within two community-organizing committees in Onaville. Young Haitian volunteers there organize weekly meetings, lead different social inclusion programs, and foster local volunteer initiatives (personal communication with Roxana Carballo, TECHO Haiti, October 25, 2013).

‡ More detailed descriptions of the students' interventions in the neighborhood of Onaville will be given in Section 17.7 From Thinking to Acting and Back.

Designing for Uncertainty

FIGURE 17.4 Ravine Lan Couline's flow channel cuts deeply into the alluvial cone. (Courtesy of Johann-Christian Hannemann, 2013.)

steeper, landslide-prone slopes, which, at times, reach an inclination of more than 40%. Topsoil degradation and the subsequent lack of higher vegetation in the watersheds have been decreasing infiltration and retention potential at least for decades—increasing the natural hazards of flooding and landslides and leaving aquifers drying, what contributes significantly to periodical major water deficit in Canaan.*

Onaville, the focus area of the research group and the most eastern part of the Canaan settlement, covers a 7-km^2-wide alluvial fan and its adjacent foothills. The neighborhood hosts—according to estimations by IOM—more than 11,000 people (IOM 2013b) and is divided from the other Canaan sections by the ephemeral river, Ravine Lan Couline (Figure 17.4), which captures runoff water of vast watersheds in the northern mountains. As the immediate consequence of local, tropical storms, hurricanes or cyclones—causing precipitation events of several hundred millimeters in a few hours—the ravine shows major flooding and erosion risk for Canaan and Camp Corail along with minor endangering of the same nature to Onaville. A GIS-based flood model for the Ravine Lan Couline was created by Valentin Heimhuber, a TUM master's student of environmental engineering, in order to estimate the flood hazard of Onaville. Using hydrologic modeling, flood hydrographs with statistical return periods of 5, 25, and 100 years were created for all relevant watersheds, which later were used in one-dimensional hydraulic modeling to perform flow simulations in the river channel. The latter revealed that large areas of the settlement are exposed

---

* Damien, the closest rain station to Canaan, lists an average annual precipitation of less than 1000 mm, which is much less than the national average. For comparison, proximate Port-au-Prince receives precipitation rates above 1300 mm (NATHAT 2010).

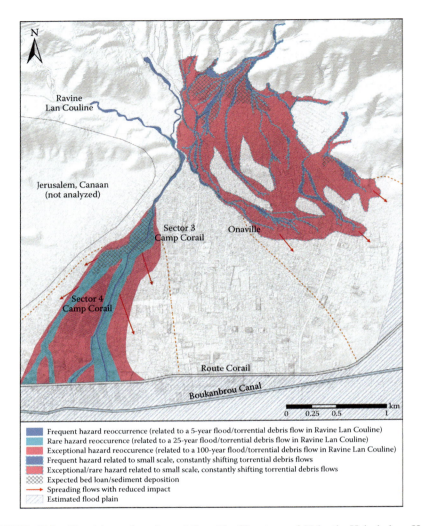

FIGURE 17.5 Flood hazard zoning of Onaville (Courtesy of Valentin Heimhuber. Hillshading based on TanDEM-X-DEM of DLR. Aerial imagery courtesy of Drone Adventures, Open Street Map and International Organization for Migration, 2013.)

to flood hazard. The artificial flow channel that was constructed along with Camp Corail only provides protection against flood events in the frequent hazard range, corresponding to a return period of 5 years. Flood runoffs with return periods of 25 and 100 years were found to exceed the capacity of the channel significantly so that devastating flood events are to be expected.* The upper part of Onaville was found to be highly vulnerable to small-scale torrential debris flows generated in four

---

* Valentin Heimhuber's hydrologic modeling of Ravine Lan Couline revealed peak flow discharges in the river channel of 110 m³ per second for a 5-year design storm event, 300 m³ per second (25-year storm), and 475 m³ per second (100-year storm).

Designing for Uncertainty

relatively small and steep watersheds north of Onaville. In order to provide an easy-to-understand overview of the flood hazard situation, the results of the hydrologic and hydraulic models were incorporated into a flood hazard map (Figure 17.5). A flood protection measure framework was developed as an outlook for further research, opening discussion for practical and feasible solutions in order to protect the settlement against a 100-year flood event.

17.5   WHAT SHOULD BE DONE IN CANAAN?

A logical result of the failure of the state and the "international community" to provide sufficient housing long before and after a major crisis, the existence of Canaan and Onaville shows that, in the current situation of unclear land ownership and stalled housing projects, "auto-construction" seems to be, despite the efforts undertaken by the Haitian government and international actors, the only way for Haiti to overcome its housing shortage in the short term. It would be prudent for the state and international actors to develop and run innovative strategies for infrastructural deficit improvement and environmental hazard mitigation.

Directed by UCLBP, recent programs in Haiti's reconstruction show the first adjustments toward a new paradigm in urban development by integrating the construction of social housing. However, budget and projected numbers of newly built housing units reveal that these measures can neither solely solve the striking deficits in the country's formal housing market nor improve situation in its precarious settlements.* Therefore, directory guidelines, combined with participatory, cooperative upgrading processes for existing, underserviced quarters, will remain as the only possibility to improve the relationship between planning authorities and residents—regaining control over Haiti's urbanization process without provoking social unrest. As Canaan's urban structure is not yet consolidated, this unique but extremely short possibility remains until urban density reaches a level at which there is simply no space left to intervene.

17.5.1   DEVELOPMENT FRAMEWORK FOR CANAAN

But so far many resources of expertise, labor, and funding still remain untapped as the planning entities struggle in Canaan with the dimension and complexity of coordination and information exchange. Residents significantly lack viable information about the current state of work; it is de facto impossible for them to participate, at some point, in the planning process concerning their neighborhood. To guide the upgrading process in the multiple hazard zone, a development framework should be

---

* In total, 3000 social housing units for earthquake-affected families with regular income will be constructed in Village Lumane Casimir in Morne-à-Cabris until the year 2015—equipped with all basic infrastructures and a sewage system (Daudier 2013). By September 2013, the government of Haiti signed a contract with the government of Chile for long-term collaboration in a sites and services program. A first phase of the program will reach about 500 families (Olivier 2013).

334 Revising Green Infrastructure

elaborated using a multi-stakeholder approach, which should bring together the following actors:

- *Important main actors from the national authorities and governmental units* concerned with urban and environmental planning, project implementation, job creation, and jurisdiction*
- *Municipal and local authorities* such as mayors and municipal planning entities as well as local police and tribunal[†]
- *Local committees, community leaders, and initiatives from civil society,* including residents and different religious leaders[‡]
- *Other stakeholders,* such as technical and legal assistants, funding and implementing bodies with backgrounds from NGOs, private economy, and academia[§]

Wrapping up, three indispensable aspects for the framework can be specified: An action plan should launch a democratic, participatory development process, which is open for all concerned stakeholders and especially includes the residents of Canaan. Project platforms should be installed on different levels in order to create an improved communication and information exchange, enabling a joint and transparent program framework for interventions. Finally, we think the key to long-term improved living conditions can only be found through integrated, intelligent local solutions within and across the disciplines of urban design, landscape architecture, environmental planning, education, and economy. Therefore, solutions need to be derived from a joint, transdisciplinary discourse with residents on the consequences of human interaction with the environment, always referring to the specific local context.

---

* *National authorities* should take over responsibility as the initiator of any development and upgrading process in Canaan (providing national rules, guidelines, and policies). The national authorities should also play the major role in the acquisition of funds.

[†] *Municipal authorities* should take over much more responsibility and strengthen their role as the initiator of municipal planning. The integration of mayors and local planning entities should be pursued by national and international actors to pave the way for future local decision-making and decentralized governance.

[‡] *Community committees and leaders* need to be integrated into the process by municipal and national planning entities and should take over responsibility through forming a Community Action Plan. A "Community Action Plan (CAP) is one of the methods used to build the capacity of community members in taking correct action in accordance with the problems, needs, and potential of the community resources" (UN-HABITAT & UNDP). If government action or initiative would fail, the CAP could ensure a certain success as community committees and leaders would be empowered with the skills to manage conflicts independently.

[§] Where existent, *NGO's* protectorate in national and regional decision making needs to be exchanged for empowerment on the national, municipal, and community level. In the context of Canaan, NGOs should search their primary role in facilitating information exchange and acting as agent between disadvantaged parts of the society, local, and national stakeholders. *Actors within the local, regional, and (inter)national economy* should be motivated much more to become part of public development programs in order to assume higher social responsibility and profit from a fair and stable social environment. *Academia* can act in an intermediary position between the community, NGOs, or other actors and authorities—mediating between technical assistance and conceptual thinking. Educational aspects and visionary, integrative concepts need to be submitted to the other stakeholders.

Designing for Uncertainty

17.5.2 Toward an Inclusive City

In the concept paper "The Global Campaign on Urban Governance," UN-Habitat defines the "Inclusive City" as "a place where everyone, regardless of wealth, gender, age, race or religion, is enabled to participate productively and positively in the opportunities cities have to offer" (UN-HABITAT 2002, 5). Referring to this statement, reality in Haitian cities can be described as rather exclusive: The majority of society is discriminated against and driven toward marginality. Whether discriminated against for their social background, precarious economic situation, gender- and age-based arguments, skin color, or handicap, people often are also spatially pushed toward marginal sites, such as ravines, steep slopes, and swamplands—segregated from the "regular city" by the lack of adequate accessibility, social, economical, educational, and sanitary infrastructure. Connected with the "formal city" often only through steep and narrow paths, entire neighborhoods remain "invisible" in the urban fabric, leaving especially their weakest inhabitants deprived of basic citizen rights and social services as well as extremely exposed to environmental hazards.

Taking this into account, Canaan can probably be seen as one of Haiti's most pressing examples of marginalized urbanization coupled with anthropogenic environmental risks. As Canaan is not yet a consolidated *bidonville*, we are convinced that the negotiation between people's aspirations and economic activity—between public (open) space, agricultural production, and environmental prudence—may reveal hidden potentials and excavate forgotten or neglected ones, paving the way for more sustainable urbanization. As it has been proven globally, the creation of income and access to jobs are key roles and the first steps for residents to begin using their earnings not only for basic necessities, such as water, food, and shelter, but for education and improvement of their neighborhood. The psychological aspect of "having work" and "earning money" may be just as important for people in order to develop self-esteem and dignity while maintaining their livelihood.*

How could such a process of social inclusion be fostered in Canaan, turning it toward being a viable and productive place to live for the many hopefuls who bet their fortunes on this desert piece of land? As elaborated in Section 17.5.1, top-down and bottom-up initiatives have to be linked in a multi-scalar way. That means a concerted and flexible strategy—always fluctuating between a generalized and a specialized point of view—seems to be one, if not the most important, key to mitigating the multiple threats of overall hazards and their social effect on people.

Aside from the mentioned social and economical threats, big parts of Canaan are exposed to the environmental hazards of inundation, landslide, and temporal drought. We think growing environmental risks caused by climate change in the tropics can

---

* For this aspect, Janice Perlman's book, *Favela—Four Decades of Living on the Edge in Rio de Janeiro* (New York: Oxford University Press, 2011) provides numerous research findings. One is that, in perception of favela residents, access to jobs and job security have always been the greatest challenge long before their living environment. Perlman refers to many of the neighborhoods' struggles after a finalization of favela retrofitting programs to the fact that the major part of investment had always been made in technical infrastructure but not in people (for example, via education, capacity building, and long-term job creation). Even if the geographical and societal context of Perlman's study is the Brazilian, the historic context of slavery and rural-urban migration might allow the transfer of certain tendencies and findings.

only be reduced if policy makers and key actors from local civil society take over responsibility, initiating strategies to direct the incremental urban growth toward more risk resistance. This means to anticipate and mitigate present threats and harsh conditions in the urban structure—using ecosystem services on site as part of engaging landscape and public open spaces. Adapted productive land use combined with ecological engineering and reforestation in the upslope watersheds of Canaan can contribute directly to flood and landslide hazard mitigation by establishing a closed vegetation cover. Once established, these "ecosystems" provide *regulating services*, such as erosion and debris flow reduction through soil stabilization, increased water infiltration, and retention potential and water and air purification as well as air cooling in urbanized surroundings. If urban sanitation and waste decomposition is achieved by composting or digestion processes, the resulting fertilizers could contribute to soil formation and nutrient cycling in the barren environment of Canaan. Beyond these agroforestry systems in upslope watersheds or tropical communities, respectively, home gardens in urbanized zones may supply important *provisioning services*, such as water, food, energy, and other raw materials, while also providing *cultural services*, such as social and educational values or recreation.*

Worldwide examples of community gardening, urban agriculture, or agroforestry in tropical, low-income contexts could serve as references on how to expand the range of local products via diversification and further processing. Both may contribute to strengthening families' nutritional and economic situation as well as a local economy by creating job opportunities, fostering solidarity networks, and formalizing Haiti's unique small-scale food-production system.† All these examples show the potential of how the integration of productive green space could act as a driver, reducing the side effects of poverty and social exclusion if focus is placed on long-term job creation and people's priorities. From the perspective of urban policy, these measures may imply an important further benefit: (Economically motivated) contributions from local cultivators and neighborhoods may significantly support private landowners and public authorities in their endeavor to stop the invasion of strategic development sites, watersheds, and environmental risk zones. The possible private-public partnership rests on individual initiative, social ties, and social control exercised by well-organized cooperation with the neighborhood.

---

* In the categorization of ecosystem services into *provisioning services, regulating services, cultural services*, and *supporting services*, we follow the Millennium Ecosystem Assessment in *Ecosystems and Human Well-Being: Synthesis* (Washington: Island Press, 2005).

† More than half of the Haitian population works in the complex organized agricultural sector between rural production, intermediate trade, and urban distribution. Livestock, crops, fruits, and vegetables are produced by small-scale farmers carrying them on their (or an intermediary's) head, back, or donkey to small markets located at local traffic nodes. There, products are sold transporting the goods with pickup trucks to the cities' central markets where they are resold to "Madame Saras" (market women) who bring the products—in the quantity they are able to carry—to the consumers on the city's streets.

Designing for Uncertainty

FIGURE 17.6 Small-scale farmers establish contour bonds with hedgerow cultures in a watershed of Petit Goâve. (Courtesy of Johann-Christian Hannemann, 2013.)

17.5.3 SHOWCASES FROM HAITI

Decades of experiences in reforestation, watershed protection, and small-scale agroforestry projects in rural Haiti (Figure 17.6) give proof of the increasing private initiative of plot-holders when confronted with higher responsibility throughout the projects' planning and implementation phases.[*] In the urban context, recent showcases from the Haitian reconstruction indicate the same tendency and show that multi-stakeholder approaches provide promising answers to deficits, such as mismanagement of projects or the lack of coordination. One of the most interesting examples is the 16/6 Projects, in which 16 underprivileged neighborhoods have been rehabilitated to facilitate the return of earthquake-affected families from six camps in Port-au-Prince. The program was initiated by the Presidential Office of Haiti (IHRC 2011) and pursues a bottom-up approach on a very high level: Based on initiative in neighborhood committees, habitants and community leaders get the possibility to become active stakeholders in a community-led reconstruction and development of their neighborhood—in time taking over responsibility inside a community action plan (Figure 17.7).[†] In the context of administration, an example of multi-stakeholder coordination needs to be mentioned:

---

[*] Personal communication with Nicolas Morand, Volmy Merise, and Benoît Jeune (HELVETAS Swiss Intercooperation Haiti, May 28, 2013).
[†] Personal communication with the architects Gerardo Gazmuri (Care/America Solidaria, April 27, 2013) and Nazanin Mehregan (Architecture for Humanity, May 19, 2013).

FIGURE 17.7 As part of the 16/6 Program, Villa Rosa (Port-au-Prince) implemented streetlights, improved pathways, and sports fields. (Courtesy of Raphaela Guin, 2013.)

The setup of six focus groups* consisting of representatives of national institutions and main stakeholders has enabled the UCLBP to exchange experiences and elaborate policy guidelines. Starting work in 2013, the focus groups so far influenced the new *Politique Nationale du Logement et de l'Habitat* (National Housing and Habitat Policy), which was presented to a public audience by October 2013[†] (UCLBP 2013a,b).

In all cases, the claimed key contents—participation, transparency, and a democratic multi-stakeholder approach—have been respected in the working methodology although to differing degrees. Nevertheless, the transfer into a new viable policy—respectively, its application to general planning reality—is not assured, especially because the budget of the Haitian government is very limited. In the case of Canaan, we think participatory elements need to be fostered from the early beginning—specifically through direct inclusion of residents, community leaders, and initiatives into local decision making. Even though bottom-up initiatives may, in some cases, require more time and trained human resources to integrate, communicate, and negotiate a consensus, they generally lead to higher efficiency in budget management as measures have residents' acceptance or—in a better case—their identification with the outcomes. Inevitably, this working method requires a well-designed

---

* Work group exists for Sites and Services, Assisted (Auto-)Construction, Canaan, Land Tenure, Standards and Guidelines, and Housing Financing (UCLBP 2013a).
† Its application will show if national and local authorities, backed by international partners, regain a leading role in urban planning, manage to upscale the promising pilot projects, and thus reduce the gap between formal and informal neighborhoods in Haiti.

Designing for Uncertainty

participatory approach in which interventions arise from fair cooperation with community leaders and residents, including *their* priorities and needs.

17.6 RESIDENTS AND ACADEMICS

Following a praxis-oriented, educative concept in fieldwork, we see the possibility of academic intervention to gear toward the integration and empowerment of residents by establishing a dual learning relationship between researchers and their local counterpart. The contact with concerned people furnishes academic research and design proposals with realism and local strategies: In community and actor meetings, students get important inside views into the realities of the site, into project management and public works in the Haitian context. In the meantime, they have the opportunity and, we think, the responsibility to bring cognizance in the form of "applied science" back to the reality of the neighborhood, partnering organizations, and authorities. Participatory diagnostics, workshop sessions, and capacity-building all incorporate the potential to become platforms on which different stakeholders can teach and learn from each other. Major problems, basic needs, and visions can be carved out; controversies across socioeconomic, cultural, and national boundaries are challenged. The result may be a community-led negotiation of directory guidelines for the local as well as for a lager scale. We think general and strategic planning issues—which are generally answered entirely by top-down planning—may develop from intensive collaboration between residents, community leaders, academics, and professionals, causing broader identification with projects and better conformation to the site's context. Praxis-oriented work in the neighborhood provides one crucial method to dissipate paternalistic structures and to kickstart local improvement of living conditions via micro interventions: People become enabled to mitigate poverty and to transform society by collective reflection, dialogue, and voluntary action.* In the following, some positions will be further elaborated using the example of interventions conducted by the TUM research group in Onaville.

17.7 FROM THINKING TO ACTING AND BACK

With site visits and stakeholder interviews, conducted by the doctoral student Gokce Iyicil in October 2012, the research project *Urban Strategies for Onaville* was launched. First calculations and analyses on potential flood and landslide zones were generated and urban structure and density analyses of different regional case studies disclosed. Development scenarios, in combination with an assessment of opportunities and threats concerning urban growth outlined initial strategies as well as minimum requirements for infrastructure, land-use, and natural resource protection. An extended research internship, completed by master of urbanism student Johann-Christian Hannemann,

---

* The working method refers, in different points, to the concept of Popular Education and Critical Pedagogy, first described by the Brazilian educator and philosopher Paulo Freire. His concept addresses lower socioeconomic classes by en-focusing social transformation—based on a dialectic learning relationship between educator and educated. By actively dismantling and defeating the origins and oppressive structures, Popular Education enables its beneficiaries to escape structural poverty (Freire 2007).

brought deeper insights into the cultural, social, and economic reality of the neighborhood. It introduced intercultural and transdisciplinary elements to fieldwork in Onaville, which was influenced by constant change. Every single day, one could witness modifications in the urban fabric: houses, kiosks, restaurants, barbershops, material stores, parochial meeting houses—often with attached schools or kindergartens—and private health care facilities are announced to be built, are built, or are closed down again. Regular meetings with two TECHO-supported community-organizing committees as well as private interviews with residents brought deep insights into motivations, driving forces, and visions behind the unhampered urbanization process, leading to a more profound comprehension of the eye-striking consequences of absent administrational authority.

17.7.1 General Assembly and First Community Workshops in Onaville

Facing the exorbitant speed of transformation, we came to the conclusion that most conventional planning approaches are doomed to fail in Canaan from the beginning as most are based extremely on labor, time, and cost-intensive bureaucratic processes.*

Thus, in this early community-creation phase, community-building and help for self-aid was set as one focal point of the academic project. Within the frame of very limited financial and temporal resources an actor-oriented planning dialogue had to be adopted. Involvement of residents and community-organizing committees demanded a high grade of organizational flexibility combined with a pragmatic, forward-oriented work process along the joint elaboration and realization of different activities. This transdisciplinary work with directly or indirectly concerned locals allowed our team of young academics and volunteers to become part of a problem-, vision-, and process-oriented cooperative process. A general assembly was designed as a kickoff event to sensitize a wide audience, promoting a further step from omnipresent social exclusion and individualism toward greater inclusion and local empowerment.† Community committees' working methodology, the intention of academic intervention, and first research findings concerning the risks of landslides, flooding, and uncontrolled urban growth were presented to a broad public, which had been informed about the assembly across different socioeconomical levels via personal invitation, distribution of flyers, and megaphone announcements. Newcomers were motivated to become part of workshops and of the neighborhood committees' weekly work.

---

* The outcomes of most regular planning processes would be outdated in the actual context long before they could reach a level of elaboration from where intervention could start.

† For further details on individualism in the Haitian society, see the chapter "The fundamental individualism of Haitian society" in Alice Corbet. *The Community-based Approach in Haiti: Clarification of the Notion of "Communities" and Recommendations* (Port-au-Prince: Groupe U.R.D., 2012), available at http://www.urd.org/IMG/pdf/Corbet_Community-based_Approach_Nov2012_ENG.pdf (accessed October 27, 2013).

Designing for Uncertainty

In the following weeks, four workshops were held in Onaville with durations limited to four hours each.* As an answer to the great distances, related long expenditures in time, and arising difficulties the residents would have to take on to participate in the workshops, it was decided to rotate locations between the community center in upper and lower Onaville. A workshop framework was designed, incorporating group formation, brainstorming and input sessions, site excursions, and a wrapping-up of the elaborated findings and issues. It was to be used in a flexible manner as its intention was not to achieve fixed results but to bring academic findings back to the people, to unite and motivate residents, and discover knowledge and priorities from the community. Participants' high level of involvement, individual inputs, and thorough feedback brought up constructive debates on the workshop topics "rainwater harvesting and water systems," "environmental risks, gardening, soil protection, and watershed protection," "sanitation," and "urban structure." Local materials, techniques, and strategies were explored and documented. The residents' visions for the future of Onaville's open spaces, urban fabric, and landscape were collected and possible solutions on individual and community levels specified. In constant change between individual, community, and regional scale, between site-specific, technical input, and cross-linking development of general correlations, joint reflection had been activated. By posing "the right questions" and referencing the identification and assessment of basic issues concerning the impacts of individual action on urban growth and environmental degradation, a disclosure of general and complex coherences had been approached, incorporating important driving factors and trends on different spatial and temporal levels. The openness of the working sessions helped essentially to make possible an open discourse while, in other moments, discussion was more directed via queries—specifying possible solutions.

This process will be put into more concrete terms by some examples from the workshop "Environmental Risks, Gardening, Soil- and Watershed Protection." After introducing all participants, environmental risks and their placement within general natural systems was reflected collectively by developing the discussed ideas and correlations simultaneously in sketches on paperboard (Figure 17.8). Inputs were given only when absolutely needed. Hazards of landslide, flooding, top soil, and riverbed erosion as well as desertification had been linked with the special role of vegetation in the water cycle (Figure 17.9).† On the basis of site description and photographs, discussion was brought back to Onaville, identifying specific risks in different areas of the settlement. By doing so, it turned out that participants remembered water only once in Onaville's riverbeds: after a summer storm in July 2011.‡ Topsoil erosion was reported to be actively promoted by illegal limestone exploitation of the upslope watersheds.

---

* Finding the right time and place to enable participation across different social and residential contexts posed a major difficulty: During the week, a large part of the interested residents are busy with work whereas on weekends many are occupied by religious activities differing in time and date.

† Of particular interest were the vegetation-provided ecosystem services improved water infiltration and retention as well as soil stabilization by deep-rooted plants.

‡ The residents' statements stand in contrast with a newspaper report stating water torrents and racing streams in the zone already by July 12, 2010, leaving destruction in camp Corail (Katz and Alvarez 2010). Nevertheless, they may be taken as an expression of the yet very low disaster awareness and preparedness in Onaville's neighborhoods as residents come from different local contexts and do not have site-specific memory or disaster experience.

FIGURE 17.8 Open discourse on the neighborhood's needs during a workshop session. (Courtesy of TECHO Haïti, 2013.)

FIGURE 17.9 Explanatory drawings on vegetation's role in flood, erosion, and landslide mitigation. (Courtesy of Johann-Christian Hannemann, 2013.)

Designing for Uncertainty

Collectively prioritized, the highest necessity for intervention was seen in raising awareness on landslide and flooding hazards as residents building in drainage channels put themselves (as well as their neighbors) at great risk. All residents addressed the need for reforestation in the hills. However, lack of clarity in ownership, growing development pressure on the surrounding mountain slopes, and the comprehensible unwillingness to invest their scarce economic resources in ventures lacking any security against eviction impeded residents' numerous ideas to become more concrete.*

The second part of the workshop contained technical input on the multiple benefits of agroforestry systems (such as avoiding total loss of harvest caused by pests or plant diseases via species diversity or allowing a big product variety via cultivation on different layers) in combination with soil-protection measures (such as contour bonds, gully plugs, and hedgerow cultures) as well as on the different characteristics of potentially site-adapted tree species (nurse tree effect, nitrogen fixation, competing effects, adaptation to pruning, and risks of invasive species).† Referring to the cultivation of fruit trees in their gardens, participants reported problems with high death rates of young seedlings they had planted.‡ As a solution for fighting high evaporation and poor soil quality, residents presented the technique of planting seedlings into old car tires, filling them with fertile soil, compost, or organic kitchen waste.§

In a concluding field trip, the species of existent trees was identified, their strengths and weaknesses reviewed, and some excellent showcases of home gardens in the neighborhood were visited. Agronomy student Jephte Chery provided practical technical feedback, introducing composting techniques and the use of Moringa (*Moringa oleifera*) leaves as a nutritional supplement while talks with the gardens' cultivators revealed other new techniques and strategies. For example, one resident showed how he irrigates his garden by using the surface water from a dirt road. Thus, the garden visits initiated knowledge transfer and an exchange of seeds and cuttings among residents. It revealed also that in May 2013, the local Church of Jesus Christ of Latter-day Saints distributed for free about 9000 tree seedlings (mostly fruit tree species) in Onaville.

To complete this short insight into the workshop sessions, some exemplary outcomes of the workshop "urban structure" will be demonstrated as well. The initial brainstorming led to an inventory of existing urban (infra) structure elements as well as a prioritization of those nonexisting but extremely needed. Predictably, this revealed affordable basic necessities, such as health care and education facilities for

---

* One participant, for example, dreamed of reforesting a big parcel in the mountain slopes with fast-growing tree species to produce charcoal for sale in the neighborhoods.

† Input content based on expert interviews, site visits of watershed reconstruction, and agroforestry projects in Petit-Goâve and Artibonite (Haiti) as well as on Joel Timyan's excellent book *Bwa yo: Important Trees of Haiti* (Washington, D.C.: South-East Consortium for International Development, 1996).

‡ Discussions with residents who successfully cultivated the same species in their gardens revealed, during the following field trip, that crucial points for reducing die-back are the planning of the right moment for planting (in respect to drought and rainy seasons) as well as regular watering of the seedlings during the first weeks after transplantation. Thus, the general lack of irrigation water, respectively, high water costs at private water kiosks seem to be major reasons for the reported loss.

§ Old car tires are available everywhere in Haiti, can be painted, are applicable even on rocky ground, and hold back humidity for longer times than if planting seedlings directly into the ground.

both children and adults, water provision, and electricity, but also unforeseen ones, including cemeteries, improved public markets, public open spaces, a representation of the municipal authorities, and local police. The input session made a city's road networks, advantages and disadvantages of dead-end streets a subject of discussion, concentrating on economic, safety and public health factors as well as on the origins of their incremental emergence in Onaville. Discourse disclosed the participants' feelings of powerlessness and skepticism regarding physical changes of their environment, surpassing their own parcel.* Frustrated participants reported their failed initiative installing a small public square on a vacant lot. It had been destroyed a short time later by an unknown external person, claiming possession of the site through the placing a pile of stones.† Starting from this, a joint reflection was launched discussing how improved communication and collaboration among neighbors could help to anticipate uncontrolled growth tendencies—leading, for example, to formation of dead-end streets—from the outset by civil courage, social control, and joint initiative.

All community meetings have shown the interesting fact that social meeting points and symbolic structures (for example, place and street name signs) get similar attention as "urgent" basic infrastructure—with the striking difference that street denomination of Onaville is only one step from implementation by the community-organizing committees.‡ Part of that project, the construction of painted road signs at Route Corail, the main access road passing the neighborhood, shows how successful initiatives can be if addressing a commonly felt need and coordinating action with municipal authorities (Figure 17.10).§ We attribute this to the fact that those gestures, on one side, strengthen the community by their intangible value but, on the other side, manifest people's right of residence.¶ Thus, open spaces and symbolic effects should not be neglected when thinking about neighborhood upgrading as both may contribute essentially to the building and well-being of the community in Onaville constituting a starting point for common action. Concluding, one can say workshops revealed many already practiced self-aid solutions but also the severe lack of communication between the very delimited, small neighborhood groups. By joining people from both subsections of Onaville, exchanging information, ideas, and techniques, the workshops served as a very first step to reduce the tendency of isolation and to build a network for collective discourse and vision making.**

---

* We refer this to the fact that social ties are still very weak in Onaville. Major efforts, such as leading a land management process either to reestablish a street connection or to negotiate rights of way, would by far exceed the neighborhood committees' capability.

† The same as in the reported case, many parcels in Canaan remain undeveloped for years. Designating land occupation by piles of stones, building foundations, or surrounding concrete walls, the "owners" wait for rising land prices.

‡ The denomination of all streets in Onaville has been planned in consensus with the municipal authority of Croix-des-Bouquets. The residents' committees are now looking for donations to implement the project.

§ Construction of the signs was carried out in December 2013 by residents and TECHO volunteers.

¶ This implies the assumption that formalized street denomination is seen among the residents as a first step toward receiving formal titles or life estate on their occupied parcels.

** Many residents participated in more than one workshop, some in all four, many discovering "the other part of Onaville" for the first time. To continue the networking, during summer 2013, all workshop sessions had been worked up as documentary manuals and handed over to the community-organizing committees promoting further activity. In addition, the reports are meant to exchange those findings derived in collaboration with the residents with other stakeholders and actors.

Designing for Uncertainty 345

FIGURE 17.10 Onaville's brand-new road sign with the hills of Canaan in the background. (Courtesy of TECHO Haïti, 2013.)

Facing the task to contribute generally to improved information exchange among stakeholders in Haiti, the research group launched an online project site providing research data and fact sheets.*

17.7.2 Second Round of Community-Based Activity

Beginning in October 2013, master's student of urban design Raphaela Guin and recent graduate of the environmental engineering program Sean Kerwin followed up work in Onaville as part of a six-week stay. Both organized a second round of practical, action-oriented workshop sessions with the special aim of fostering the "Inter-Haitian" knowledge exchange between the communities and TECHO volunteers. A first workshop addressed "public health," explaining simple hygiene techniques based on four small, practical low-cost experiments: fly traps and a "Tippy Tap" for hand washing were constructed (Figure 17.11); water treatment with locally available seeds of Moringa and Solar Disinfection (SODIS) were executed. The second workshop "smart household layout and design" focused on the theoretical components, including thermic ventilation techniques, grey water reuse by home garden irrigation, foresight concerning plot organization, and earthquake-proof construction techniques. Both sessions had been held with residents and the community-organizing

---

* The project site can be accessed on the open-source platform Open Architecture Network: http://open architecturenetwork.org/projects/urban_strategies_ona.

FIGURE 17.11 Workshop participants construct and test the Tippy Tap for correct functioning. (Courtesy of Mariana Aramayo Donoso, 2013.)

committees in Onaville as well as for TECHO volunteers in Port-au-Prince, so they can further develop and repeat the workshops in other Haitian communities.

17.7.3 Outlook of Academic Action

The research group will continue consolidating enhanced local networking among different initiatives as we strongly believe that experiences and expertise collected in self-organized and -managed neighborhood projects may bring the most crucial inputs for successful community-driven interventions. Joint capacity building sessions between TECHO volunteers and TUM students, therefore, will be one focus of collaboration as the Haitian volunteers may play a leading role in cross-linking neighborhood initiatives of particular interest, for example, from other TECHO-supported communities.* We are convinced this knowledge transfer would affect the neighborhood Onaville in a voluntary and thus highly effective way to take over responsibility for their living environment and conduct their own projects. A resulting higher level of autonomous exchange of information and better practice then would allow academics and professionals to focus again on providing technical and design assistance while facilitating collaboration with public and private actors. In this context, further research visits are planned for 2014, adding the aspect of engaging

---

* Gariche Prince, a community with already tightened social ties, gives proof of what can develop from community initiative: Happiness Plaza—a public square—has been planned and designed along all phases by the community committee and TECHO volunteers. In the last planning phase, the group was reinforced by a young Haitian architect. To get financial support for the project's implementation costs over estimated US$10,000, the working group chose to go in unconventional ways: By September 2013, a crowdfunding campaign was accomplished, raising $1900 in two weeks runtime. (Personal communication with Roxana Carballo, TECHO Haiti, October 25, 2013; Indiegogo 2013.)

Designing for Uncertainty 347

midterm and final feedback sessions on the students' research. The overall goal will remain the mediation between technical, academic findings, scenario building, and problem-oriented, practical workshops in Onaville—exchanging gained knowledge with local and national actors and authorities.

17.8 PRELIMINARY CONCLUSION

Based on the current state of reconstruction in Haiti, different questions arise, especially when facing the projected modus operandi for Canaan: Could community action, with residents in a key role in decision making, reduce poverty and improve the living conditions by real development cooperation and handmade "Urban Acupuncture"—in which the latter arises from social and creative urban practice? If embedded in a multi-stakeholder framework, how could commonly accepted planning tools and strategies be identified to create, on one hand, more affordable and risk-resilient housing and, on the other, to anticipate further occupation of natural hazard zones and open spaces holding essential productive, social, and ecological functions? Would joint initiative and the neighborhood-led planning approach be capable of promoting alternatives to nonrenewable resource exploitation—for example, via intelligent ecosystem restoration, in which mutual benefits, higher risk resilience, and new business opportunities could lead toward a better equilibrium in the relationship between men, city, and environment?

We see the great opportunity that neighborhood-driven initiatives may become starting points of a societal change reducing socioeconomic exclusivity and anti-democratic tendencies of today's Haitian cities. Obviously, this transformation cannot be solved solely by architects, planners, and engineers. Embedded in a wider frame of solid social, educational, cultural, and local economic interventions, we think that spatially operating pilot projects may provide catalyst effects on a community's social development. By opening eyes for the changeability of reality and manifesting communities' existence, they may become a resource of hope. Thereby, handmade urban interventions may create a sense of common responsibility and ownership toward infrastructure and space, in the long run determining a project's success or failure. Encouraging further self-initiative thus may be a key to change the quality of life in Onaville, Canaan, and other Haitian settlements—especially when the most poor and excluded parts of society are equally included in the process (Figure 17.12). But when all comes back to political will and money, the question arises if the Haitian government will be a main promoter of this societal change and if it will be able to raise the necessary funds and long-term loans to build public-private partnerships to kickstart such a development process in Canaan. Simultaneously, the national planning entities would need to foster development in the country's rural areas to decrease migration pressure on the metropolitan area of Port-au-Prince, which is caused by socioeconomic mono-centrality and people's dream of a better life in the city. It would be desirable that the country's striking deficits in urban and environmental development will be attacked as soon as possible. Various promising tools and strategies have been developed in the aftermath of the earthquake or even before. The next years will have to answer the question if the current and subsequent governments—backed by civil society, diaspora, and the so-called "international

FIGURE 17.12 Handmade public square in upper Onaville. (Courtesy of Johann-Christian Hannemann, 2013.)

community"—are willing and able to adopt a courageous, transparent, and proactive course to ensure livelihood, improved access to basic services, and disaster preparedness of the entire Haitian population.

REFERENCES

Books and Papers

Farmer, P. (2012). *Haiti After the Earthquake.* New York: Public Affairs.

Forsman, Å. (2010). *Strategic Citywide Spatial Planning: A Situational Analysis of Metropolitan Port-au-Prince, Haiti.* Nairobi: United Nations Human Settlements Programme (UN-HABITAT). http://www.unhabitat.org/pmss/listItemDetails.aspx?publicationID=3021 (accessed October 27, 2013).

Freire, P. (2007). *Pedagogy of the Oppressed.* New York: Continuum.

Hartberg, M., Proust, A., and Bailey, M. (2011). From relief to recovery. Supporting good governance in post-earthquake Haiti. Oxfam briefing paper 142. Oxford: Oxfam GB, January 2011. http://www.oxfam.org/sites/www.oxfam.org/files/bp142-relief-to-recovery-130111-en.pdf (accessed October 26, 2013).

International Finance Corporation. (2011). *Integrated Economic Zones in Haiti: Site Assessment.* Washington, DC: International Finance Corporation. https://www.wbginvestmentclimate.org/advisory-services/investment-generation/special-economic-zones/upload/IEZs-in-Haiti-Sites-Assessment-English.pdf (accessed October 27, 2013).

International Organization for Migration (IOM). (2013a). *Displacement Tracking Matrix (DTM) v.2. DTM_v2.0_30_June_2013.kmz.* Last modified: June 2013. http://iomhaitidataportal.info/dtm/downloads.aspx?file=~/downloads\15.DTM%20Report_Jun%2713/DTM_v2.0_30_June_2013.kmz (Login required for download; Accessed October 29, 2013).

Designing for Uncertainty 349

International Organization for Migration. (2013b). *Displacement Tracking Matrix (DTM) v.2. DTM Report-English.* September 2013. http://iomhaitidataportal.info/dtm/downloads .aspx?file=downloads/16.DTM%20Report_Sep%2713/DTM_V2_Report_Sep_2013 _English.pdf (Login required for download; Accessed October 29, 2013).

Jean-Louis, M., and Hilaire, E. (2011). *Debris Removal Situation, Context and Recommendations.* Port-au-Prince: Development Innovations Group, January 2011. http://haiti-ecap .com/Websites/dig/images/H-Debris-SF3.pdf (accessed October 31, 2013).

Kristoff, M. and Panarelli, L., United States Institute of Peace (USIP). (2010). Haiti: A Republic of NGOs? *USIP Peace Briefs no. 23.* Washington, DC: United States Institute of Peace. http://www.usip.org/sites/default/files/PB%2023%20Haiti%20a%20Republic%20 of%20NGOs.pdf (accessed October 31, 2013).

Millennium Ecosystem Assessment. (2005). *Ecosystems and Human Well-Being: Synthesis.* Washington: Island Press. http://millenniumassessment.org/documents/document.356 .aspx.pdf (accessed February 25, 2014).

Perlman, J. (2011). *Favela—Four Decades of Living on the Edge in Rio de Janeiro.* New York: Oxford University Press.

Ramachandran, V., and Walz, J. (2012). Haiti: Where has all the money gone? *CGD Policy Paper 004.* Washington, DC: Center for Global Development. http://www.cgdev.org /content/publications/detail/1426185.

Richener, N. (2012). *Reconstruction et environnement dans la région métropoli-taine de Port-au-Prince: Cas de Canaan ou la naissance d'un quartier ex-nihilo.* [Reconstruction and the Environment in the Metropolitan Region of Port-au-Prince: Canaan — A Neighbourhood Built from Scratch]. Port-au-Prince: Groupe U.R.D. http://www .urd.org/IMG/pdf/ReconstructionetEnvironnement_Rapport_Canaan_Nov2012.pdf (accessed October 27, 2013).

Richener, N. (2013). *Reconstruction et environnement dans la région métropolitaine de Port-au-Prince: Morne-Hôpital ou l'histoire d'un déni collectif.* [Reconstruction and the Environment in Haiti: Morne L'Hôpital—A Case of Collective Denial]. Port-au-Prince: Groupe U.R.D. http://www.urd.org/IMG/pdf/Groupe_URD_Rapport_Morne_Hopital _Mars_2013-2.pdf (accessed October 27, 2013).

UN-HABITAT. (2002). *The Global Campaign on Urban Governance—Concept Paper.* Nairobi: United Nations Human Settlements Programme (UN-HABITAT) http://www .unhabitat.org/pmss/listItemDetails.aspx?publicationID=1537 (accessed February 18, 2014).

UN-HABITAT & UNDP. ANSSP Guidelines. Volume 2. CAP (Community Action Planning) & Village Mapping. *Aceh-Nias Settlements Support Program.* UN-HABITAT & UNDP, Year n/a. http://www.sswm.info/sites/default/files/reference_attachments/UNDEP%20 ny%20CAP%20Community%20Action%20Plan%20and%20Village%20Mapping.pdf (accessed October 27, 2013).

Unité de Construction de Logements et de Bâtiments Publics (UCLBP). (2013a). *Réunion des parties prenantes de la Table sectorielle du logement.* [Stakeholder meeting of the Sectoral Table of Housing]. Port-au-Prince: UCLBP. http://uclbp.gouv.ht/pages/34-reunion-des -parties-prenantes-de-la-table-sectorielle-du-logement.php (accessed October 27, 2013).

Unité de Construction de Logements et de Bâtiments Publics (UCLBP). (2013b). *Politique nationale du logement et de l'habitat (PLNH).* [National Housing and Habitat Policy]. Port-au-Prince: UCLBP. http://uclbp.gouv.ht/download/pnlh-resume-executif.pdf (accessed October 27, 2013).

United Nations Development Program (UNDP). (2012). *Fiche de Projet. Projet appui au Secteur de la Reconstruction et du Logement.* [Project fiche. Reconstruction and Housing Sector Support Project]. Port-au-Prince: UNDP, December 2012. http://www .undp.org/content/dam/haiti/docs/fiche%20de%20projet%20-%202013/relevement /UNDP-HT-Fic-proj-rel_APPUI%20UCLBP_jan2013.pdf (accessed October 31, 2013).

United Nations Office of the Special Envoy for Haiti. (2011). *Has Aid Changed? Channelling Assistance to Haiti Before and After the Earthquake.* New York: United Nations Office of the Special Envoy for Haiti. http://www.lessonsfromhaiti.org/download/Report_Center /has_aid_changed_en.pdf (accessed October 31, 2013).
United Nations Office of the Special Envoy for Haiti. (2012). *New York Conference Recovery Pledge Status and Modalities as of December 2012 in USD millions.* New York: United Nations Office of the Special Envoy for Haiti, December 2012. http://www.lessonsfrom haiti.org/download/International_Assistance/6-ny-pledge-status.pdf (accessed October 31, 2013).
World Bank. (2007). *Social Resilience and State Fragility in Haiti. World Bank Country Study.* Washington, DC: World Bank. http://documents.worldbank.org/curated/en/2007 /01/8341103/social-resilience-state-fragility-haiti (accessed October 31, 2013).

Journal Article
Saffache, P. (2002). La Plaine du Cul-de-Sac (République d'Haïti): des dégradations à un aménagement raisonné.' *Ecologie et Progrès* 2: 90–97.

Websites
Haiti Grassroots Watch (HGW). (2013a). Controversy over Camp Corail. Last modified: June 17, 2013. http://haitigrassrootswatch.squarespace.com/31controverseEng.
Haiti Grassroots Watch (HGW). (2013b). Reconstruction's Massive Slum Will Cost "Hundreds of Millions." Last modified: June 17, 2013. http://haitigrassrootswatch.squarespace. com/haiti-grassroots-watch-engli/2013/6/17/reconstructions-massive-slum-will-cost -hundreds-of-millions.html (accessed October 25, 2013).
Indiegogo. Happiness Plaza: Changing the life of a Haitian community. Last modified: October 09, 2013. http://www.indiegogo.com/projects/happiness-plaza-changing-the -life-of-a-haitian-community?c=activity.
Interim Haiti Reconstruction Commission (IHRC). 16 Neighborhoods - 6 Camps Project. Last modified: August 18, 2011. http://reliefweb.int/report/haiti/16-neighborhoods-6-camps -project (accessed October 25, 2013).
International Organization for Migration (IOM). IOM Assists in Relocation of Displaced from Petionville Golf Club to Corail Cesselesse Site. Last modified: April 9, 2010. http:// reliefweb.int/report/haiti/haiti-iom-assists-relocation-displaced-petionville-golf-club -corail-cesselesse-site (accessed October 25, 2013).
NATHAT. Haiti Monthly Precipitation (Pluviométrie mensuelle). Last modified: May 2010. http://haitidata.org/data/geonode:hti_hazardclimate_monthprecipitationstation _point_052010 (accessed October 31, 2013).
United Nations Peacekeeping (UN). MINUSTAH: United Nations Stabilization Mission in Haiti. Last modified: n/a. http://www.un.org/en/peacekeeping/missions/minustah/ (accessed October 26, 2013).
UN Office of the Secretary-General's Special Adviser on Community-based Medicine & Lessons from Haiti. Key statistics. Last modified: n/a. Link: http://www.lessonsfrom haiti.org/lessons-from-haiti/key-statistics/ (accessed October 31, 2013).
U.S. Geological Survey (USGS). Earthquake Information for 2010. Last modified: December 12, 2011. http://earthquake.usgs.gov/earthquakes/eqarchives/year/2010/ (accessed October 27, 2013).

Online Newspapers
Daudier, V. (2013). Le Village "Lumane Casimir" inauguré. [Village "Lumane Casimir" inaugurated.] *Le Nouvelliste*, May 16, 2013. http://www.lenouvelliste.com/article4 .php?newsid=116901 (accessed October 30, 2013).

Katz, J., and Alvarez, M. (2010). Haiti: Summer storm floods "safe" refugee camp. *The Seattle Times*, July 12, 2010. http://seattletimes.com/html/nationworld/2012341617_apcbhaiti homelesscamp.html (accessed March 01, 2014).

Olivier, L.-J. (2013). Le Chili en soutien à la politique du logement en Haïti. [Chile supports housing policy in Haiti.] *Le Nouvelliste*, September 3, 2013. http://lenouvelliste.com /article4.php?newsid=120790.

Valencia, M. (2013). Comenzará con 470 sitios: US$ 2 millones aporta Chile a primera política habitacional de Haití. [It will start with 470 sites: Chile contributes 2 million US$ to Haiti's first housing policy.] *El Mercurio Nacional*, September 04, 2013. http://impresa.elmercurio.com/pages/detail-view.htm?enviar=%2FPages%2FNewsDetail .aspx%3Fdt%3D2013-09-04%26NewsID%3D0%26BodyID%3D3%26PaginaId%3D8 %26SupplementID%3D0 (accessed October 30, 2013).

Images (DVD, Video or Film)

Assistance Mortelle (Fatal Assistance). (2012). [TV Screening] Directed by Raoul Peck. France/Haiti/USA/Belgium: Velvet Film France, In co-production with Arte France and Radio Télévision Belge Francofone (RTBF).

Section IV

Applied Design

18 Water-Sensitive Design of Open Space Systems
Ecological Infrastructure Strategy for Metropolitan Lima, Peru

Eva Nemcova, Bernd Eisenberg, Rossana Poblet, and Antje Stokman

CONTENTS

18.1 Lima: Urban Landscape Development of a City in a Desert 356
18.2 Urban Water Cycle and Open Space 356
    18.2.1 Decorative Approach to the Design of Open Space 359
    18.2.2 Open Space with High Water Demand 359
    18.2.3 Open Space with High Demand on Water Quality 361
    18.2.4 Disconnection between the Local Ecology, Open Space System, and Hydrological Infrastructure 361
18.3 Lima Ecological Infrastructure Strategy (LEIS) 361
    18.3.1 Theoretical Framework and Adaptation 362
    18.3.2 Methodology of LEIS 362
        18.3.2.1 LEIS Principles 362
        18.3.2.2 LEIS Tool 363
        18.3.2.3 LEIS Manual 363
18.4 Water-Sensitive Urban Design Approach 365
18.5 Design Method for Integrated Solutions 366
18.6 Saving Water by Design of Open Space 367
18.7 Treating Polluted Water by Design of Open Space 367
18.8 Prototypical Water-Sensitive Urban Design for Different Hydro-Urban Typologies 369
    18.8.1 Polishing Parks and Corridors for Insufficiently Treated Wastewater 371
    18.8.2 Domestic Wastewater Treatment Parks and Corridors for Settlements Not Connected to WWTP 371
    18.8.3 Irrigation Channel Water Treatment Parks and Corridors 374

356   Revising Green Infrastructure

18.9   Strategic Projects in the Lower Chillon River Watershed 379
        18.9.1 Toward a Water-Sensitive Future of the Lower Chillon River
              Watershed 379
        18.9.2 Chillon River Linear Park 381
18.10 Conclusion 383
References.. 384

18.1 LIMA: URBAN LANDSCAPE DEVELOPMENT OF A CITY IN A DESERT

The Peruvian capital of Lima is located in the coastal desert on the Pacific Ocean. The water-scarce Pacific basin possesses only 1.8% of the total water resources of the country, but it hosts 62.40% of its population. (PNUD 2009, 40 part II). According to the 2007 National Census, metropolitan Lima, including the Callao Province, had 8.5 million inhabitants, and its population was projected to reach 9,585,636 in 2013 and 9,886,647 in 2015 (INEI 2012, 32). Due to these characteristics, Lima is considered the second most extensive city in the world built in a desert after Cairo. In Lima, however, there is much less water than in Cairo: It only has an average of 9 mm of rainfall per year (compared to 35 mm in Cairo), and its three rivers, Rimac, Lurin, and Chillon—which are mainly seasonal torrents of water—have an average monthly flow of 39 m^3/s (compared to the Nile River, which has an average flow of 2830 m^3/s). Large infrastructural projects were put in place to collect and transfer rainwater from the Andean mountains to feed the rivers supplying Lima. Additionally, the Transandean tunnel was constructed to transfer rainwater from the Amazon basin to the Pacific basin to cover the water deficit of the city. Because Lima's water supply is dependent on water transfers, rain, and glaciers in the Andes that are quickly melting due to climate change, Lima is considered one of the most vulnerable cities in the world to the effects of climate change. Additionally, the water vulnerability is increased by a lack of awareness about water saving and efficient technologies by stakeholders and the population (Zucchetti et al. 2010). Although there is almost no rain, Lima's climate is very humid due to the effects caused by the interface of the cold Humboldt current and the hot equatorial current over the Pacific Ocean in the west and the high Andes mountains in the east. Therefore, from June through December, a thick fog covers the city and turns some of the desert hills into temporarily green, blooming meadows called lomas (Figure 18.1).

18.2 URBAN WATER CYCLE AND OPEN SPACE

The fast population growth, lack of implementation of urban and regional planning instruments, economic crisis, and other factors have led to a vast expansion of informal settlements, including the occupation of vulnerable areas on steep hills and river flood zones, lomas, and wetlands. Fast land-use change is resulting in the disappearance of agricultural land in the river valleys (Figure 18.2). Many settlements in the peri-urban areas suffer from poor housing standards, high levels of pollution, and a lack of basic urban services, such as water supply and wastewater infrastructure. Over one million inhabitants, living mainly in the hilly and peri-urban areas, are not

Water-Sensitive Design of Open Space Systems

FIGURE 18.1 Seasonal biotope lomas, Villa María del Triunfo district, metropolitan Lima. (From Marius Ege, 2012.)

FIGURE 18.2 Rimac River, metropolitan Lima. (From Evelyn Merino Reyna, evelynmerinoreyna@limamasarriba.com.)

Water-Sensitive Design of Open Space Systems

connected to the public water supply networks. They receive drinking water, often of very poor quality, from private water vendors at high prices.

There is no green without irrigation in Lima. Besides the seasonal biotope loma, wetlands, and sporadic green in the river corridors, all other green areas would not survive without irrigation. Although there is a limited amount of water resources and many people lack access to safe water, the potable water or groundwater is still widely used for irrigation of urban green where it can be afforded. For example, water consumption data from state water utility Sedapal for May 2011 and May 2012 show that more than 3000 parks were irrigated with potable water (Sedapal 2012, 2013).

Moreover, in 2011, only 17% of the total treated wastewater was reused, and the remaining wastewater, often of poor quality, was discharged into the rivers and ocean (Moscoso Cavallini 2011). The last monitoring of the water company Sedapal in 2011 (Salazar 2011) shows that the effluent from only four out of the 18 wastewater treatment plants (WWTP) operated by Sedapal meet the water-quality standards for the irrigation of green areas. Even though most wastewater is collected in central WWTPs, there are numerous peri-urban areas not connected to the wastewater network. Instead, silos or latrines are used or wastewater is discharged directly into rivers or irrigation channels. With the opening of La Taboada WWTP in 2013 and La Chira WWTP (currently under construction), it is expected that most of the wastewater produced in the city will be treated at least with primary treatment and the outflow discharged trough submarine outfalls into the ocean without reuse in the city.

18.2.1 Decorative Approach to the Design of Open Space

García Calderon et al. (2011) states that the planning of green areas and the ecological protection in Lima have been led by two separate disciplines: urban planning and ecology. While urban planning has focused on the green and recreational spaces within the city and urban sectors, ecology has focused on the ecosystems and biotopes outside of the urban core. The lack of integration of these two disciplines has led to the creation of decorative green patches lacking consideration of the local ecology, especially the climate, seasonal aspects, water cycle, and ecological processes (Figure 18.3). Because the design approach to urban green systems does not acknowledge the fact that Lima is a city located in and expanding into the desert, the implementation and maintenance of urban parks is very costly. At the same time, the ecological approach of protecting urban ecosystems without actively developing for human benefit cannot resist the huge development pressures on unbuilt land.

18.2.2 Open Space with High Water Demand

The merely decorative approach to the design of open spaces results in a high water demand of green areas within the city. The most frequent vegetation cover is grass (specifically Stenotaphrum secundatum), and the most commonly planted tree species are Ficus benjamina and Eucalipto grandis. Grass requires roughly three times more water (2100 L/m^2/year) than the little-used xerophytic lawn (630 L/m^2/year), which can provide similar visual qualities. Ficus and Eucalipto require roughly three times more water (1260 L/m^2/year) than the common native species Schinus molle

FIGURE 18.3 Decorative, water-intensive green areas in the desert, Carabayllo district, metropolitan Lima, 2012. (From Eisenberg, E. et al., Lima Ecological Infrastructure Strategy, Integrated Urban Planning and Design Tools for a Water Scarce City, Institute of Landscape Planning and Ecology, University of Stuttgart, 2014.)

Water-Sensitive Design of Open Space Systems 361

(530 L/m²/year) (Ascencios Templo 2012). Additionally, the water demand of green areas increases due to the installation of inefficient irrigation technologies and poor irrigation practices leading to losses by infiltration or evaporation.

18.2.3 Open Space with High Demand on Water Quality

The decorative design characterized by extended grass areas requires a high water quality standard for irrigation. Available non potable water sources often do not meet such water quality standards. Rivers with polluted water are channelized and decorative parks created along them, which have to be irrigated with potable water. Even though, in metropolitan Lima, there are experiences using treated water from irrigation channels (e.g., Surco district using water from Surco channel), in most urbanizing areas, the irrigation channels are closed and not considered as a potential water source because the polluted water does not meet irrigation water-quality standards.

The increasing water demand for the irrigation of green areas and the increasing cost of drinking water have led to the search for alternative solutions. The wastewater infrastructure is under pressure to provide treated wastewater of a quality that is sufficient for irrigation of green areas. Many district municipalities and administrative bodies started to build small wastewater treatment plants often without coordinating their actions with Sedapal or across district boundaries. These facilities have often failed to function due to the lack of knowledge and management capacity.

18.2.4 Disconnection between the Local Ecology, Open Space System, and Hydrological Infrastructure

In the recent open space projects of the municipality of Lima the vegetation covers with less water demand, efficient irrigation schemes and alternative water sources are used. Treated wastewater is considered as the main alternative water source. However, the treatment and reuse of polluted water or domestic wastewater is dealt with as a solely technological/engineering issue to be solved by water engineers. So far, the focus has been on improving the treatment technology to provide a sufficient amount and quality of treated wastewater for irrigation. The potential of integrated, system-based planning and design of open space and water infrastructure has not yet been explored. Lima's hydrological infrastructures and open space design need radical rethinking to make urban and natural systems perform in concert with one another and keep up with the increasing water and green area demand of a growing, livable city.

18.3 LIMA ECOLOGICAL INFRASTRUCTURE STRATEGY (LEIS)

In order to reverse or at least mitigate unsustainable urban development processes in cities like Lima, an urban water paradigm shift is needed. Within the Federal Ministry of Education and Research (BMBF) megacity project "Sustainable Water and Wastewater Management in Urban Growth Centres Coping with Climate Change—Concepts for Metropolitan Lima (Perú)" (LiWa), a new planning approach called "Lima Ecological Infrastructure Strategy" (LEIS) was developed.

The ecological infrastructure is based on the following hypothesis: The water cycle facilitates the creation of open spaces; it provides water and promotes diverse habitats. On the other hand, a well-established open space system can help to improve and protect the water cycle.

LEIS is a new approach towards sustainable urban development, which integrates water and wastewater management with open space planning and design. It stresses the need for adapting them to the desert city context, considering the city as a water source and the open space as a catchment area, promoting the reuse of nonpotable waters, contributing to closing the urban water cycle in the city and increasing green areas in a sustainable way, creating resilience in coping with climate change (Eisenberg et al. 2014).

18.3.1 THEORETICAL FRAMEWORK AND ADAPTATION

The LEIS draws upon the concepts of tackling challenges, such as uncontrolled urban growth and the overuse of resources with the focus on water in today's megacities, including the green infrastructure (GI) and water sensitive urban design (WSUD). The two concepts were integrated and adapted for their use in the arid climatic conditions of metropolitan Lima with the emphases on the urban water cycle and the open space system. The ecological infrastructure, consisting of natural and man-made ecosystems, considers the urban water cycle, readapts ecological processes and increases essential environmental services in the city. It is assumed that the coordinated designation of multifunctional open spaces tackles the urban development challenges more efficiently than conventional land-use planning approaches.

18.3.2 METHODOLOGY OF LEIS

LEIS is an integrative, interdisciplinary, and multi-scalar approach elaborated in three main parts: LEIS principles leading to policy recommendations; LEIS tool for citywide GIS-based analyses and localization of potentials and challenges; and LEIS manual, including water-sensitive design recommendations. The strategy was tested and applied in the demonstration area in the Lower Chillon River Watershed. To support the research project activities in the demonstration area, two academic design studios for strategic landscape planning and two international and interdisciplinary summer schools for building and testing of water-sensitive design solutions on-site were organized by the Institute of Landscape Planning and Ecology of the University of Stuttgart.

18.3.2.1 LEIS Principles

Among the main obstacles for integrated planning is the lack of collaboration of urban planners, ecologists, and water engineers. Therefore, a multidisciplinary key actor working group was set up, composed of urban policy makers; water and environmental engineers; urban planners and architects; and members of the state water utility Sedapal, Metropolitan Planning Institute (IMP), Metropolitan Park Authority

Water-Sensitive Design of Open Space Systems

(SERPAR), National and Local Water Authority (ANA-ALA), Lima and Callao regional governments, and San Martin de Porres District Municipality, among others. The principles defined within this group were harmonized with the Lima Regional Concerted Development Plan 2012–2025 (PRDC). The participatory process helped to define LEIS principles as future policies for integrative urban planning and water management to be included in planning instruments securing future implementation at different scales. The defined principles are

- Protect, develop, and implement ecological infrastructure considering availability and integral management of water resources
- Protect and consolidate agricultural land and add value to improve ecosystem performance
- Transform high-risk areas as part of the ecological infrastructure
- Promote water-sensitive urban development that considers water catchment, saving, treatment, and reuse of water in the city and develop water-sensitive urban design according to water sources
- Integral and sustainable city management, including city vision for a water-sensitive urban development with a sustainable and resilient approach

18.3.2.2 LEIS Tool

A geographical information system (GIS)–based tool was developed to store, analyze, and synthesize data layers of open space, water sources, and socioeconomic data. Satellite imagery analyses were carried out to localize and quantify green areas in the city. The tool provides meso-scale spatial analyses for localization and quantification of green areas—water demand and water sources—water supply, with the focus to localize spatial potentials. The results of the green areas study contributed to the city green inventory completed in 2012 by local authorities. The tool contains and provides the information for the spatial framework of the ecological infrastructure (Figure 18.4).

18.3.2.3 LEIS Manual

The LEIS manual contains water-sensitive urban design recommendations with the focus on water saving and water and wastewater treatment and reuse. Interactive design tools were developed to facilitate the consideration of water demand and treatment early in the design process. Finally, in order to bridge the detailed design recommendations with a larger-scale and citywide perspective, prototypical application of water-sensitive urban design for eight different urban typologies was developed. The recommendations were conceived in a multidisciplinary working group of experts in urban and landscape planning and design, urban ecology, and water treatment and reuse, supported by the Metropolitan Park Authority (SERPAR) and other institutions from which survey data were obtained. The LEIS manual will be examined in more detail in this chapter, therefore, those who provided major contributions are here mentioned and include Ing. Julio César Moscoso Cavallini, Prof. Rosa Maria Miglio Toledo de Rodriguez, and PhD candidate Maribel Zapater Pereyra.

FIGURE 18.4 Open space framework of the ecological infrastructure of metropolitan Lima, elaborated 2013. (From Eisenberg, E. et al., Lima Ecological Infrastructure Strategy, Integrated Urban Planning and Design Tools for a Water Scarce City, Institute of Landscape Planning and Ecology, University of Stuttgart, 2014.)

18.4 WATER-SENSITIVE URBAN DESIGN APPROACH

The green inventory delivers the first citywide study assessing the distribution and the percentage of "green" within the different types of registered green areas. The results show that many of the registered green areas are, in reality, partially or fully dry or built-up. The deficit of green areas in 2012, based on the ratio of 4 m²/inh, is up to 3 m² in large areas of metropolitan Lima (Figure 18.5). Based on these city analyses

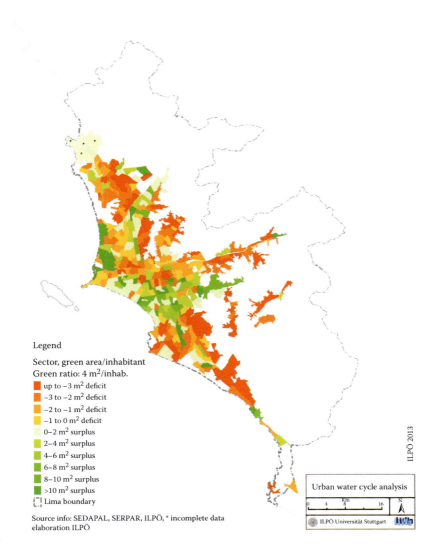

FIGURE 18.5 Deficit of green areas in metropolitan Lima in 2012. (From Eisenberg, E. et al., Lima Ecological Infrastructure Strategy, Integrated Urban Planning and Design Tools for a Water Scarce City, Institute of Landscape Planning and Ecology, University of Stuttgart, 2014.)

and the population growth rate, it can be assumed that the demand for green areas will increase in the future in order to cover the deficit of green in the existing settlements as well as in the urban expansion areas. To keep up with the increasing water demand for green areas and limited water resources of metropolitan Lima, there is a need for a shift from image-based design toward a system-based design of open spaces and hydrological infrastructure. Additionally, metropolitan Lima is characterized by an uneven distribution of socioeconomic classes over the city and uneven distribution of water infrastructure and green areas. Water-related problems and opportunities vary greatly from place to place due to different natural and urban contexts and require different solutions to integrate the urban water cycle and open space.

18.5 DESIGN METHOD FOR INTEGRATED SOLUTIONS

In order to develop site-specific water-sensitive urban design solutions, synergies have to be found and elaborated between water infrastructures, open space, and the ecology of the site to combine functions and develop new qualities and aesthetics while following safety regulations. Such an integrated approach of water-sensitive design is often difficult to realize due to the sectoral division and different working methods and approaches of the different disciplines involved, such as urban and landscape planning and design, urban ecology, and sanitation, among others. In order to overcome the different working methods of the disciplines that inhibit them from reaching integrated design solutions, a suitable working method had to be developed.

An interdisciplinary working group was set up as part of the LiWa research project that carried out a survey of 20 different open-space typologies to understand the current challenges and potentials. The selected cases represent different hydro-urban characteristics (geomorphology, water sources, state of water infrastructure, state of consolidation process, open-space typologies). On the one hand, they include different linear typologies of open space: river corridors, irrigation channel corridors within settlements, electrical lines, transit areas along metro and median strips, and hill areas without connection to water infrastructure. For the urban patches, on the other hand, the following typologies were selected: district and zonal parks, university campus, loma biotope, and wetland in the context of the built structure. The survey contained questions focusing on the following aspects: general information about the area, technical information about the open space (land use, type of vegetation, irrigation demands in dry season and wet season, type of irrigation, source of nutrients, etc.), existing natural and man-made water sources (water quality, quantity), and also aesthetic and design aspects of the water sources (visibility, accessibility, river bank structure, and if and how water is reflected in the design of the open space). Further questions were included about maintenance costs and institutional responsibilities. To prove the accuracy of the obtained data on the sites was not possible. Some of the numerical values showed little consistency and could not be used for exact water-demand calculations. However, the results provided a better understanding of the urban water system, the use of water sources, and qualitative observations. The collected information was complemented with secondary data sources, maps, aerial photographs, and photographs taken by the working group members.

Water-Sensitive Design of Open Space Systems

The collected information was the base for the design workshops of the working group. During the workshops, each surveyed case was analyzed with the focus on the interrelations of open space, hydrological process, and people, and intermediate design proposals were drawn, discussed, calculated, and reflected. Those design proposals, together with findings from the planning and design work in the demonstration area of the LiWa project, were the basis for the definition of water-sensitive design recommendations. Drawing on the potentials of the ecological processes, combined with technical solutions and spatial considerations, new solutions were proposed, which perform water saving and water-quality improvement. The results were structured in three main groups giving responses to three guiding questions:

- How to save water by design of open space?
- How to treat polluted water by design of open space?
- How to apply water-sensitive urban design in different urban settings of metropolitan Lima?

These three parts formed the structure of the LEIS Manual, which aims to facilitate an interdisciplinary process for water-sensitive urban design. Each of the three parts will be described in a separate section.

18.6 SAVING WATER BY DESIGN OF OPEN SPACE

There is a large potential to reestablish the connection between people and the arid landscape and to lower the water demand of green areas by a context-considered— water-sensitive—selection of vegetation cover and irrigation schemes. Even though in metropolitan Lima there are new projects that utilize more sustainable vegetation schemes, the challenge remains how to integrate such knowledge into the daily design practice as well as development of programs and projects. The manual provides fundamental knowledge about water demand of vegetation, impact of irrigation systems on water demand, and water-quality requirements for irrigation of green areas for the use of designers and planners. An Excel-based *"Design testing and water demand calculation tool"* was developed that includes predefined information and can be used for fast calculation of the water demand of planned green areas with different vegetation covers, irrigation schemes, and it indicates automatically the water-quality requirements. This tool serves for quick and simple testing of the feasibility of a design proposal and its variations in terms of water demand. It aims to contribute to water saving and the improvement of the local ecology by providing information on native and context-considered species for the design of open space.

18.7 TREATING POLLUTED WATER BY DESIGN OF OPEN SPACE

Water sources other than potable water, e.g., domestic wastewater, insufficiently treated domestic wastewater, or surface water, are often polluted and cause environmental degradation when disposed of into the environment. They often do not meet water-quality standards and cannot be used for irrigation of green areas without treatment.

Many different treatment systems are available, ranging from low-tech, low-cost technologies to high-tech, costly technologies. The most common low-tech technology used in metropolitan Lima is facultative lagoon (10 cases) (Moscoso Cavallini 2011). The lagoons require large amounts of space because part of the treatment is done by solar radiation; therefore, low extra energy input is necessary, and accordingly, the operational costs remain low. Constructed wetland is another low-tech technology used in Metropolitan Lima (two cases) (Moscoso Cavallini 2011). Low-tech technologies require low energy input but a great deal of space for treatment. Space is a limited resource in the cityscape and is, therefore, a limiting factor in the implementation of low-cost ecological treatment technologies.

Increasingly, high-tech ecological treatment technologies are being used in metropolitan Lima. The high-tech technologies require a small area for treatment in comparison to the low-tech technologies. The consequence of the limited space demand is the need for high input of energy for the treatment process, resulting in considerably high operational cost. The most common high-tech technology in metropolitan Lima is activated sludge (14 cases); less used are trickling filters (two cases) (Moscoso Cavallini 2011). The operational and maintenance costs of the activated sludge technology are up to three times higher than the operational and maintenance costs of facultative lagoons or constructed wetlands (Moscoso Cavallini 2009). Despite the high costs of the high-tech technologies, they are favored and increasingly implemented in parks and other green areas in order to provide water for irrigation. One of the arguments supporting their implementation is that they take little space away from green and recreational spaces (Eisenberg et al. 2014).

For water-sensitive urban design, low-tech treatment systems, including treatment lagoons, constructed wetlands, and treatment reservoirs are recommended. The treatment lagoons stabilize the organic matter and reduce the pathogens contained in the wastewater through natural processes. The constructed wetland is composed of macrophytes, a bed media, microorganisms, and wastewater. The microorganisms provide a biological treatment, and additionally, the macrophytes create green area that is irrigated by the wastewater that is being treated. Finally, the treatment reservoirs improve the water quality mainly by reducing the concentration of human pathogens (bacteria, viruses, and parasites). Water surface or vegetated surface has a high potential to be integrated into the urban surroundings.

The manual provides design guidelines that facilitate the integration of natural water flow, water infrastructure, and treatment system with the open space design. The *"Design testing and treatment space demand estimation charts"* were developed to estimate the space required for the treatment system so that it can be integrated into the design proposal from the initial stage of the design process. The understanding of the processes of the treatment systems and their technical components enables designers to shift from creation of decorative green areas with high water consumption to the creation of open spaces that treat water or wastewater and create new ecology of the site and new aesthetics of the open space. Such new treatment landscapes are very distinct from the decorative green areas the inhabitants are accustomed to, so to gain the people's acceptance of such landscape, awareness-raising process are required.

18.8 PROTOTYPICAL WATER-SENSITIVE URBAN DESIGN FOR DIFFERENT HYDRO-URBAN TYPOLOGIES

The identification of so-called hydro-urban typologies is a method to assess the distinct conditions of metropolitan Lima in terms of water characteristics and the urban landscape characteristics, which then allow for the development of site-specific water-sensitive urban design recommendations. The typologies are defined by similar characteristics of geomorphology (slope), hydrology (water sources, state of water infrastructure), and urban structure (level of consolidation, open-space typologies, and their condition). From areas studied within the working group, nine hydro-urban typologies were identified (Figure 18.6). Because the identified typologies are based on the study of a limited number of areas, they do not represent the whole territory of metropolitan Lima. Also, in the scope of the research project, it was not possible to develop recommendations for all identified typologies. The focus is on the treatment and reuse of polluted waters in the open space, no recommendations were developed for areas that have the potential for harvesting water from fog.

For the selected typologies, mappings were produced to illustrate the current hydro-urban characteristics and the water-sensitive urban development. Additionally, for each typology, a map illustrates the potential for the large-scale application of the proposed solution in the context of the whole city.

The proposed measures integrate the open-space system with the water treatment and reuse system. The solutions for treatment of domestic wastewater and grey water consider the existing wastewater infrastructure and build upon it in a number of ways:

- Addition to the existing system: Polishing parks and corridors for insufficiently treated wastewater
- Complement to the existing system in areas where it is not properly functioning, nonexistent, or difficult to implement: Domestic wastewater treatment parks and corridors for settlements not connected to WWTP, grey water treatment parks, and corridors for settlements in areas on steep hills without drinking water infrastructure
- Combined with the existing centralized system: Domestic wastewater treatment parks and corridors in areas connected to WWTP
- Independent from the existing system: Grey water treatment and reuse in large facilities in a large open space

Solutions for surface water sources include river water treatment parks and corridors and irrigation channel water treatment parks and corridors.

The integration of the open space with the treatment function and provision of green areas can strengthen the resistance of open spaces (green areas) against informal occupation, especially flood zones and areas along main streets, below electrical lines, or around archaeological sites. When the water-sensitive design recommendations are applied on a large scale in different urban contexts, a system of water-sensitive open spaces can be created, functioning as an ecological infrastructure of

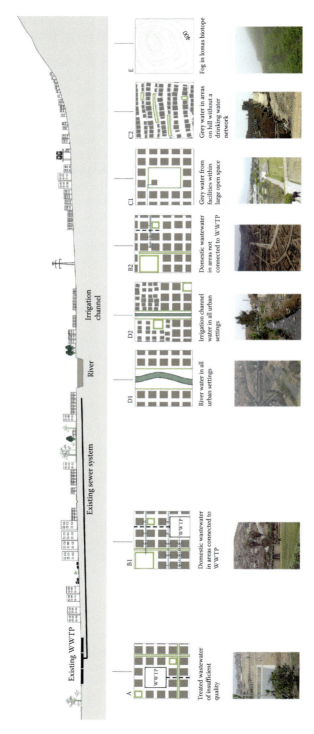

FIGURE 18.6 Overview of identified hydro-urban typologies. (From Eisenberg, E. et al., Lima Ecological Infrastructure Strategy, Integrated Urban Planning and Design Tools for a Water Scarce City, Institute of Landscape Planning and Ecology, University of Stuttgart, 2014.)

Water-Sensitive Design of Open Space Systems 371

the city and providing essential ecosystem services. This has to be assured by integrating the proposed solutions into both water infrastructure development schemes and urban development plans.

In Sections 18.8.1 to 18.8.3, the article describes three different hydro-urban typologies, their current hydro-urban characteristics, water-sensitive design development, and the potential areas for the large-scale implementation in the city.

18.8.1 POLISHING PARKS AND CORRIDORS FOR INSUFFICIENTLY TREATED WASTEWATER

At the beginning of the chapter, it was mentioned that wastewater, even after the treatment in the WWTP, often does not yet meet standards for reuse for irrigation of green areas. Such an urban situation is portrayed in which water treated in an existing WWTP is not sufficiently clean for reuse, and it is discharged into river and ocean, causing significant environmental damage. Open spaces are either dry or are irrigated with groundwater or potable water, which increases the irrigation costs. The WWTPs are commonly located within the cityscape where space is in high demand, and therefore, it is difficult to dedicate more land for an additional treatment stage (Figure 18.7a).

The water-sensitive urban design solution introduces an additional treatment step. This treatment step is here referred to as "polishing of treated wastewater." It is integrated in large parks or large linear open spaces, for example, median strips or electrical lines, which are located in the proximity of the WWTP, and therefore, no additional space has to be fenced up for the sole use of the treatment facility. The recommended technologies treatment reservoirs or constructed wetlands (for certain types of water quality) can remove the remaining pollutants from the wastewater so that it can be reused for irrigation of green areas (Figure 18.7b). The Figure 18.8 shows the locations of the WTTPs in metropolitan Lima operated by Sedapal and their outflow quality and suitability for reuse for irrigation of green areas, based on the water quality monitoring in 2011.

18.8.2 DOMESTIC WASTEWATER TREATMENT PARKS AND CORRIDORS FOR SETTLEMENTS NOT CONNECTED TO WWTP

In metropolitan Lima, about 330,000 households are not connected to a proper wastewater collection and treatment network (COOPI and IRD 2011). Households with drinking water supply dispose of their backwater into septic tanks, rivers, or irrigation channels. Households not connected to drinking water dispose their feces in latrines or dry toilets. Small amounts of grey water from washing and cooking are usually disposed of by dumping it on the street to keep the dust down on the unpaved streets. Most open spaces (e.g., district parks, median strips, electrical lines) are dry due to the lack of water (Figure 18.9a).

For such areas with a high number of households not connected to an existing WWTP, an alternative system for the treatment and reuse of domestic wastewater is proposed. A semi-decentralized approach is proposed, in which large volumes of

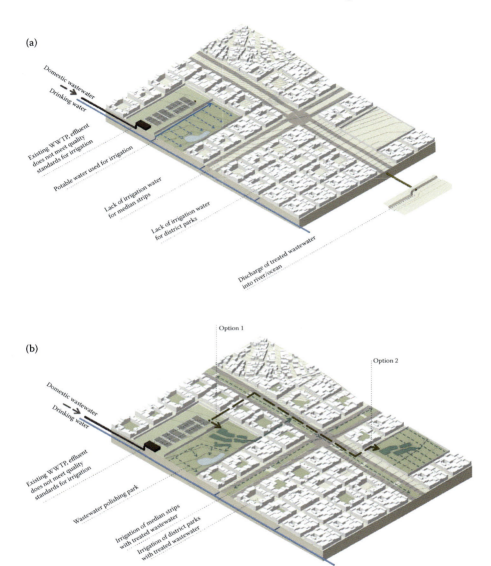

FIGURE 18.7 Hydro-urban typology: treated wastewater of insufficient quality in settlements in its proximity. (a) Portrait of current hydro-urban characteristics: treated wastewater currently does not reach quality sufficient for irrigation and is discharged into surface watercourses, instead potable water is used to irrigated green areas. (b) Portrait of water sensitive urban development: insufficiently treated wastewater is treated by ecological treatment systems integrated in recreational open space which so create wastewater polishing parks or corridors. Potable water is replaced by treated wastewater for irrigation. (From Eisenberg, E. et al. Lima Ecological Infrastructure Strategy, Integrated Urban Planning and Design Tools for a Water Scarce City, Institute of Landscape Planning and Ecology, University of Stuttgart, 2014.)

Water-Sensitive Design of Open Space Systems 373

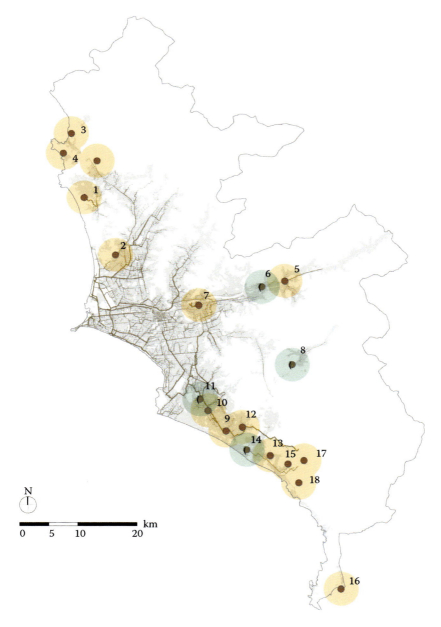

FIGURE 18.8 Location of WWTP's operated by Sedapal with indication of the effluent quality in 2011.

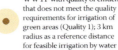

No.	District	Name of WWTP	Average total flow, July 2011 (l/s)	Effluent FC (MPN/100 mL)	BOD$_5$ (mg/L)	TSS (mg/L)
1	Ventanilla	WWTP Ventanilla	226	7.80E+05	70	23
2	San Martin de Porres	Puente Piedra	536	5.30E+06	126	315
3	Acon	WWTP Ancon	27	9.05E+06	122	65
4	Santa Rosa	WWTP Santa Rosa	8	9.00E+06	27	15
5	Ate Vitarte	WWTP Carapongo	441	9.00E+05	48	84
6	Luricagncho Chosica	WWTP San Antonio de Carapongo	18	1.80E+01	4	4
7	El Agustiono	WWTP Nueva Sede Atarjea	1	2.00E+03	65	76
8	Cieneguilla	Cienegilla	82	1.80E+01	7	3
9						
10	Villa El Salvador	WWTP Huascar / Parque 26	82	2.20E+03	82	58
11	San Juan de Miraflores	WWTP San Juan	385	1.80E+01	19	29
12	Villa Maria	WWTP José Galvez	75	2.20E+06	61	52
13	Lurín	WWTP San Pedro de Lurín	25	4.90E+06	87	60
14	Lurín	WWTP J.C.Tello	26	1.80E+01	10	47
15	Lurín	WWTP Nuevo Lurín	72	–	–	–
16	Pucusana	WWTP Pucusana	26	5.40E+06	76	153
17	Lurín	PTAR San Bartolo	821	2.90E+03	46	45
18	Punta Hermosa	Punta Hermosa	18	5.70E+06	68	109

FIGURE 18.8 (Continued) Location of WWTP's operated by Sedapal with indication of the effluent quality in 2011. (From Eisenberg, E. et al., Lima Ecological Infrastructure Strategy, Integrated Urban Planning and Design Tools for a Water Scarce City, Institute of Landscape Planning and Ecology, University of Stuttgart, 2014.)

wastewater should be collected and treated in one WWTP on the level of a district/group of neighborhoods, which can be then reused close to the place of its production. The wastewater is collected by an underground pipe system and is treated in treatment lagoons or constructed wetlands located in large open space: Domestic wastewater treatment parks or corridors are created (Figure 18.9b). Figure 18.10 shows areas with potential for such development.

18.8.3 IRRIGATION CHANNEL WATER TREATMENT PARKS AND CORRIDORS

The river valleys of metropolitan Lima have been characterized by agricultural fields with an extended network of irrigation channels. But as a consequence of the expansion of urban settlements and decreasing value of agricultural production, the agricultural land is disappearing together with the irrigation channels. However, in many consolidated areas, as well as peri-urban settlements that lack basic services and receive their potable water from water trucks at high prices, the irrigation channels are the main source of water to irrigate public parks. One of the main problems that hinders the coexistence of the irrigation channels with the city is their high water pollution caused by industrial wastewater, domestic wastewater, and solid waste discharge. The self-treatment function of the irrigation channels does not keep up with

Water-Sensitive Design of Open Space Systems

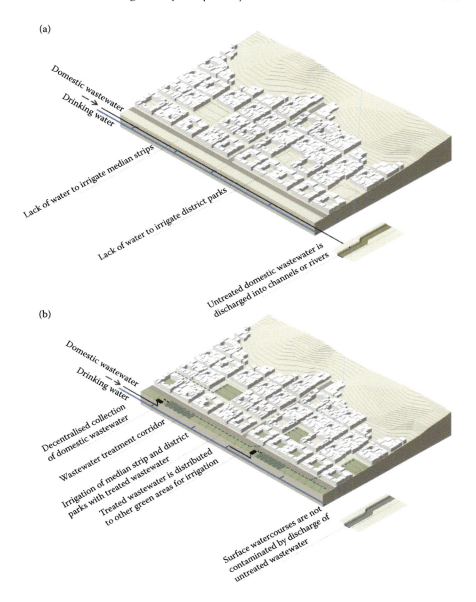

FIGURE 18.9 Hydro-urban typology: raw domestic wastewater in settlements not connected to WWTP. (a) Portrait of current hydro-urban characteristics: untreated domestic wastewater is discharged into surface watercourses. There is a lack of green public spaces. (b) Portrait of water sensitive urban development: raw domestic wastewater is collected in a decentralized manner; it is treated in ecological wastewater treatment technologies and reused for irrigation close to the area of its production. Treatment technologies are integrated into the open spaces system and create domestic wastewater treatment corridors or parks. (From Eisenberg, E. et al., Lima Ecological Infrastructure Strategy, Integrated Urban Planning and Design Tools for a Water Scarce City, Institute of Landscape Planning and Ecology, University of Stuttgart, 2014.)

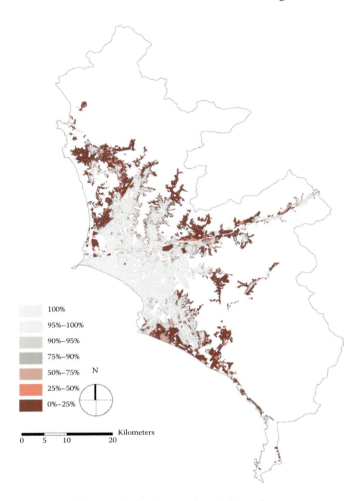

FIGURE 18.10 Potential areas for implementation of domestic wastewater treatment parks and corridors: the map shows connectivity of households to wastewater network. High potential for implementation of proposed measure is in areas without connection to the wastewater treatment shown in dark red (max. 25% households within a block) and in pink (20%–50% households within a block). (Elaboration, ILPÖ, 2013; Based on data from INEI, 2007; Connections, Coperazione Internazionale [COOPI] and Institut de Recherche pour le Développement [IRD], Estudio SIRAD: Recursos de respuesta inmediata y de recuperación temprana ante la ocurrencia de un sismo y/o tsunami en Lima metropolitana y Callao, Lima, 2011.) (From Eisenberg, E. et al., Lima Ecological Infrastructure Strategy, Integrated Urban Planning and Design Tools for a Water Scarce City, Institute of Landscape Planning and Ecology, University of Stuttgart, 2014.)

the high level of pollution, and they turn into smelly and dangerous places unwanted by the local population. Once the channels are closed, the parks previously irrigated by the water from the channels become dry (Figure 18.11a).

A water-sensitive urban design prototype is developed in which the constructed ecology—wetland or reservoir—replaces the natural treatment function of the

Water-Sensitive Design of Open Space Systems

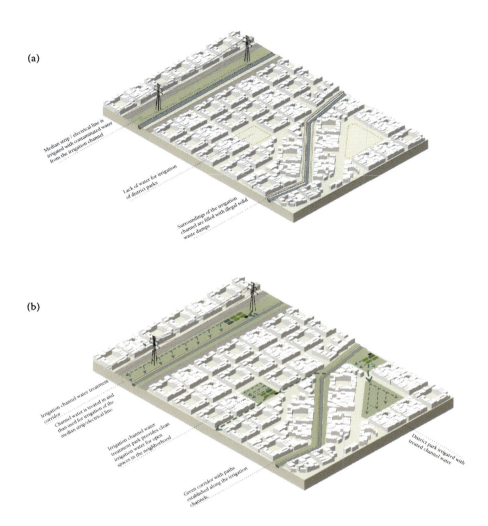

FIGURE 18.11 Hydro urban typology: irrigation channel water in settlements in its proximity. (a) Portrait of current hydro-urban characteristics: on the formally agricultural land occupied by new settlements the irrigation channels receive the informal discharge of wastewater and solid waste. They become neglected places in the urban space triggering social problems and are step by step channelized or fully closed. The previously green corridors and parks dry out with the closures of the channels. (b) Portrait of water sensitive urban development: the irrigation channels are the source of water for irrigation of green open spaces. The polluted water is treated by ecological treatment technologies integrated into open space and so Irrigation channel water treatment parks or corridors are created. (From Eisenberg, E. et al., Lima Ecological Infrastructure Strategy, Integrated Urban Planning and Design Tools for a Water Scarce City, Institute of Landscape Planning and Ecology, University of Stuttgart, 2014.)

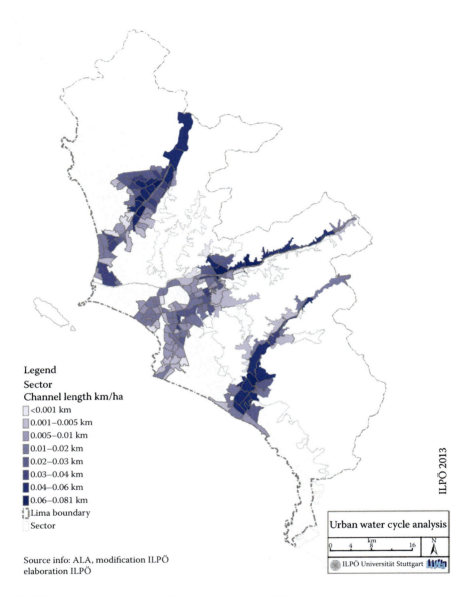

FIGURE 18.12 Potential areas for implementation of irrigation channel water treatment parks and corridors: The map shows the length of irrigation channels per spatial unit (water sector). Elaboration, ILPÖ, 2013, based on data from ALA: Surco y Nana 2003–2005; Chuquitanta, 2009. (From Eisenberg, E. et al., Lima Ecological Infrastructure Strategy, Integrated Urban Planning and Design Tools for a Water Scarce City, Institute of Landscape Planning and Ecology, University of Stuttgart, 2014.)

Water-Sensitive Design of Open Space Systems

damaged ecosystem; it provides water of quality sufficient for irrigation of parks with inedible plants, a cleaner environment, and space for recreation in the shade of trees. The irrigation channels are reconnected with the urban fabric as a multifunctional open space system with treatment and irrigation functions combined with continuous paths and recreational spaces with shade areas (Figure 18.11b). Figure 18.12 shows the potential areas in metropolitan Lima for implementation of this measure.

18.9 STRATEGIC PROJECTS IN THE LOWER CHILLON RIVER WATERSHED

The Lower Chillon River Watershed was chosen as a study area by the LiWa research project to demonstrate the water-sensitive urban development approach. The Chillon River is characterized by a challenging seasonal water regime: Between May and December, it carries no water and becomes a torrent between January and April. Poorly treated outflow from a wastewater treatment plant is flowing directly into the river. Irrigation channels diverted from the polluted Chillon River irrigate the disappearing agricultural fields in the valley. So far, planning concepts have failed to address the environmental degradation and rapid land-use change, resulting in the occupation of vulnerable areas in the flood zone, the loss of agricultural land and river flood plains and in the occupation and neglect of archaeological sites. To reverse the current development, the proposal focuses on water-sensitive urban development and its potential to serve as a framework for the sustainable future development of the Lower Chillon Watershed.

18.9.1 Toward a Water-Sensitive Future of the Lower Chillon River Watershed

A Strategic Landscape Framework Plan for the Lower Chillon River Watershed was developed, integrating landscape planning with the perspective of improving the urban water management, including social, economic, and legal aspects to guide the implementation of a water-sensitive demonstration area (Figure 18.13). This plan was presented to the local planning authorities of metropolitan Lima with the aim of integration into the formal planning instruments, such as the Land Management Plan (POT) and the Urban Development Plan for Metropolitan Lima 2035, which is currently under development.

During two international summer schools in 2012 and 2013, water-sensitive design prototypes for water saving and treatment and reuse of grey water were designed, built, and tested in an effort to promote their development on a larger scale (Figure 18.14). In parallel, the research project team developed a conceptual design proposal for a linear river park in order to demonstrate future possibilities of modern urban development while improving the natural environment, cultural landscape, and archaeological heritage. The proposal was accepted in 2012 by the Metropolitan Park Authority (SERPAR). The research project supported the elaboration of a pre-investment study, which is necessary for allocation of municipal funds.

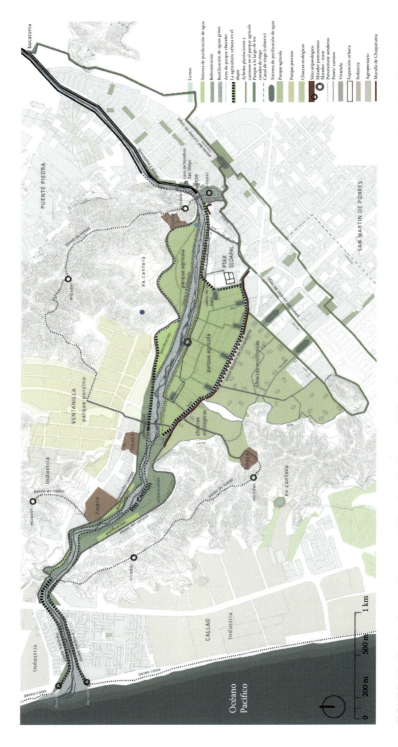

FIGURE 18.13 Landscape Framework Plan for the Lower Chillón River Watershed, metropolitan Lima, elaborated in September 2012. (From Eisenberg, E. et al., Lima Ecological Infrastructure Strategy, Integrated Urban Planning and Design Tools for a Water Scarce City, Institute of Landscape Planning and Ecology, University of Stuttgart, 2014.)

Water-Sensitive Design of Open Space Systems

FIGURE 18.14 Water-sensitive design prototype for grey water treatment and reuse, result of academic summer school in San Martin de Porres, metropolitan Lima. (From Stokman, A. et al., Lima Beyond the Park. International Summer School, Institute of Landscape Planning and Ecology, University of Stuttgart, 2012.)

The study was approved in November 2013. Additionally, the LiWa research project has designed and implemented the irrigation channel water treatment park in the area of Chuquitanta, also located in the Lower Chillon River watershed. The park was inaugurated in August 2014.

18.9.2 CHILLON RIVER LINEAR PARK

The river park design covers 300 m and an area of 11.5 ha of the south bank of the river Chillon. A new dike system is proposed, which is designed for the extreme seasonal variation from dry to medium to high water flow. The flood protection dike creates a recreational space that can be used during different water seasons. The river park integrates the treatment and reuse of the available water sources in the area (Figure 18.15). The majority of the linear park is designed as a purification corridor for polishing the effluent of the WWTP of insufficient quality. The low-tech treatment technology of constructed wetland is adjacent to the dike, which provides treatment of the polluted river water, green areas, and recreational space (Figure 18.16). At the same time, the park integrates the low-tech technology called treatment reservoirs to improve the quality of the irrigation channels in the agricultural land and serve for recreation.

FIGURE 18.15 River park design, Chillon River, metropolitan Lima. (From ILPÖ, flux Dieterle Landschaftsarchitektur, 2012 and Eisenberg, E. et al., Lima Ecological Infrastructure Strategy, Integrated Urban Planning and Design Tools for a Water Scarce City, Institute of Landscape Planning and Ecology, University of Stuttgart, 2014.)

Water-Sensitive Design of Open Space Systems 383

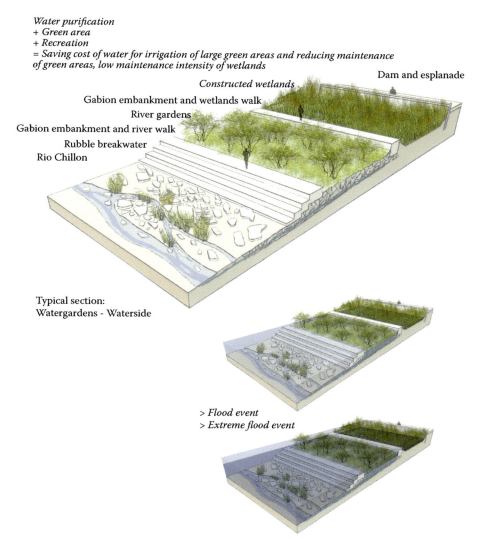

FIGURE 18.16 Polishing of insufficiently treated wastewater in the river park. (From ILPÖ, flux Dieterle Landschaftsarchitektur, 2012 and Eisenberg, E. et al., Lima Ecological Infrastructure Strategy, Integrated Urban Planning and Design Tools for a Water Scarce City, Institute of Landscape Planning and Ecology, University of Stuttgart, 2014.)

18.10 CONCLUSION

While designing an urban area, urban water flows are altered. Urban and landscape design that is based on the understanding of the urban water cycle and natural processes initiates new natural processes with new ecosystem services for the city. Such places can provide services, such as water and wastewater purification, and thus create an essential and regenerative infrastructure for the city: an ecological infrastructure. For the desert megacity Lima, an ecological infrastructure strategy was developed. Tools and methods for understanding the city as a water source and for linking the urban water cycle with open space in different urban settings were presented and applied to different contexts. Prototypical water-sensitive urban scenarios show how different types of open spaces can integrate different low-tech wastewater treatment technologies (e.g., water channel purification corridors, wastewater purification parks, etc.) and thus proactively improve the urban water cycle. A system of such water-sensitive design prototypes should create an innovative urban ecological infrastructure for Lima's future.

REFERENCES

Ascencios Templo, D. (2012). Demanda de agua en áreas verdes urbanas en Lima, University of Stuttgart, LiWa.

Coperazione Internazionale (COOPI) and Institut de Recherche pour le Développement (IRD) (2011). Estudio SIRAD: Recursos de respuesta inmediata y de recuperación temprana ante la ocurrencia de un sismo y/o tsunami en Lima metropolitana y Callao, Lima.

Eisenberg, E., Nemcova, E., Poblet, R., and Stokman, A. (2014). Lima Ecological Infrastructure Strategy, Integrated Urban Planning and Design Tools for a Water Scarce City, Institute of Landscape Planning and Ecology, University of Stuttgart.

García Calderon, J., Torres, V., Odiaga, L., and Vasquez Rodriguez, R. (2011). Bases para el Sistema Metropolitano de Áreas Verdes, Recreativas y de Reserva Ambiental de Lima.

INEI (2012). Perú: Estimaciones y Proyecciones de Población Total por Sexo de las Principales Ciudades, 2000–2015. Boletín Especial N° 23 200902389, p. 50.

Marius, E. (2011). Parque Lomas, Diplomarbeit, Institut für Landschaftsplanung und Ökologie, 2012.

Moscoso Cavallini, J. C. (2011). Estudio de opciones de tratamiento y reuso de aguas residuales en Lima Metropolitana, LiWa.

Programa de las Naciones Unidas para el Desarrollo (PNUD) (2009). Informe sobre Desarrollo Humano Perú 2009: Por una densidad del Estado al servicio de la gente, Parte I: las brechas en el territorio, Perú.

Salazar, J. (2011). Promoción de reúso de agua residuals tratadas en el riego de áreas verdes, Taller estratégico de LIWA, Sunass.

Sedapal (2012). Area-Verde-Total: Water consumption in SEDAPL serviced public green areas, Consumption from May 2011–May 2012, Lima.

Sedapal (2013). Facturación Por Agua Por Tipos De Tarifa—Sector: 1—803 (Excel-File): Total Red—No Grandes Y Grandes Clientes—Uso De La Red, September, October, November, December 2012, March 2013.

Stokman, A., Poblet, R., and Nemcova, E. (2012). Lima Beyond the Park. International Summer School, Institute of Landscape Planning and Ecology, University of Stuttgart.

Zucchetti, A., Guerrero Barrantes, J. A., Amiel Bermúdez, C., del Carmen, A., Pareja, E., Gómez Lavi, A. X., Jurado Sotelo, M. Y., Loro Ocampos, A. C., Soberón Roberts, F. M., and Ríos Pereyra, A. (2010). Reporte Ambiental de Lima y Callao 2010: Evaluación de avances a 5 años del Informe GEO, Percy Encinas Carranza (UCSUR), Fondo Editorial Universidad Científica del Sur, Lima.

19 Green Infrastructure
Performance, Appearance, Economy, and Working Method

Paulo Pellegrino, Jack Ahern, and Newton Becker

CONTENTS

19.1 Introduction 385
19.2 Challenges for Green Infrastructure Acceptance 387
    19.2.1 Performance..388
    19.2.2 Appearance 389
    19.2.3 Economy 389
    19.2.4 Working Method..389
19.3 Adaptive Design and Designed Experiments 390
19.4 Green Infrastructure, Climate Change, and the Increase in Extreme
    Rainfall in the State of São Paulo, Brazil..392
19.5 The Santo André Green Infrastructure Study: How Much Could Green
    Infrastructure Help with Urban Flood Management in a São Paulo
    Neighborhood?..393
19.6 Interdisciplinary Green Infrastructure Field Experiment, University of
    São Paulo 396
    19.6.1 Discussion of Water Quality Analysis...399
19.7 Toward an Adaptive Design and Planning Research Agenda for Green
    Infrastructure 400
References...402

19.1 INTRODUCTION

As the 20th-century infrastructure of the "developed" world degrades, rusts, and decays, it needs a more sustainable, resilient replacement (Bélanger 2013). The 20th-century modernist-functionalist urban infrastructure reflects single-purpose efficiency thinking and depends on energy-intensive materials and performance (National Academy Press 1986). If the predominately urban world of the 21st century embraces the global aspiration for sustainability, replacement of 20th-century urban infrastructure should not rely on such modernist, specialized single-purpose

385

Infrastructure type	Engineered/grey	Hybrid/landscape	Greenways/ecosystems services–oriented
References	Novotny et al., 2010; National Academy Press 1986	Margolis and Robinson 2010; Bélanger 2013; European Commission 2013; FHWA 2006, UNEP 2013	Benedict and McMahon 2006; Maryland Dept Nat Resources; The Conservation Fund 2013; Sustainable Sites Initiative 2009
Construction cost	High	High-moderate	Low
Resilience	Inflexible-rigid	Variable, including rigid and regenerative	Regenerative, living systems–based
Planning strategies	Offensive	Opportunistic, offensive	Protective
Spatial genesis	Transportation utility corridors, property ownership/easements	Interstitial, complementary	Physiographic, topographic, dendritic
Functionality	Limited/single function	Explicitly multifunctional, ecosystem services–based	Ecosystem services–based, wildlife/conservation corridors

Urban infrastructure continuum

FIGURE 19.1 An urban infrastructure continuum.

Green Infrastructure

infrastructure (Benedict and McMahon 2006; The Conservation Fund 2014). Rather it should employ a "green infrastructure" that provides a diverse and broad suite of biophysical and cultural ecosystem services and contributes to urban sustainability, assuring that infrastructure redundancy builds urban resilience—the ability to absorb disturbance and recover without changing to a different state. This green infrastructure—built on a multifunctional, performance-based foundation—holds the potential to reshape and redefine an aesthetic character that will define the cultural identity of future cities and urban landscapes. If green infrastructure can provide an "immersive, aesthetic experience, it can lead to recognition, empathy, love, respect, and care for the environment" (Meyer 2008). Another infrastructure is also needed in the emerging urban metropolis of the developing world. In the world's most rapidly growing cities and urban regions, urbanization often precedes infrastructure development of any type (Davis 2006). The world's emerging and developing cities therefore have the unique opportunity to "get it right the first time" and "leapfrog" the modernist/industrial phase of monofunctional, low-performance, unsustainable industrial infrastructure—and start de novo with a multifunctional ecosystem services–oriented green infrastructure. Landscape intervention experiments in Brazil and Argentina explored a range of working methods with residents of informal urban developments meaningfully engaged and local materials and building practices employed (Beardsley and Werthman 2008). In the context of developed and developing world infrastructure, urban infrastructure can be understood as a continuum from conventional/grey to green or ecosystem services–based (Figure 19.1). Engineered infrastructure is the status quo of the developed world. It is single purpose, expensive to build and maintain, inflexible and rigid to disturbance, and monofunctional. Because it is built for single functions, its spatial genesis is anthrocentric and employs an offensive planning strategy. Hybrid or landscape infrastructure integrates engineered with ecosystem-based systems to deliver multiple functions (FHWA 2006). Its cost can vary, depending on context and function. It often occupies interstitial urban space guided by an opportunistic or offensive planning strategy. Hybrid/landscape infrastructure seeks innovative approaches to add essential ecosystem services to conventional and familiar urban forms, including roads, parking lots, and buildings. It can be understood as the infrastructure of urban sustainability. Greenways or ecosystem-based infrastructure is a protective or pre-urban development approach that is appropriate for peri-urban and urban fringe landscapes to retain functional ecological integrity (Benedict and McMahon 2006). The urban infrastructure continuum is proposed as a rubric to explain or understand infrastructure in an ecosystem services and physical form context. It is not intended as an argument for a particular typology on the continuum.

19.2 CHALLENGES FOR GREEN INFRASTRUCTURE ACCEPTANCE

Green infrastructure is defined as "spatially and functionally integrated systems and networks of protected landscapes supported with protected, artificial and hybrid infrastructures of built landscapes that provide multiple, complementary ecosystem and landscape functions to the public, in support of sustainability" (Ahern 2007). Articulating the ecosystem services provided by green infrastructure is an emerging research theme (Center for Neighborhood Technology 2014; Landscape Architecture Foundation 2014; Sustainable Sites Initiative 2009). Green infrastructure addresses the "imperative to act" to make future urban environments more sustainable through "learning by doing" in a context of rapid urban expansion or redevelopment. Green infrastructure delivers measurable ecosystem services and benefits that are fundamental to the concept of the sustainable city (Ahern 2013; Habitat U.N. 2006). Because green infrastructure is, by definition, multifunctional, the ecosystem services concept is a useful concept to apply to explicitly identify and assess its multiple functions. Landscape architecture has an unprecedented opportunity to lead in planning and designing green infrastructure in both developed and developing urban contexts. To capture this opportunity, landscape architects will need to address challenges, including performance, appearance, economy, and working method for green infrastructure.

The introduction of the term "green infrastructure"—a counterpoint to engineered infrastructure—has raised a series of questions and inspired a set of experiments, as shown here, that promise to deepen our understanding from "ecology in the city to ecology of the city" and thereby providing the tools needed to have cities that function economically but also that provide the ecosystem services that the city needs to be sustainable (Pickett et al. 2001).

But is green infrastructure technically effective? Is it cost-effective? Is it valued by the public as attractive? Can a network of protected and adapted open spaces be assessed and measured as other infrastructure networks can? Could such assessments aid decisions that concern the best solutions for urban form to support both urban resilience and development? For replacing old, single-purpose infrastructure or as a new conception? These are the challenges that a vision of landscape as infrastructure must overcome to be truly viable and applicable in different urban contexts and stages of development.

19.2.1 PERFORMANCE

The multiple functions provided by green infrastructure can be understood as ecosystem services (Habitat U.N. 2006). Although these functions and ecosystem services are conceptually well understood, empirical measurements of specific functional performance are not familiar to most design professionals nor are they routinely collected or analyzed. Two recent initiatives in the United States promote monitoring of green infrastructure performance. The Landscape Performance Series is a program sponsored by the Landscape Architecture Foundation to document and assess the ecological/ecosystem performance of specific landscape projects. The series provides online resources and tools for designers to evaluate ecosystem performance and make the case for sustainable landscapes (Landscape Architecture Foundation 2014). The LAF series classifies benefits into seven categories: land, water, carbon energy and air quality, habitat, economic, materials and waste, and social. The Sustainable Sites Initiative is a "green certification" program that rates and classifies landscape projects for their overall ecological and environmental performance system for sustainable landscape certification. The initiative provides a set of economic, environmental, and social arguments for the use of sustainable landscape practices

Green Infrastructure 389

(Sustainable Sites Initiative 2009). In answer to this challenge for performance measures, this chapter presents an original case study with empirical testing and monitoring for storm water treatment alternatives in São Paulo, Brazil.

19.2.2 APPEARANCE

Green urban infrastructures are challenged to provide a recognizable visual, experiential expression of urban sustainability and resilience. As the modernist infrastructures manifested an industrial aesthetic and ideology, green infrastructure may give form and meaning to sustainable urban landscapes. To be accepted by the public and to realize long-term public support, sustainable cities of the future need to have both sound ecosystem-service performance and provide high-quality aesthetic and cultural character that is recognized, understood, and valued by the general public and by local resident stakeholders (Musacchio 2011; Nassauer et al. 2009). Elizabeth Meyer (2008) asks, can landscape increase the sustainability of the biophysical environment through the experience it affords?

Green infrastructure employs living systems, sharing the biotic capacity to process information and learn. This intrinsic regenerative ability derives from both abiotic and biotic processes (Margolis and Robinson 2010). When this biotic/regenerative capacity of green infrastructure is legible, it contributes to a new urban aesthetic of sustainability in which performance is integral with appearance.

19.2.3 ECONOMY

The new green infrastructure must function efficiently in economic terms if it is to be sustainable. Infrastructure must support each city in realizing its economic potential. Green infrastructure is often less expensive than conventional engineering (European Commission 2013). If green infrastructure is economically inefficient, it will likely fail as a passing fashion or green indulgence. Both the Sustainable Sites Initiative and the Landscape Performance Series in the United States emphasize explicit measures of green infrastructure economic performance, including cost comparisons with conventional infrastructure systems (Sustainable Sites Initiative 2009; LAF 2014; City of New York Parks and Recreation 2013). Cities that have actively pursued sustainability through green infrastructure practice cost accounting to monitor economic cost and performance.

19.2.4 WORKING METHOD

To achieve high-performance infrastructure that is supported by the public and contributes to cultural identity will require new "transdisciplinary" working methods engaging all design professions with stakeholders and decision makers—so that all contribute to and are invested in the GI solutions. Without legitimate, tested, and verified performance, green infrastructure may ultimately fail in the realm of public and political acceptance. A truly transdisciplinary approach engages design professionals, scientists, decision makers, and stakeholders and integrates an inclusive diversity of human perspectives and values throughout the design or planning process. In an interdisciplinary working method, design professionals may consult with stakeholders at the "scoping" stage of a project and again to consider alternative proposals. In a transdisciplinary approach, knowledge and information is generated and flows multilaterally and continuously among the participants (Tress et al. 2005). Consequently, a transdisciplinary approach is more likely to be understood and valued by the urban community over time.

19.3 ADAPTIVE DESIGN AND DESIGNED EXPERIMENTS

Adaptive design is a transdisciplinary method for planning and design that addresses uncertainty and promotes innovation through a "learning by doing" approach in which designs and plans are conceived as experiments that can adapt if the results fail or to learn new approaches and practices if the experiments succeed (Ahern 2011; Felson and Pickett 2005; Jansson and Lindgren 2012; Nassauer and Opdam 2008). Adaptive design tests, in situ, new solutions and combinations of functions. And it is well suited to "safe-to-fail" design experiments in which innovations are implemented and monitored in an experimental but fine-scale mode in which the possibility of failure is understood and acceptable. Because these small-scale designs are true experiments, the possibility of failure is real, but the risks of failure are explicitly understood and accepted by decision makers and stakeholders (Ahern 2011). Urban design experiments are defined as collaborations among scientists, planners, and designers collaborating to insert experiments into the urban mosaic, balancing ecological goals with context, aesthetics, amenity, and safety (Felson and Pickett 2005; Lister 2007; Musacchio 2011; Pickett et al. 2004; Rottle and Yocum 2011).

Adaptive design starts by articulating an explicit set of planning or design goals or objectives, presented as specific, desired ecosystem services, for example, hydrology, climate mitigation, habitat provisioning, or cultural/social services. These goals are then prioritized in a transdisciplinary process, including scientists, design professionals, decision makers, and stakeholders. The prioritization decision incorporates social concerns and economic and scientific feasibility (Dickinson et al. 2012). With the goals prioritized, designs are structured as "safe-to-fail" experiments to test alternative spatial configurations and treatments. Indicators and metrics are identified to measure the performance of the respective design experiments in terms of the specific, prioritized ecosystem services. These metrics may include, among others, water-quality parameters, storm water infiltration rates, observations of social use and activity, biodiversity indicators, and economic cost-effectiveness. At this point, the data obtained from monitoring of design experiments can support an adaptive design process in which designs or management practice are modified, "adapted" based on the performance of the design experiment. Because specific goals were preestablished for particular ecosystem services, the monitoring results will show to what extent the built experiments produced the desired results. If the results meet expectations, the design experiment is confirmed, and perhaps an innovative approach has been validated. If the results do not meet the desired goals, the design experiment is not a failure but rather an example of "learning by doing." Figures 19.2 and 19.3 illustrate monitoring methods and systems to monitor the ecosystem service performance of green infrastructure for water-quality improvement and biodiversity support.

Green Infrastructure

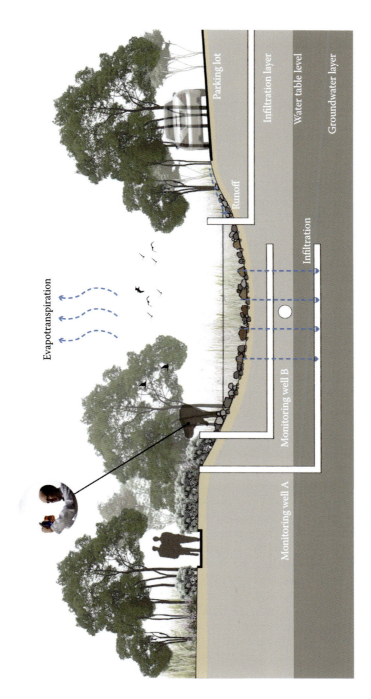

FIGURE 19.2 Monitoring water quality with monitoring wells installed at multiple depths to test the performance of rain-garden soils to improve storm water quality. (Image courtesy of Zhangkan Zhou.)

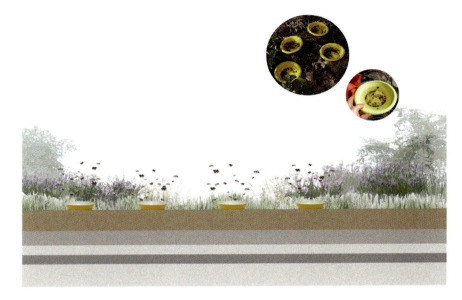

FIGURE 19.3 Monitoring insect pollinators in green infrastructure. The pan-trap method attracts insect pollinators with yellow-colored pans filled with soapy water. The number and diversity of species can be monitored at regular intervals to document the pollinating ecosystem service, for example, from a green roof. (Image courtesy of Bin Liu.)

19.4 GREEN INFRASTRUCTURE, CLIMATE CHANGE, AND THE INCREASE IN EXTREME RAINFALL IN THE STATE OF SÃO PAULO, BRAZIL

The most common motivation for and the application of urban green infrastructure is to manage storm water quantity and water quality (White 2010). This water-centric trend is likely to continue because hydrology is a fundamental physical and ecological process: All life depends on water; government regulations address water resources, water transports materials and nutrients, and cities are increasingly facing challenges to manage larger amounts and frequencies of extreme rainfall (Novotny et al. 2010).

The pattern of rainfall in southeastern South America analyzed by Re and Barros (2009) from 1959 to 2002 found an increasing trend in annual rainfall and increases in the frequency of high-precipitation events, defined as 50–150 mm/day. A specific study for the state of São Paulo, Brazil, from 1950 to 1999 identified a significant increase in the number of days with rainfall above 20 mm and of the maximum rainfall in five-day periods (Dufek and Ambrizzi 2008). Future climate change scenarios for South America developed by the Intergovernmental Panel on Climate Change (Alley et al. 2007) and Grimm (2011) predict a significant increase in extreme precipitation events. In the state of São Paulo, the concurrences between simulations and recent empirical observations of extreme precipitation events reinforce the likelihood of increased intensity and frequency of extreme climatic events in the future (Marengo et al. 2009). Projections by the Brazilian National Institute for Space Research (INPE) for the Metropolitan Area

Green Infrastructure 393

of São Paulo (MASP) between 2070 and 2100 predict a doubling in the number of days with over 10 mm of rainfall (Nobre et al. 2010). There is therefore a clear consensus regarding the increase in the frequency and intensity of the heaviest rains in the state and metropolitan region of São Paulo. This trend provides a strong rationale to reevaluate existing storm water management policies and practices and to consider a green infrastructure approach. Here an adaptive approach to green infrastructure planning and design could be an important part of the solution—promoting "learning by doing" and design experiments to promote a culture of experimentation and innovation.

The current scenario of rapid urban development and urban sprawl in a context of increasing heavy precipitation has favored conventional solutions to urban drainage. Because these conventional, monofunctional engineered solutions have provided effective flood control until recently, they have gained respect and acceptance and have been widely implemented (Marengo et al. 2009). The promise of control and precise predictability supports a flood-management philosophy favoring rapid conveyance to detention reservoirs without prior treatment. However, these systems have already reached or exceeded their capacity without consideration for expected increases in extreme precipitation associated with climate change. This conveyance-based, monofunctional flood control infrastructure causes significant impacts including non-point source water pollution and sedimentation in the receiving rivers. In addition, this conventional infrastructure provides few, if any, of the collateral ecosystem services that are provided by green infrastructure.

19.5 THE SANTO ANDRÉ GREEN INFRASTRUCTURE STUDY: HOW MUCH COULD GREEN INFRASTRUCTURE HELP WITH URBAN FLOOD MANAGEMENT IN A SÃO PAULO NEIGHBORHOOD?

The 45 ha Santo André sub-basin is located within the São Paulo metropolitan area. Flood control in this basin is provided by detention reservoirs that were built in 1991. Although the reservoirs have functioned well for their intended flood-control function to date, their effectiveness in a context of increased climate change–related precipitation increases, as discussed previously, is uncertain. Analysis of the water quality of the detention reservoirs revealed the presence of pollutants, including heavy metals and high concentrations of total coliform bacteria. In addition, the detention reservoirs have a bad odor and a general unpleasant aspect. Analysis also confirmed that the reservoir and stream pollution worsens with rainfall events, transporting a myriad of non-point source pollutants to accumulate in the reservoir basins. The limitations of the monofunctional flood control approach are becoming significant and recognized by the local population and by São Paulo city officials.

To explore green infrastructure–based flood-control alternatives for Santo André, multiple bio-retention and infiltration strategies were simulated and modeled for the neighborhood. The potential volume of storm water storage was simulated for all local streets inside the catchment area for the reservoirs. In the simulations, runoff was reduced by a proposed narrowing of the paved surface of the streets and using traffic-calming practices (Figures 19.4 and 19.5). To

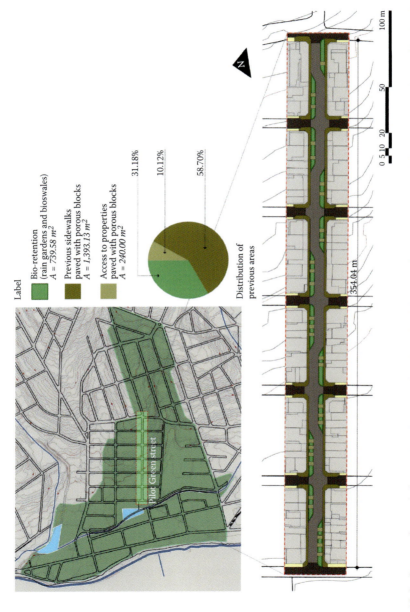

FIGURE 19.4 Proposed green infrastructure demonstrating the potential to introduce pervious green infrastructure to the neighborhood.

Green Infrastructure

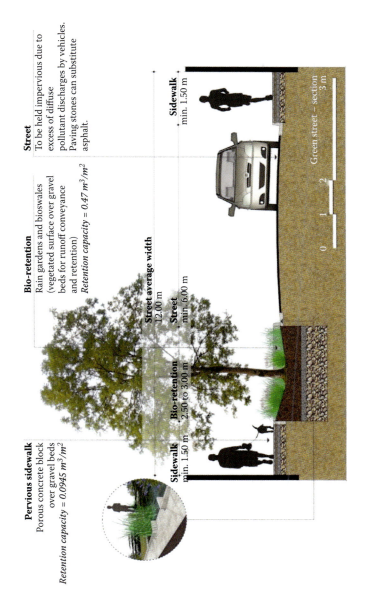

FIGURE 19.5 Proposed modifications for Rua Santarem to enhance infiltration and reduce runoff in the Santo André neighborhood, São Paulo.

increase infiltration, simulations were made for subsurface gravel, rain gardens, bioswales, and permeable paving on sidewalks and building lots. The areas of these bioretention and infiltration elements were summed, and the percentages identified were replicated for other streets through all the urban catchment area. The retention volume was calculated for each retention element and pavement construction. Even without taking into account the natural terrain infiltration and evapotranspiration and planting substrate to be held by the vegetation cover, the resulting volume of runoff that can be stored by this landscape infrastructure could add more than 42% to the capacity of the traditional detention reservoir techniques. In addition to the intended storm water management services, this landscape infrastructure would provide a broad suite of ecosystem services, including mitigating the urban heat island effect, providing wildlife habitat, and improving the aesthetic character of the neighborhood.

With the once-extraordinary climate events now occurring more frequently, it's becoming clear that cities need to prepare. One option is to use green infrastructure to clean water, defend from floods, cool the weather, enhance the ecosystems, and define neighborhood character. The challenge is to employ natural systems and processes for their ecosystem-services benefit and to integrate these natural processes with the urban fabric to make the urban landscape a sustainable and sheltering garden.

19.6 INTERDISCIPLINARY GREEN INFRASTRUCTURE FIELD EXPERIMENT, UNIVERSITY OF SÃO PAULO

A field test was designed and installed in 2012 at the University of São Paulo to test the performance of alternative rain garden/bio-retention plantings for water-quality retention and improvement. This test was part of a research collaboration between the programs of landscape and environment and hydraulic and environmental engineering at the University of São Paulo. The experiment was designed to address the specific climatic conditions in São Paulo to apply regionally familiar construction materials and practices and to examine the differential performance of native versus exotic plants in the rain garden (Dufek and Ambrizzi 2008; Moura et al. 2008). The experiment includes electronic, real-time, continuous monitoring of 21 water-quality parameters.

This test evaluated storm water runoff samples collected in its curb inlet and paired outlets to measure the efficiency of alternative treatments to improve water quality (Figure 19.6). The experiment consists of two hydrologically independent vegetated plots connected to the street gutter through a concrete channel. Each plot has its own spillway, where "outlet" samples are collected for laboratory analysis and compared to "input" street gutter runoff samples. During periods of high runoff, the surplus is directed to a drainage channel located immediately adjacent to the plots. The plots are installed in hydrologically sealed concrete basins and equally structured with 60 cm of crushed stone at the base, overlaid by 15 cm of gravel, geotextile fabric, then 5 cm of coarse sand, and, finally, 45 to 75 cm of planting substrate with side slopes (Figure 19.7). The soil surface is covered with an organic mulch. The identical substrates in each plot support two types of plantings:

Green Infrastructure

FIGURE 19.6 Inlet to the University of São Paulo bio-retention experiment.

FIGURE 19.7 Paired bio-retention cells, University of São Paulo bio-retention experiment, mixed planting (M) on right and lawn (L) on left.

- Plot 1–mixed garden (M): ground covers with a predominance of native shrubs and herbaceous vegetation distributed in wetter (bottom) and drier (side slopes) zones. Diversification and proximity among species is aimed at stimulating competition, densification, biomass increase, and complete rooting throughout the planting substrate.

- Plot 2–lawn (L): covered only with emerald grass carpet (Zoysia japonica), which has been extensively used for lawns throughout Brazil. This planting followed the same shape as in the mixed garden with side slopes and bottom.

Three sets of samples were collected during precipitation events in the following sequence: gutter (G), mixed garden (M), and lawn (L). In each monitored rain event, an average of four to six samples for each point were collected with a total of 12 to 18 samples. The interval between collections and between each set of samples is the shortest possible, estimated at approximately five minutes due to the time required to fill and pack the set of sampling vials. One hundred sixteen samples have been collected during seven rain events from March 2012 to March 2013 (Tables 19.1 and 19.2).

TABLE 19.1

Water-Quality Results from Sampling at Gutter (G), Mixed-Planted Cell (M), and Lawn-Planted Cell (L), (March 27, 2012) Sets 1 and 2

		Set 1			Set 2		
Sample Collection Point		G	M	L	G	M	L
Water Quality Parameter	**Unit**	**16:05**	**16:15**	**16:25**	**17:05**	**17:12**	**17:19**
Total alkalinity	mg/L	13	61	112	15	65	107
pH		8.64	8.70	8.87	8.91	8.57	8.77
Conductivity	µS/cm	173.00	253.00	160.00	47.00	250.00	154.00
Hardness	mg/L	18.111	38.381	67.072	17.961	38.709	65.987
Calcium (Ca)	mg/L	5.04	11.51	20.53	5.05	11.54	20.09
Magnesium (Mg)	mg/L	1.59	2.78	4.56	1.54	2.85	4.56
Iron (Fe)	mg/L	10.42	0.321	0.331	10.02	0.198	0.393
Chrome (Cr)	mg/L	0.020	0.013	0.010	0.020	0.005	<0.006
Zinc (Zn)	mg/L	0.267	0.026	0.021	0.252	0.070	0.028
Copper (Cu)	mg/L	0.046	0.008	0.006	0.046	0.005	0.015
Cadmium (Cd)	mg/L	0.003	0.0002	0.0002	0.003	0.0003	0.0001
Chloride	mg/L	2.21	2.56	2.59	2.86	2.61	2.66
Sulfide	mg/L	0.047	0.032	0.012	0.038	0.040	0.018
Fluoride	mg/L	0.32	0.40	0.62	0.30	0.53	0.60
Biochemical oxygen demand	mg/L	10	<2	2	30	<2	<2
Nitrate (NO_3)	mg/L	0.65	<0.07	0.07	0.58	0.35	<0.07
Nitrite (NO_2)	mg/L	0.12	<0.02	<0.02	0.12	<0.02	<0.02
Total organic carbon	mg/L	18.10	2.10	2.90	28.40	2.70	2.80
Oils and greases	mg/L	8	<5	2	11	2	<5
Total suspended solids	mg/L	17	<5	18	17	2	7
Total dissolved solids	mg/L	104	152	96	28	150	92

Source: Operator Environment. Analysis Bulletin No. 4166/2012-1.0.

Green Infrastructure 399

TABLE 19.2

Comparison of Results of Water-Quality Samples Showing Increases and Decreases in the Parameters Analyzed (G_0–M_f = gutter to mixed planting, G_0–L_f gutter to lawn planting) (March 27, 2012)

Indicator	G_0–M_f	G_0–L_f
Total alkalinity	415.385%[a]	707.692%[a]
pH	−3.009%[b]	−4.630%[b]
Biochemical oxygen demand	−90.000%[b]	−80.000%[b]
Total organic carbon	−86.740%[b]	−85.635%[b]
Nitrate (NO_3)	−89.231%[b]	−89.231%[b]
Nitrite (NO_2)	−83.333%[b]	−83.333%[b]
Oils and greases	−37.500%[b]	−25.000%[b]
Iron (Fe)	−97.466%[b]	−95.633%[b]
Chrome (Cr)	−85.000%[b]	−80.000%[b]
Zinc (Zn)	−91.386%[b]	−91.011%[b]
Copper (Cu)	−80.435%[b]	−73.913%[b]
Cadmium (Cd)	−43.333%[b]	−96.667%[b]
Conductivity	−10.405%[b]	46.821%[a]
Hardness	126.774%[a]	266.015%[a]
Calcium (Ca)	137.897%[a]	300.595%[a]
Magnesium (Mg)	101.887%[a]	188.050%[a]
Chloride	4.977%[a]	38.009%[a]
Sulfide	740.426%[a]	−59.574%[b]
Fluoride	65.625%[a]	87.500%[a]
Total suspended solids	−52.941%[b]	−94.118%[b]
Total dissolved solids	−10.577%[b]	46.154%[a]

[a] Increase.
[b] Decrease.

19.6.1 Discussion of Water Quality Analysis

Both lawn (L) and mixed garden (M) treatments showed very significant reductions (mostly above 80%) in biochemical oxygen demand, total organic carbon, nitrate, nitrite, iron, chrome, zinc, and copper. This pattern of improvement in the quality of storm water runoff remained in the other monitored rain events, attesting to the efficiency of the best management practices (BMPs) under evaluation for most of the parameters. However, there were significant increases in total alkalinity, hardness, calcium, magnesium, and fluoride from both plots.

Considering the water-quality parameters selected, overall, the mixed garden (M) was more efficient than the lawn (L) in reducing the concentration of pollutants.

However, the lawn plot (L) was also effective in reducing pollution concentrations and outperformed the mixed garden (M) for reducing total suspended solids and sulfide.

As it is difficult to measure precisely the amount of pollutants from runoff, it is difficult to determine the effects of best management practices in the field. Most of the experiments and research on this issue have used laboratory-controlled environments. In the field experiment described here, the conditions of the urban realm are analyzed with no artificial interference. We can assume that similar positive findings would occur if the results of the experiment were adopted as a best management practice.

This experiment has provided scientific answers that demonstrate the effectiveness of the performance of both lawn and mixed vegetated surfaces to retain and treat storm water. With proper experimental design, this model could be routinely tested in green infrastructure projects as "designed experiments." This research should complement, not replace, conventional methods of urban drainage in local cities. Green infrastructure may be integrated with other types of infrastructure as illustrated in Figure 19.1, an urban infrastructure continuum. Changing the current storm water management paradigm in Brazilian urban areas demands a transdisciplinary alliance between private and public interests, scientists, engineers, and designers to test innovative practices of runoff retention and treatment. In a transdisciplinary approach, neighborhood residents and decision makers would be involved from the initial planning to the evaluations of alternative treatments to the detailed design decisions, importantly, to learn their impressions and opinions of the built projects over time. The collateral ecosystem services should also be monitored to learn if such green infrastructure provides multiple benefits, for example, urban biodiversity, local urban climate, and public preferences among alternatives.

In Brazil, national surface water-quality regulations only address point sources of pollution. Non-point sources from diverse sources of water pollution are not yet regulated but are known to cause substantial degradation of urban water quality. Recently published São Paulo storm water management manuals describe bio-retention typologies in isolation, focusing on their importance as elements of engineering but not of the urban landscape. For green infrastructure to be accepted as a practice to improve non-point urban water pollution as well as to provide other ecosystem services in São Paulo, it needs to be researched further with monitoring of its expected ecosystem services.

19.7 TOWARD AN ADAPTIVE DESIGN AND PLANNING RESEARCH AGENDA FOR GREEN INFRASTRUCTURE

This chapter has proposed an adaptive approach to planning and designing urban green infrastructure in which urban infrastructure, and "designed experiments" can be deployed to explore and measure its performance in public, visible locations under local ecological and climatic conditions. It argues for a redefinition of infrastructure as an emerging convergence of resources, processes, and ecosystem services that support 21st-century urbanization (Bélanger 2013). Water is the focus of much attention here because water is understood as an equally life-sustaining and erosive-hazardous force (Margolis and Robinson 2010). The acceptance of urban green infrastructure as an alternative system faces multiple challenges. The performance that green infrastructure claims to provide must be rigorously monitored over time. A culture of monitoring needs development as does an "ecosystem services assessment toolbox" with an appropriate set of indicators, methods, and monitoring protocols. Acceptance of green infrastructure depends on its provision of multiple ecosystem services, including public acceptance and aesthetic preference. To be sustainable, green infrastructure needs to provide humans with meaning, beauty, and delight (Meyer 2008). These goals warrant and deserve appropriate research methods and discussion. As economics is one of the three pillars of sustainability, so it must be included in measures of green infrastructure performance. The European Commission has adopted a strategy to promote green infrastructure citing its economic benefits as a justification (EU 2013). Adapted and new economic cost-benefits are needed to compare the lifetime costs of alternative infrastructure versus conventional. Green construction certification programs and professionally sponsored research are starting to address this challenge (Landscape Architecture Foundation 2014; Sustainable Sites Initiative 2009).

If there are other ways to equip cities, other kinds of open space design and open space networks that can somehow give rise to more resilient and pleasant cities, then we may be more inclined to ask what is so special about the existing infrastructure in our cities that cannot be a transition for a new one? Instead of a traditional centralized solid infrastructure, which is inherently fragile, can we now imagine something different, something that can synthesize nature and the city, ecology and design?

Because green infrastructure aspires to provide multiple ecosystem services, no single discipline can dominate it. A transdisciplinary approach is appropriate in which scientists, engineers, designers, planners, decision-makers, and local stakeholders are all involved, not only as consumers of knowledge, but as contributors and collaborators. Transdisciplinarity is arguably the appropriate modus operandi for sustainability but remains a rudimentary aspiration in most contemporary practice. The empirical field experiment in São Paulo presented here compared the performance of two green infrastructure options for removing specific contaminants from storm water runoff. Because the experiment was robust in its design, data collection, and analysis, the results are both useful and defensible but only for water-quality performance. Other experiments could be designed to measure and compare green infrastructure performance for economic cost-benefits, biodiversity support, climatic effects, or public preference. The São Paulo field experiment provides a model for interdisciplinary research on green infrastructure performance. To be truly transdisciplinary, green infrastructure experiments such as this should involve local stakeholders and public decision makers from the conception of what to measure, who to involve in the experiment, and how the results will be used—and by whom?

For the concept of green infrastructure itself to be sustainable, it will need to demonstrate and document its ecosystem service performance and economic value continuously and repeatedly—in every specific location that it is employed, including the cultural and social benefits associated with human use and aesthetic appreciation for the urban environment.

REFERENCES

Ahern, J. (2007). Green infrastructure for cities: The spatial dimension. In *Cities of the Future: Towards Integrated Sustainable Water and Landscape Management*, V. Novotny and P. Brown (Eds.), 267–83. London: IWA.

Ahern, J. (2011). From fail-safe to safe-to-fail: Sustainability and resilience in the new urban world. *Landscape and Urban Planning 100 no. 4*:341–3.

Ahern, J. (2013). Urban landscape sustainability and resilience: The promise and challenges of integrating ecology with urban planning and design. *Landscape Ecology 28 no. 6*:1203–12.

Alley, R.; Berntsen, T.; Bindoff, N. L.; Zhenlin, C. et al. (2007). Summary for Policymakers: Contribution of Working Group I to the Fourth Assessment Report of the Intergovernmental Panel on Climate Change. Geneva: IPCC/UNEP/WMO.

Beardsley, J. and Werthman, C. (2008). Dirty work. *Topos 64*:36–43.

Bélanger, P. (2013). *Landscape Infrastructure: Urbanism Beyond Engineering*. Wageningen: Wageningen Universiteit en Researchcentrum.

Benedict, M. A. and McMahon, E. T. (2006). *Green Infrastructure: Linking Landscapes and Communities*. Washington: Island Press.

Center for Neighborhood Technology (2014). The Value of Green Infrastructure: A Guide to Recognizing Its Economic, Environmental and Social Benefits. http://www.cnt.org /repository/gi-values-guide.pdf (accessed February 5, 2014).

City of New York Parks and Recreation (2013). New York City High Performance Landscape Guidelines: 21st Century Parks for NYC. http://www.nycgovparks.org/sub_about/go _greener/design_guidelines.pdf (accessed February 5, 2014).

The Conservation Fund (2014). Green Infrastructure. http://www.conservationfund.org/our -conservation-strategy/focus-areas/green-infrastructure/ (accessed February 5, 2014).

Davis, M. (2006). *Planet of Slums*. London: Verso.

Dickinson, J. L.; Shirk, J.; Bonter, D. et al. (2012). The current state of citizen science as a tool for ecological research and public engagement. *Frontiers in Ecology and the Environment 10 no. 6*:291–7.

Dufek, A. S. and Ambrizzi, T. (2008). Precipitation variability in São Paulo State, Brazil. *Theoretical and Applied Climatology 93*:167–78.

European Commission (2013). Environment: Investing in green infrastructure will bring multiple returns to nature, society and people. Press Release, Brussels. http://europa.eu /rapid/press-release_IP-13-404_en.htm (accessed May 6, 2013).

Felson, A. J. and Pickett, S. T. A. (2005). Designed experiments: New approaches to studying urban ecosystems. *Frontiers of Ecology and the Environment 3 no. 10*:549–66.

Grimm, A. M. (2011). Interannual climate variability in South America: Impacts on seasonal precipitation, extreme events, and possible effects of climate change. *Stochastic Environmental Research and Risk Assessment 25*:537–54.

Habitat, U. N. (2006). *State of the World's Cities. The Millennium Development Goals and Urban Sustainability*. London: Earthscan.

Jansson, M. and Lindgren, T. (2012). A review of the concept 'management' in relation to urban landscapes and green spaces: Toward a holistic understanding. *Urban Forestry & Urban Greening 11 no. 2*:139–45.

Landscape Architecture Foundation, LAF (2014). Landscape Performance Series. http://www .lafoundation.org/research/landscape-performance-series/ (accessed February 5, 2014).

Lister, N. M. (2007). Sustainable large parks: Ecological design or designer ecology? In *Large Parks*. G. Hargreaves and J. Czerniak, (Eds.), 35–54. New York: Princeton Architectural Press.

Green Infrastructure

Marengo, J. A.; Jones, R.; Alves, L. M.; and Valverde, M. C. (2009). Future change of temperature and precipitation extremes in South America as derived from the PRECIS regional climate modeling system. *International Journal of Climatology 29 no. 15*:2241–55.

Margolis, L. and Robinson, A. (2010). *Living Systems. Innovative Materials and Technologies for Landscape Architecture*. Basel: Birkhäuser Verlag.

Meyer, E. K. (2008). Sustaining beauty: The performance of appearance, a manifesto in three parts. *Journal of Landscape Architecture no. 1*:6–23.

Moura, N. C. B.; Pellegrinio, P. R. M.; and Martins, J. R. S. (2008). Transição em infraestruturas urbanas de controle pluvial: Uma alternativa de adaptação às mudanças climáticas." 11° ENEPEA, Campo Grande, MS.

Musacchio, L. R. (2011). The grand challenge to operationalize landscape sustainability and the design-in-science paradigm. *Landscape Ecology 26 no. 1*:1–5.

Nassauer, J. I. and Opdam, P. (2008). Design in science: Extending the landscape ecology paradigm. *Landscape Ecology 23 no. 6*:633–44.

Nassauer, J. I.; Wang, Z.; and Dayrell, E. (2009). What will the neighbors think? Cultural norms and ecological design. *Landscape Urban Planning 92*:282–92.

National Academy Press (1986). *Engineering Infrastructure Planning and Modeling*. Washington, D.C.

Nobre, C. A.; Young, A. F.; Saldiva, P. et al. (2010). Vulnerabilidade das Megacidades Brasileiras às Mudanças Climáticas: Região Metropolitana de São Paulo—Sumário Executivo. INPE/UNICAMP/USP/IPT/UNESP. Julho, 2010. http://mudancasclimaticas .cptec.inpe.br/~rmclima/pdfs/publicacoes/2010/SumarioExecutivo_megacidades.pdf (accessed May 1, 2011).

Novotny, V.; Ahern, J.; and Brown, P. (2010). *Water-Centric Sustainable Communities: Planning, Retrofitting, and Building the Next Urban Environment*. Hoboken, NJ: John Wiley & Sons.

Pickett, S. T. A.; Cadenassso, M. L.; and Grove, J. M. (2004). Resilient cities: Meaning, models, and metaphor for integrating the ecological, socio-economic, and planning realms. *Landscape and Urban Planning 69 no. 4*:369–84.

Pickett, S. T. A.; Cadenassso, M. L.; Grove, J. M. et al. (2001). Urban ecological systems: Linking terrestrial ecological, physical, and socioeconomic components of metropolitan areas. *Annual Review of Ecological Systematics 32*:127–57.

Re, M. and Barrros, V. R. (2009). Extreme rainfalls in SE South America. *Climatic Change 96*:119–36.

Rottle, N. and Yocom, K. (2011). *Ecological Design*. Lausanne: AVA Publishing.

Sustainable Sites Initiative (2009). The Case for Sustainable Landscapes. http://www.sustainable sites.org/report (accessed February 5, 2014).

Tress, B.; Tress, G.; and Fry, G. (2005). Integrative studies on rural landscapes: Policy expectations and research practice. *Landscape and Urban Planning 70 no. 1–2*:177–91.

United Nations Environment Programme (2013). Sustainable Urban Planning in Brazil. http:// www.unep.org/greeneconomy/SuccessStories/SustainaibleUrbanPlanninginBrazil /tabid/29867/Default.aspx (accessed December 8, 2013).

U.S. Federal Highway Administration, FHWA (2006). Eco-logical: An Ecosystem Approach to Developing Infrastructure Projects. http://www.environment.fhwa.dot.gov/ecological /ecological.pdf (accessed December 8, 2013).

White, I. (2010). *Water and the City: Risk, Resilience and Planning for a Sustainable Future*. New York: Routledge.

20 The Caribbean Landscape Cyborg
Designing Green Infrastructure for La Parguera, Puerto Rico

José Juan Terrasa-Soler, Mery Bingen, and Laura Lugo-Caro

CONTENTS

20.1 Introduction 405
20.2 Setting...407
20.3 Landscape Context..409
20.4 Methodology and Analysis 410
20.5 Proposed Strategies 413
    20.5.1 Mangrove/Wetland Restoration and Buffer Systems 414
    20.5.2 Vegetated Swales/Bio-Retention Cells 414
    20.5.3 Wet Ponds 414
    20.5.4 Rain Gardens 415
    20.5.5 Filtering Practices/Sand Filter 415
    20.5.6 Rainwater Harvesting 415
    20.5.7 Permeable Pavement/Road Stabilization...415
20.6 Green Infrastructure in Tropical Areas 416
    20.6.1 Site 1: Green Street Bio-Retention Cell..419
    20.6.2 Site 2: Sand Filter...419
    20.6.3 Site 3: Mangrove Restoration and Forebay.....................................421
20.7 Conclusion 424
Acknowledgments..425
References...425

20.1 INTRODUCTION

"Green infrastructure" is a term that has acquired fad status as it is manipulated and misused to "brand" interventions in the landscape. Certain groups, mainly in the engineering camp, consider it to be simply the new term for storm water management. The U.S.

Environmental Protection Agency (USEPA), for example, defines green infrastructure mainly in terms of storm water management (USEPA 2010). In this view, green infrastructure is undistinguishable from "low-impact development" or LID, an approach focused on the hydrologic cycle and hydrologically sound strategies for site development (PGC 1999; Schueler and Holland 2000; Schueler 2005; Schueler et al. 2007).

A more holistic view of green infrastructure, however, has evolved within the discipline of landscape architecture. This view is rooted in the tradition of the multifunctional landscapes of Europe and North America. The infrastructural functions of landscape are at the heart of Frederic Law Olmsted's famous works, from Central Park to the Emerald Necklace, as well as in the "water landscapes" of The Netherlands, just to reference a few examples (Meyer 1997; Berrizbeita 1999). In these cases, landscape is called upon to deal with storm water and even sewage in urbanizing or urban contexts; it is intended to do work and to solve an immediate problem that conventional infrastructure would normally solve. At the same time, however, the landscape is designed to provide a slew of other ecological and social benefits, including aesthetic experience. For landscape architects, green infrastructure is a new term for what, in one form or another, has historically been conceived as designed landscapes that perform a variety of functions, from the aesthetic to the recreational to the hydrological to the ecological.

Contemporary understanding of green infrastructure has also been influenced by authors like Strang (1996), who pointed out that infrastructure is inherently part of landscape and its functioning should be clear and experiential. Both the view of "landscape as infrastructure" and Strang's view of "infrastructure as landscape" are included and blurred together in our current notion of green infrastructure, at least as used in landscape architecture. Although Olmsted might have started from the idea of landscape as leisure space and ended up with a multifunctional landscape that also managed storm water, many contemporary practitioners start from the challenge of integrating infrastructure with natural processes while providing meaningful spaces for humans and ecologically functioning landscapes for wildlife. Green infrastructure, therefore, could be conceptualized as a polyvalent strategy or bundle of strategies to harness the power of designed nature for the mutual benefit of human society and wildlife. Or put in another way, green infrastructure could be thought of as a strategic "alliance" between humans and the rest of nature for their mutual benefit and mediated through landscape design.

Green infrastructure is perhaps a clear example of what Elizabeth Meyer has called the landscape cyborg (Meyer 1997). The landscape cyborg—half machine and half biological entity—brings together mechanical and biological processes to achieve a variety of objectives. The landscape cyborg performs a multiplicity of functions for both humans and wildlife. It is a multifunctional machine that thrives because of human agency to the benefit of both human and nonhuman. The landscape cyborg goes well beyond the site to include regional ecological networks that provide a framework for the life-sustaining relationship envisioned between humans and landscapes (Terrasa-Soler 2006).

Many recent projects point to this polyvalent, landscape-cyborg view of green infrastructure. The "Rising Currents" exhibition at the Museum of Modern Art in New York City in 2010 is a great example (Bergdoll 2011). Even though this exhibition offered a proposal rather than an actual built project, it reflects the broad view of green

The Caribbean Landscape Cyborg

infrastructure as designed nature doing work. The plan that the "Rising Currents" exhibition presented is full of biologically or ecologically based strategies to protect New York's harbor from climate change and its consequences. Coastal ecologies that include designed clam beds and other marine life are recruited to provide physical protection from storm waves; geotextile-contained marine sediments are used to "soften" rather than "harden" the coast and make it more resilient to the increased force of storms under climate change scenarios. The "Rising Currents" exhibition also highlighted the multidisciplinary nature of the contemporary urban design project and the importance of landscape as urbanizing medium (Brown 2007). This medium is the fundamental set of ecological processes that sustain both human and nonhuman life and that green infrastructure takes advantage of to achieve its multiple objectives.

A recent review of urban ecology by Pickett et al. (2011) highlights the many benefits of designed nature in the city. Beyond the obvious benefits to water quality in urban streams, green infrastructure in the city helps to reduce the heat-island effect, increase biodiversity, reduce noise, and improve air quality. The emerging picture is one of the intermingled nature of human and wildlife in the city through the integration of spontaneous and designed landscapes. As more and more of humanity lives in cities and urbanized areas, the opportunities for integrating designed nature and conventional infrastructure as green infrastructure offer a new frontier for landscape architects and environmental designers.

Beyond its ecological benefits, green infrastructure is often used by landscape architects as a poetic device. That infrastructure can be poetic and provide an aesthetic experience of the city or the "urban condition" should not be a surprise. Many designers have specifically exploited the different intertwined layers of landscape and infrastructure (the landscape cyborg) to reveal a different experience of the environment. See, for example, the landscapes of George Descombes or the recent projects by Michael Van Valkenburgh (Descombes 1999; Cooper 2010). In these projects, simple devices or landscape operations offer new ecological functions full of experiential opportunities for humans.

In the project that we present in this chapter, we strived to show that green infrastructure in the tropics could do more than simply manage storm water better. We tried to incorporate the polyvalent nature of true green infrastructure in interventions that are based on local community needs coupled with a broader need to conserve tropical biodiversity. The resultant landscape cyborgs are intended to satisfy these multiple demands that we put on the landscape today through the integration of the biological with the purely infrastructural.

20.2 SETTING

La Parguera, a coastal town of 26,000 people in southwestern Puerto Rico (Figure 20.1), is home to some of the more diverse coral reef ecosystems in Puerto Rico and the Caribbean (Miller and Lugo 2009). The La Parguera area has the highest number of shallow-water stony coral species (38) in any reef system in Puerto Rico (Miller and Lugo 2009) and contains significant stands of both *Acropora palmata* and *Acropora cervicornus*, species listed as threatened under the U.S. Endangered Species Act. The diversity of La Parguera is due in part to its extensive coastal

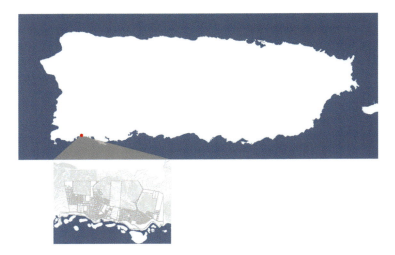

FIGURE 20.1 Location plan showing the fishing village of La Parguera in southwestern Puerto Rico. Topography and street layout are shown.

shelf (13 km) and historically clear waters. A nearby bioluminescent bay (*Bahía Fosforescente*) adds to the richness of this Caribbean coast.

La Parguera entertains a thriving tourism industry associated with snorkeling, diving, boating, and the bioluminescent bay and receives about 100,000 visitors annually. However, storm water runoff from the hillside town and streets flows directly into the tidal waters and mangroves, carrying high levels of pollutants that are damaging the near shore coral reefs. The La Parguera ecosystem was subject to significant coral bleaching in 1987 and 1990 (García et al. 1998) and continues to be threatened by the extensive use of its coastal resources and storm water pollution. The U.S. National Oceanic and Atmospheric Administration (NOAA) has monitored PAH (polycyclic aromatic hydrocarbons, many of which are toxic and carcinogenic compounds associated with combustion of fossil fuels) levels and heavy metals in the near shore sediments, and the data show a clear plume from the town and its impervious surfaces. Further analysis has determined that automobiles, as opposed to boat engines, are the primary source of these PAHs, implying that storm water runoff from the heavily developed town is the main contributor of these pollutants (Pait et al. 2007) (Figure 20.2).

Studies have shown that the storm water from La Parguera also carries significant loads of nitrogen, sediment, and bacteria as well as other pollutants harmful to coral reefs (Pait et al. 2007). Research has also indicated that unpaved roads and disturbed soils from development or road clearing on the La Parguera hillsides are significant sources of sediment in storm water runoff. The coral *Acropora palmata* is known to be particularly sensitive to sedimentation, which affects both reproduction and growth rates.

Our project addressed land-based sources of pollution by developing a comprehensive and integrated green infrastructure plan consisting of landscape interventions that also addressed improvements in community space, urban landscape character, and ecosystem functioning along the coast.

The Caribbean Landscape Cyborg

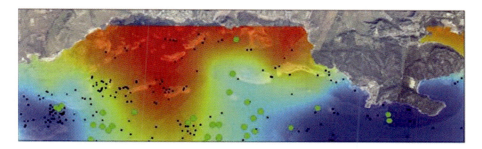

FIGURE 20.2 Polycyclic aromatic hydrocarbon (PAH) plume in La Parguera Bay (NOAA).

20.3 LANDSCAPE CONTEXT

La Parguera is nestled between the dry coastal hills and the ocean in southwestern Puerto Rico. A series of parallel, small, and steep watersheds form the hydrological context (Figure 20.3). Average rainfall is around 1118 mm per year, and storms are usually short and intense. Water flows to the coast rapidly and with little chance of interception (due to sparse vegetation cover) or infiltration (prevalence of impermeable surfaces). Streams are small and ephemeral, but seasonal, intense rainfall (May and September–October) can produce localized flooding.

La Parguera is a hub for year-round tourism because of the beauty of the coastal and reef ecosystems, local Caribbean cuisine, the bioluminescent bay, and boating and fishing opportunities, among others. However, La Parguera displays significant problems beyond coastal pollution. There is a lack of coherent transportation routes and connections, fragmented ecosystems, poor infrastructure, lack of urban landscape identity, and poor quality of public space.

After an initial analysis, we concluded that well-designed green infrastructure could contribute to solve the challenges confronted by La Parguera, both environmental

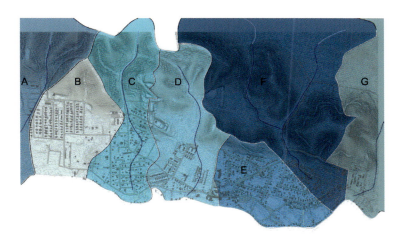

FIGURE 20.3 Coastal sub-watersheds in La Parguera.

and in terms of urban landscape quality. The challenge was to turn storm water treatment infrastructure and strategies (which have been mostly developed for temperate climates; see, for example, PGC 1999; Schueler 2005; Schueler et al. 2007) into urban space strategies that would work also as green infrastructure for a tropical, coastal, Caribbean town. The ultimate goal, of course, is to enhance the social, economic, and environmental outlook of La Parguera and allow for the sustainable development of what already is an enchanting Caribbean coastal fishing village.

20.4 METHODOLOGY AND ANALYSIS

Sub-watershed land use and hydrological analysis, along with visual landscape analysis, were the basis for developing conceptual interventions designed to reduce near coastal pollution. The conceptual design strategy was to improve environmental and hydrological performance by using the improved urban landscape as water treatment infrastructure and new habitat, combining mechanical and biological processes that work synergistically (the landscape cyborg).

Two main tools were used to gather landscape and hydrological system information needed for design: A comprehensive and fieldwork-intensive documentation of landscape conditions and a multifactor visual analysis. Fieldwork was carried out by Center for Watershed Protection* (CWP) personnel, the authors, and other partners

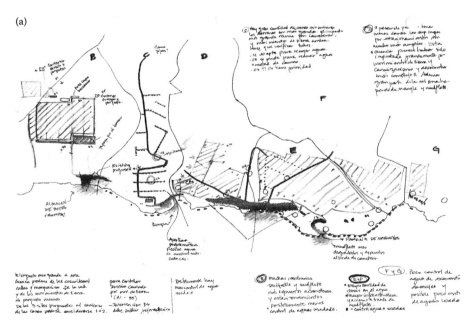

FIGURE 20.4 Landscape analysis for La Parguera, Puerto Rico. The analysis included conceptualization of the landscape (a).

---

* Center for Watershed Protection, 8390 Main Street, 2nd Floor, Ellicott City, MD 21043 USA, http://www.cwp.org.

The Caribbean Landscape Cyborg

in August 2010. After conceptualizing and delineating the sub-watersheds that form the La Parguera coastal plain (Figure 20.3), teams described, in the field, each of the sub-watersheds (photographs, drawings, field notes, maps) and identified opportunities to improve hydrological functioning and reducing risk of storm water pollution. Each team also indicated on annotated aerial photographs the main storm water flow paths and the location of potential green infrastructure interventions. In addition, the local community was consulted during fieldwork to obtain its opinion on landscape values and environmental problems in La Parguera.

In the studio, drawings representing different layers of information were constructed to scale and analyzed visually (Figure 20.4). For example, different layers contained all of the impermeable surfaces, the active sediment sources, the topography, and the main hydrological flow paths. These layers together provided a model for the movement of sediments in each of the sub-watersheds. Other layers

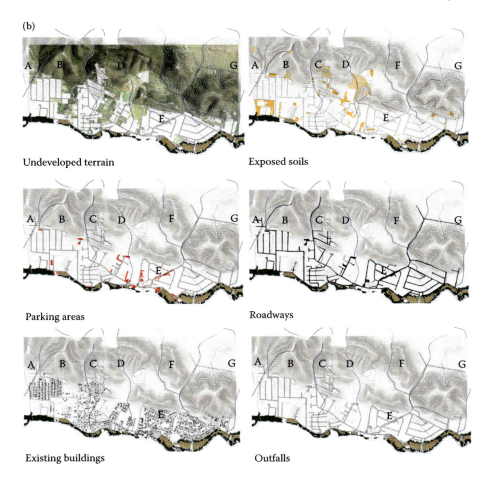

FIGURE 20.4 (Continued) Landscape analysis for La Parguera, Puerto Rico. The analysis included a site inventory (b).

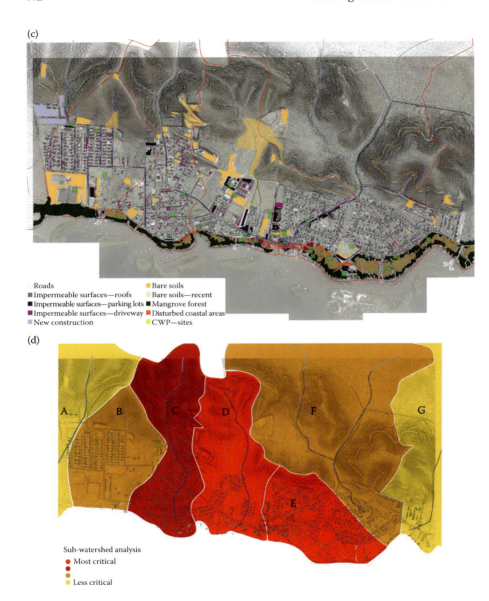

FIGURE 20.4 (Continued) Landscape analysis for La Parguera, Puerto Rico. The analysis included a multilayer visual analysis (c) and prioritization of sub-watersheds (d). In (c), sub-watershed boundaries are shown as red lines, main streams/drainage network are blue lines, mangrove forest is shown in dark green, salt flats in brown, bare soils in yellow, new construction in blue, large impermeable surfaces in black, and disturbed coastal areas in red. Fieldwork was carried out in August 2010.

documented the regional ecological networks that occur at La Parguera. We also put together layers of information to identify the tensions or conflicts in land use, visual impacts to the landscape, hydrological stressors, and other environmental problems. Some of the conflicts and problems identified were lack of coherent connections, fragmented ecosystems, erosion on steep slopes, high proportion of impermeable surfaces, sparse vegetation cover, deforested or impacted coastal habitat, uncontrolled storm water flows, steep and long flow paths, poor modulation and quality of urban space, poor infrastructure, and lack of identity. The final step in the analysis was to prioritize each of the sub-watersheds for design interventions.

20.5 PROPOSED STRATEGIES

The project proposes to address land-based sources of pollution by retrofitting best storm water management practices at specific discharge points and sediment source areas in order to capture and treat overland flows before they enter the tidal waters and near shore environment. Several intervention strategies were identified as appropriate for the landscape of La Parguera, and a palette of interventions was developed using the concept of locally adapted landscape cyborgs. The design work included the application of these interventions to specific locations within La Parguera and the retrofitting of each general strategy to accommodate tropical materials and desired urban quality. Figure 20.5 presents the assembled strategies or "master plan." The palette included the interventions described in the following sections.

FIGURE 20.5 Green infrastructure strategies for La Parguera, Puerto Rico. Landscape interventions include mangrove restoration (shown in red), freshwater wetland restoration/creation (dark blue), wet ponds (light blue), sand/gravel filters (brown), bio-swales and bio-retention cells (light green and aqua), permeable pavement (black), and road stabilization (orange).

20.5.1 Mangrove/Wetland Restoration and Buffer Systems

Wetland restoration (mangrove, freshwater swamp, and freshwater herbaceous wetlands) is intended to increase vegetation cover along the coast, improve coastal habitat quality, serve as final treatment of storm water, and protect the coast from storm surges. It includes forebays for the treatment of contaminated or nutrient-enriched storm water and to receive discharges from storm water collection systems and other outfalls. This intervention type makes use of natural wetland processes (plants, soil, and associated microorganisms) and is intended to reduce flow velocity, capture suspended sediments, and adsorb contaminants. The cost is proportional to the number and sizes of treatment cells required (US$9–37 per square meter for constructed treatment wetlands or about 50% to 90% less than conventional treatment techniques). This set of interventions is marked in red on Figure 20.5.

20.5.2 Vegetated Swales/Bio-Retention Cells

Vegetated swales and bio-retention cells are open, shallow channels or pits with thick vegetation covering the side slopes and bottom and that collect and slowly infiltrate or convey storm water to downstream discharge points. These interventions are best suited for residential, industrial, and commercial areas with low flow and smaller populations. The feasibility of installation depends on the area, slope, and the contributing watershed. General limitations include that they are impractical in areas with very flat grades, steep topography, or poorly drained soils. The selection of plants, as well as the design of the soil, is crucial for their success. La Parguera is well suited to these interventions except where very steep catchment areas exist. This set of interventions is marked in light greens on Figure 20.5 (dark green is existing mangroves). The increased evapotranspiration that vegetated swales and bio-retention cells provide also helps reduce the heat-island effect in urbanized areas.

20.5.3 Wet Ponds

Generally, wet ponds are a combination of a permanent pool and an extended detention or shallow marsh equivalent to the entire treatment volume (usually the common storm, a 10–25 mm storm). A sediment forebay is important for maintenance and longevity of a storm water treatment pond. It includes a pond buffer that extends 8 m outward from the maximum water surface elevation of the pond. Woody vegetation may not be planted or allowed to grow within 5 m of the toe of the embankment. This set of interventions is marked in light blue on Figure 20.5. The old treatment pools of an abandoned sewage treatment plant in La Parguera could be converted to wet ponds to help treat sediment-laden flows in the western sub-watersheds (where recent construction activities have occurred). Wet ponds are also opportunities to increase habitat for aquatic birds when the shoreline is well designed and when appropriate plants are selected.

The Caribbean Landscape Cyborg

20.5.4 Rain Gardens

Rain gardens are excavated shallow depressions planted with specially selected native vegetation to capture and treat storm water. They are typically underlain by a sand or gravel infiltration bed and are best suited for areas with at least moderate permeability (more than 25 mm/hour). Rain garden cost is about US$177–247 per cubic meter. Surface area is dependent upon storage volume requirements but should not exceed a maximum loading ratio of 5:1. Water accumulation depth should not exceed 15 cm and should empty within 48 hours. This set of interventions is marked in light green on Figure 20.5. Rain gardens have the potential to modulate urban space because they are generally small interventions appropriate at the street scale. They can also function as wildlife refuge in urban areas.

20.5.5 Filtering Practices/Sand Filter

Landscape filters are intended to capture and store, temporarily, the treatment volume and pass it through a filter bed of sand, organic matter, soil, or other media. Filters are generally applied to land uses with a high percentage of impervious surfaces. Sediment should be cleaned out when it accumulates to a depth of 15 cm or more. This set of interventions is marked in brown on Figure 20.5. In La Parguera, they are recommended for the more dense and intensive use areas of the town where other interventions are impractical because of space limitations. The landscape filter usually includes "finishing" stages that depend on biological processes and form the landscape cyborg. Microbial processes may also occur within the filter medium itself, adding another layer of biological functioning to the cyborg.

20.5.6 Rainwater Harvesting

Rainwater harvesting refers to the capture, diversion, and storage of rainwater for a number of different purposes. It is practical only when the volume and frequency of rainfall and size of the catchment surface can generate sufficient water for the intended purpose. Given the relative scarcity of rainfall in La Parguera, rainwater harvesting might be useful for maintenance of the local urban landscape and, therefore, it would work as a form of detention. This set of interventions is marked in aqua on Figure 20.5.

20.5.7 Permeable Pavement/Road Stabilization

Permeable pavement allows rainwater to infiltrate through it, to be filtered, and to percolate into the ground as groundwater. The approximate cost of this type of pavement is US$75–160 per square meter, including the underground infiltration bed. It can reduce overall project cost by eliminating the need for traditional storm water infrastructure. It is ideally situated on shallow slopes above soils with permeability rates greater than 60 mm/hour.

416   Revising Green Infrastructure

20.6   GREEN INFRASTRUCTURE IN TROPICAL AREAS

The application of the landscape cyborg concept to La Parguera, guided by the results of our analysis, resulted in the green infrastructure plan shown in Figure 20.5. The plan explores ways to adapt to the tropics through appropriate material selection and systems design, green infrastructure concepts developed for temperate climates. This plan will serve as a guide for the future implementation of key projects to decrease nutrients, sediment, and other pollutants in the storm water from La Parguera, ultimately improving the quality of the receiving tidal waters, coral reefs, and the resilience of the coastal ecosystem. The plan is also a framework to guide the implementation of green infrastructure with a coherent urban design language. Local partners are already developing specific construction projects emanating from this plan.

Three interventions included in the green infrastructure plan were developed further to explore materiality, spatial quality, systems integration, and cost of construction. They also served to explore the condition of tropical landscape cyborg. These three interventions, which will likely be built in the near future, are described here. Table 20.1 describes the tropical, vernacular, materials proposed for each of the three interventions.

TABLE 20.1

Summary of Tropical Materials for the Interventions Proposed for Sites 1 through 3

Site 1: Green Street Bio-Retention and Site 2: Sand Filter between Plaza and Building Preliminary Materials

Plant	Common Name	General Description
Bucida buceras	Úcar, Black olive tree	Medium to large tree 9–18 m tall
Conocarpus erectus	Mangle botón, Button mangrove	Small evergreen tree up to 6 m tall
Pennisetum—variety	Fountain grass	Medium grass
Sansevierias trifasciata—variety	Lengua de vaca, Snake Plant	Small herbaceous perennial
Cyperus polystachyos	Manyspike flatsedge	Annual or perennial sedge. Stems thin, triangular. Very common in moist and grassy places at lower and middle elevations, ditches and roadsides
Coccoloba uvifera	Uva playera, Seagrape	From shrubs suitable for hedging to small trees reaching up to 9 m tall

Other Materials

Limestone rocks	Different sizes	White to cream native stone
Sand		
Topsoil		

(continued)

The Caribbean Landscape Cyborg

TABLE 20.1 (Continued)
Summary of Tropical Materials for the Interventions Proposed for Sites 1 through 3

Site 3: Mangrove Restoration and Forebay
Preliminary Materials

Plant[a]	Common Name	General Description
Rhizophora mangle	Mangle rojo, Red mangrove	Evergreen tree. Distinguished by its erect and aerial roots that form a dense brush. Forms pure colonies when in direct contact with seawater on quiet coasts. Reaches heights of 10 m or more and trunks of 30 cm in diameter or more
Laguncularia racemosa	Mangle blanco, White mangrove	Evergreen tree; reaches up to 20 m high, produces pneumatophores. Mangrove swamps in coastal areas; usually found on the landward edge. Provides shelter for wildlife
Avicennia germinans	Black mangrove	Evergreen tree. Trunks up to 30 cm in diameter; 16 m high. More salt tolerant than other mangrove species and produces aerial root pneumatophores. Mangrove swamp forests and sometimes on rocky and sandy shores; estuarine systems. Provides shelter for wildlife
Conocarpus erectus	Mangle botón, Button mangrove	Evergreen tree. Generally up to 3–5 m high but it can reach 20 m. Mangrove swamp forests and sometimes on rocky and sandy shores; estuarine systems. Provides shelter for wildlife
Thespesia populnea	Emajagüilla, Portia tree, Spanish cork	Shrub/tree of coastal woods, up to 9 m tall. Trunk of 20 cm diameter; dense crown
Hibiscus pernambucensis	Emajagua	Shrub/tree up to 15 m tall. Coastal woods and thickets and borders of mangrove swamps
Sansevierias trifasciata—variety	Lengua de vaca, Snake Plant	Small herbaceous perennial
Coccoloba uvifera	Uva playera, Seagrape	From shrubs suitable for hedging to small trees reaching up to 9 m tall

Other Materials

Limestone rocks	Different sizes	White to cream native stone
Sand		
Topsoil		

[a] Mangrove species listed following distribution according to distance from seashore inland. See Ashton (1989), Francis and Lowe (2000), and DNER (2001) for additional information.

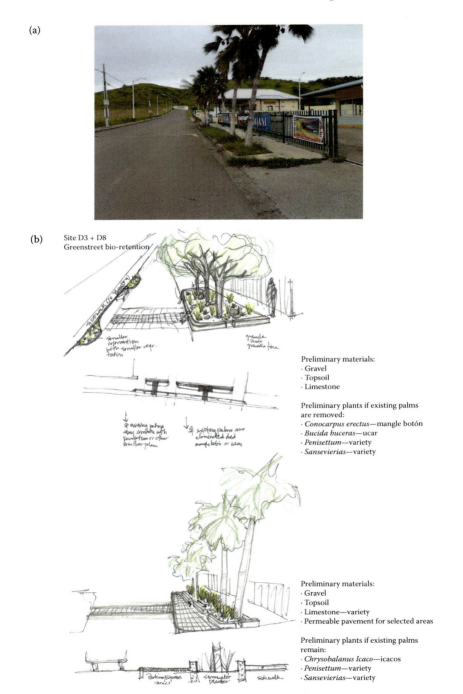

FIGURE 20.6 Site 1. Existing conditions (a) and sketch of proposed green street bio-retention cell intervention (b).

20.6.1 SITE 1: GREEN STREET BIO-RETENTION CELL

The purpose of this intervention is to reduce the load of pollutants that flow from a long road (one of the main access routes to the coast) and commercial area. The storm water flow path is long and steep over impermeable surfaces. The bio-retention cell is intended to allow infiltration and evapotranspiration of storm water from common storms (10–25 mm of rainfall). The location was selected to break the existing flow path and to be near potential sources of pollution (small commercial area). The palette of materials includes limestone, locally available gravel and sand, *Conocarpus erectus* (Button mangrove), *Bucida buceras* (Úcar), *Cyperus polystachyos* (Manyspike flatsedge), *Pennisetum* spp., and *Sansevieria* spp. The constructed intervention also aspires to modulate the urban space, dominated by a long, straight, length of road and a parking lot, and to serve as a habitat stepping stone for migrating birds and other wildlife (Figure 20.6).

20.6.2 SITE 2: SAND FILTER

This intervention intends to abate the impact of polluted storm flows at the center of town. The filter is needed at this location because there is little space to do a plant- and soil-based intervention, and the topography concentrates polluted urban storm flow at this location. The material palette includes locally available sand, gravel, and limestone fragments as well as *Bucida buceras* (Úcar) and *Sansevieria* spp. for the secondary planting strips in the plaza and kiosks area. The secondary planting strips will receive part of the effluent from the sand filter before it is discharged to the sea (Figure 20.7). The planting strips are also intended to soften the hard edge of the shore at this location and to modulate one of the most popular spaces of town. In this intervention, the coupling of the constructed, mechanical, component (sand filter) with the biological component (planting strips) has to be effective in order

FIGURE 20.7 Site 2. (a) Existing conditions for La Parguera's busiest area.

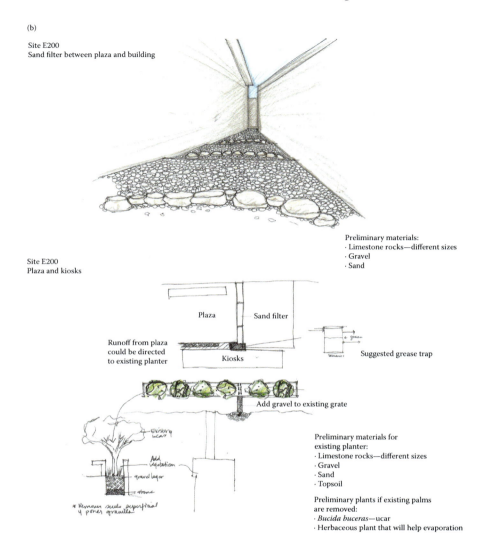

FIGURE 20.7 (Continued) Site 2. (b) Sketch of sand filter strategy for La Parguera's busiest area.

for the mixed intervention (cyborg) to work. If the flows from the sand filter constantly exceed the capacity of the planting strips, for example, the "finishing" effect expected of the planting strips will be lost. Likewise, if the sand filter is not maintained properly, it could become clogged and the water diverted directly to the shore instead of flowing through the filter and planting strips. The precise coupling of flow volumes between the mechanical and biological components of this landscape cyborg is crucial to its proper functioning.

20.6.3 SITE 3: MANGROVE RESTORATION AND FOREBAY

The site for this intervention has been used as informal parking, impacting the previously existing salt flats and mangrove stand. There is a concentrated storm water discharge point at this location that collects water from a broad sub-watershed and that also has impacted the mangrove stand. For this reason, a vegetated forebay is proposed to mechanically reduce the energy of the flows and to pretreat the storm water before entering the mangrove restoration area (Figure 20.8). The planting palette includes (from upland to seashore): *Stahlia monosperma* (Cóbana negra, a threatened species), *Hibiscus pernambucensis* (Emajagua), *Thespesia populnea* (Emajagüilla), *Conocarpus erectus* (Button mangrove), *Avicennia germinans* (Black mangrove), *Laguncularia racemosa* (White mangrove), and *Rhizophora mangle* (Red mangrove). The intervention is also intended to restore coastal habitat along a dense area of town to prevent shoreline erosion and increase the resilience of the coastal ecosystem. The sequence of species proposed is usually found at other coastal locations throughout Puerto Rico. The idea is that the primary mechanical component of the cyborg (the forebay) produces lower-energy water flows appropriate to the salt flat–mangrove coastal ecosystem type. If the energy of the storm flow is not

FIGURE 20.8 Site 3. (a) Existing conditions near town parking lot.

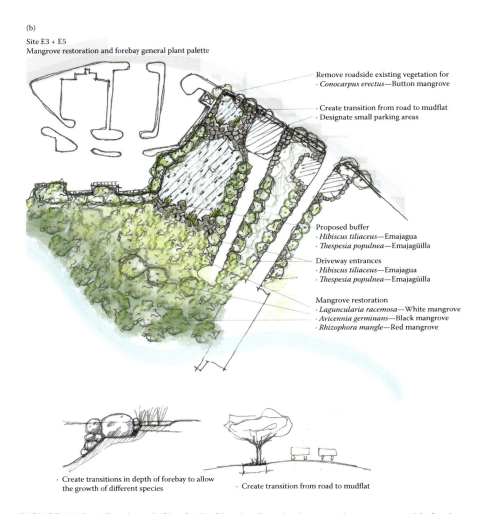

FIGURE 20.8 (Continued) Site 3. (b) Sketch of wetland restoration strategy with forebay general plant palette near town parking lot.

effectively reduced or the water quality is not sufficiently improved by the forebay, the mangrove restoration (and primary biological component of this cyborg) will not prosper. Again, the precise coupling of the mechanical and biological components of the landscape cyborg is crucial for the effective functioning of this designed landscape in La Parguera. It is important to note that both the forebay and the mangrove restoration have biological and mechanical functions; however, the forebay has primary mechanical functions while the mangrove restoration represents the primary biological functions of the cyborg.

The Caribbean Landscape Cyborg

(c)

Mangrove distribution according to distance from sea to land

Roadside, parking area and mangrove entrance trees

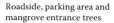

Rhizophora mangle—Red mangrove
Growing habit—Evergreen tree. Distinguished by its erect and aerial roots that forms a dense brush. Forms pure colonies when in direct contact with sea in quiet coasts. Reaches heights of 10 m or more and 30 cm in diameter or more.

Habitat—Common in mangrove swamps around coastal areas, near river mouths. Provides shelter and nesting for aquatic birds. Estuarine systems.

Avicennia germinans—Black mangrove
Evergreen tree. Stems up to 30 cm of diameter, 16 m long. More salt tolerant than other mangrove species. Habitat—Coastal lagoons and swamps, estuarine systems. Provides nesting and shelter of wildlife.

Thespesia populnea—Emajaguilla (Portia tree, Spanish cork, Otaheita)
Growing habit—Shrub/tree of coastal woods, up to 10 m high. Trunk of 20 cm diameter; dense crown. Habitat—Coastal woods, thickets and mangrove swamps.

Hibiscus pernambucensis—Emajagua
Growing habit—Shrub or tree; to 15 m tall.
Habitat—In brackish swamps and inner margins of mangrove, ascending to the humid mountains.

Laguncularia racemosa—White mangrove
Growing habit—Evergreen tree; reaches up to 20 m high, produces pneumatophores. Habitat—Mangrove swamps in coastal areas. Provides shelter for wildlife.

Conocarpus erectus—Button mangrove
Growing habit—Evergreen tree; generally up to 3–5 m high but it can reach 20 m. Habitat—Mangrove swamp forests and sometimes on rocky and sandy shores. Estuarine systems. Provides shelter for wildlife.

Stahlia monosperma—Cobana negra
Growing habit—Evergreen tree. Stems up to 20 m high. Habitat—Coastal woodlands and borders of mangroves. Threatened with extinction.

FIGURE 20.8 (Continued) Site 3. (c) Description of the selected plant palette.

20.7 CONCLUSION

The authors sought to test in La Parguera the idea that a landscape cyborg, or the coupling of mechanical and biological components mutually supporting each other and operating synergistically, which coincides with the contemporary landscape architectural notion of "green infrastructure," is the most effective way to deal with the environmental and spatial problems of this Caribbean fishing village. A detailed, field-based assessment was conducted that, coupled with the analysis of other spatial data, provided an interpretation of this landscape and a diagnosis of significant problems. From the analysis, we developed a palette of appropriate green infrastructure strategies, conceived as landscape cyborgs in which both mechanical and biological components work synergistically. Three of the proposed interventions were developed in more detail with the intention of testing tropical materials and eventually building pilot projects to calibrate the interaction among the cyborgs' components.

What might be unique about the development of a coordinated green infrastructure plan for La Parguera is the multidisciplinary, systematic approach that was part of the project's planning and execution and the unique landscape setting of a Caribbean coastal watershed. The project encompasses detailed fieldwork, scientific analysis, community consultation, landscape architectural design, and the future construction and scientific evaluation of demonstrative landscape interventions contained in the plan. How well these demonstration projects perform might determine if funding is secured for the complete construction of all of the interventions contained in the plan. The La Parguera Plan has brought together the local community, the scientific community, and the design community to achieve the common goal of restoring and protecting a magnificent landscape and cultural hub through green infrastructure.

La Parguera's unique landscape, enormous ecological value, and touristic attractions will continue to draw many locals and visitors alike (Figure 20.9). The green infrastructure strategies for La Parguera described herein are an effort to make this

FIGURE 20.9 Partial view of La Parguera's coastline from the village plaza.

The Caribbean Landscape Cyborg

landscape a sustainable and resilient landscape in the face of climate change through green infrastructure and for the enjoyment of all. The strong sense of community in La Parguera is definitively a key ingredient in the potential success of this project.

At the end, green infrastructure is part of the living landscape and depends on the care and sense of ownership that the local residents and visitors have about it. It is not merely some piece of equipment that is installed; it has to mean something to the people and complement the character of the place and its supporting web of ecological interactions. The degree to which residents of and visitors to La Parguera understand that a well-functioning landscape is essential for the survival of this wonderful place will be the degree to which we might be able to enjoy La Parguera for years to come.

ACKNOWLEDGMENTS

This project was made possible by a grant from the U.S. National Oceanic and Atmospheric Administration's Coral Reef Program, through the Center for Watershed Protection, Ellicott City, Maryland. Field data collection was done by the authors, the Center for Watershed Protection team, and local partners. The support of the Polytechnic University of Puerto Rico and The Office of Marvel & Marchand Architects, both in San Juan, Puerto Rico, is also kindly acknowledged.

REFERENCES

Ashton, P. M. S. (1989). *Forester's Field Guide to the Trees and Shrubs of Puerto Rico.* 2nd Edition. Tropical Resource Institute, School of Forestry and Environmental Studies, Yale University, New Haven, CT.

Bergdoll, B. (2011). *Rising Currents: Projects for New York's Waterfront.* The Museum of Modern Art, New York.

Berrizbeita, A. (1999). The Amsterdam Bos: The modern public park and the construction of collective experience. In *Recovering Landscape: Essays in Contemporary Landscape Architecture*, J. Corner, ed., 187–204. New York: Princeton Architectural Press.

Brown, J. (2007). Landscapes as complex adaptive systems. *Harvard Design Magazine* 27 (Fall 2007/Winter 2008): 77–79.

Cooper, J. E. (2010). Brooklyn Bridge Park: The evolution of a new New York tradition. *Topos* 72: 88–93.

Descombes, G. (1999). Shifting sites: The Swiss way, Geneva. In: *Recovering Landscape: Essays in Contemporary Landscape Architecture*, J. Corner, ed., 79–86. New York: Princeton Architectural Press.

DNER [Puerto Rico Department of Natural and Environmental Resources]. (2001). *Guide to Identify Common Wetland Plants in the Caribbean Area: Puerto Rico and the U.S. Virgin Islands.* Editorial de la Universidad de Puerto Rico, San Juan, Puerto Rico.

Francis, J. K. and Lowe, C. A., eds. (2000). *Bioecología de Árboles Nativos y Exóticos de Puerto Rico y las Indias Occidentales.* U.S. Department of Agriculture, Forest Service, International Institute of Tropical Forestry, San Juan, Puerto Rico.

García, J. R., Schmitt, L., Heberer, C., and Winter, A. (1998). La Parguera, Puerto Rico. In: Kjerfve, B., ed. *Caribbean Coral Reef, Seagrass and Mangrove Sites.* United Nations Educational, Scientific, and Cultural Organization, Paris.

Meyer, E. K. (1997). The expanded field of landscape architecture. In *Ecological Design and Planning*, G. F. Thompson and F. R. Steiner, eds., 45–79. New York: John Wiley & Sons.

Miller, G. L. and Lugo, A. E. (2009). *Guide to the Ecological Systems of Puerto Rico.* Gen. Tech. Rep. IITF-GTR-35. U.S. Department of Agriculture, Forest Service, International Institute of Tropical Forestry, San Juan, Puerto Rico.

Pait, A. S., Whitall, D. R., Jeffrey, C. F. G., Caldow, C., Mason, A. L., Christensen, J. D., Monaco, M. E., and Ramirez, J. (2007). *An Assessment of Chemical Contaminants in the Marine Sediments of Southwest Puerto Rico.* NOAA Technical Memorandum NOS NCCOS 52. Silver Spring, MD.

PGC [Prince George's County, Maryland]. (1999). *Low-Impact Development Design Strategies: An Integrated Design Approach.* Prince George's County, Department of Environmental Resources, Largo, MD.

Pickett, S. T. A. et al. (2011). Urban ecological systems: Scientific foundations and a decade of progress. *Journal of Environmental Management* 92: 331–362.

Schueler, T. (2005). *An Integrated Framework to Restore Small Urban Watersheds* (version 2.0; Urban Subwatershed Restoration Manual No. 1). Center for Watershed Protection, Ellicott City, MD.

Schueler, T., Hirschman, D., Novotney, M., and Zielinski, J. (2007). *Urban Stormwater Retrofit Practices* (version 1.0; Urban Subwatershed Restoration Manual No. 3). Center for Watershed Protection, Ellicott City, MD.

Schueler, T. R. and Holland, H. K. (2000). *The Practice of Watershed Protection.* Center for Watershed Protection, Ellicott City, MD.

Strang, G. (1996). Infrastructure as landscape. *Places* 10(3): 8–15.

Terrasa-Soler, J. J. (2006). Landscape change and ecological corridors in Puerto Rico: Towards a master plan of ecological networks. *Acta Científica* 20(1–3): 57–62.

USEPA [U.S. Environmental Protection Agency]. (2010). *Green Infrastructure Case Studies: Municipal Policies for Managing Stormwater with Green Infrastructure.* EPA-841-F-10-004. USEPA, Office of Wetlands, Oceans and Watersheds, Washington, DC.

21 Forests and Trees in the City
Southwest Flanders and the Mekong Delta

Bruno De Meulder and Kelly Shannon

CONTENTS

21.1 Introduction 427
21.2 Nutshell History: Urban Forests and Trees... 429
21.3 Shifting Paradigms 430
21.4 Southwest Flanders 431
21.5 Mekong Delta... 439
21.6 Forests and Trees in Delta Cities 447
References... 448
Suggested Reading... 449

21.1 INTRODUCTION

Throughout the history of urbanism, there has almost always been an interweaving of structures of plantation with urban armatures and tissues. In Europe and Asia, forests have traditionally formed the counter-figure of the city (from Karlsruhe to the Barcelona plan of Cerda—although that plan is usually represented without the unrealized forest), embedded the city (from Versailles to Suzhou), or complemented the city (from Paris to Dalat). In an attempt to locate Islamabad in a "natural environment," its foundation (plan of Doxiadis, 1960) was developed hand in hand with the systematic afforestation of the Margalla Hills, a previously barren mountain chain. In Europe, the Forêt de Soignes in Brussels, Epping Forest in London, Bois de Boulogne or de Vincennes in Paris, and Wiernerwald (Vienna Woods), among others, bear witness to how forests have become an essential identity of the city (Figure 21.1) (Forrest and Konijnendijk 2005). In many parts of Asia, cities historically developed in relation to a worldview that included geomancy (*feng shui*) and divination, which limited the activities of man; beliefs in land gods, river kings, and forest spirits were both practical and mystical (Cuc 1999; Yu et al. 2006). The central highlands Vietnamese city of Dalat is known as the "city of thousands pine trees" (De Meulder and Shannon 2012), Suzhou, known as "Venice of the East" and famous for its canals and garden, yet it is embedded in an equally awe-inspiring forest setting. At the same time, lines

FIGURE 21.1 Bois de Boulogne, Paris: extension of nature. Jean-Charles Adolphe Alphand—as head of "Service des Promenades et Plantations de Paris," working for Louis Napoleon III and his prefect, Georges-Eugène Haussmann III—designed and developed the park to the west of the urban core (also a large one to the east of the city, Bois de Vincennes) as ownership passed from the Crown to the City of Paris. Existing woodlands were extended and included new plantations, lakes, and multiple buildings. Throughout the city, an entirely new nature was created via promenades and tree-lined boulevards. (From Alphand 1867–1873, Paris.)

Forests and Trees in the City   **429**

of trees planted on public spaces—including promenades, alleés, and boulevards; along walls; on ramparts and moats; beside canals; in city squares; and parallel to city streets and railroads—have, for millennia in Asia and centuries in Europe, been systematically planned and constructed on national scales and directly organized first by emperors and later by central and municipal governments.

Forests can be considered as self-regenerating ecosystems (and for that reason as well, a renewable and eternal resource), which nonetheless do not restrain mankind from managing them. For centuries, forests have been planned and systematically exploited and maintained; often this management was more extensive and sophisticated than the town planning of the same time (Bridel 1798). Forests, finally, not only appear as an early planning instrument as a counter-figure of the city, but have also always been a counter-model when it comes to (urban) development: self-regeneration versus continuous self-destruction, (self-)renewable resource versus continuous consumption of production goods. In a certain way, forests stand for the ecological archetype of sustainability. The introduction of forests within cities, as counter-figure or otherwise, is consequently often related to an idea of naturalization of the (artificial) city. The city, in itself an unsustainable element, is embedded within a (sometimes artificially created) nature.

21.2 NUTSHELL HISTORY: URBAN FORESTS AND TREES

The written history, in English, of urban forests is more prevalent for Europe than Asia. Wealthy private landowners, monastic orders, monarchs, nobility, and municipal authorities of medieval Europe owned forests as well as communities for whom forests were part of the commons. Uses range from wood pastures to cultivation for timber, to hunting grounds, to wildlife preserves, as protectors of water resources, and for public amenities. Over time, the forests became managed, often reforested and completely designed landscapes. Because they were sources of recurrent income, forest planning gained importance. In Asia, *feng shui* forests are sacred forests, often patches of native woodlands, also choreographed into a city-nature worldview fitting to the national culture of the moment. Yet the industrial revolution saw the felling of forests worldwide while others were planted for timber production. Anyhow, throughout the 20th century, urban forests became increasingly used for recreation and recognized as embodying social ("spiritual and bodily hygiene" and reduced crime) and economic benefits (high return on investment, increased property values, community appeal) in addition to the known environmental qualities. The more industrialization and mass migration colored cities grey, dusty, and monotonous, the more urban forests were on the agenda and assigned the function of "green lungs."

In antique times, plane trees (*Platanus*), which offer ample shade, were common in public spaces and promenades. From the 17th century onward, nature was fundamentally transformed by designers in Europe; promenades of trees were planted as urban parks and developed by the likes of John Nash in London, Jean-Charles Adolphe Alphand in Paris, Peter Josef Lenné in Berlin, and Holger Bolm in Stockholm (Da Costa Meyer 2013; Forrest and Konijnendijk 2005). Defunct military ramparts and bastions were planted with trees in the 18th century (excluding those in Lucca planted two centuries earlier). Promenades for the flâneur were complemented

430 Revising Green Infrastructure

by those for the horse-drawn carriage. The advent of the car led to parkways. The first evidence of state support for tree planting is from as early as 158 AD; in Ancient Greece, there were "in the form of vines, figs, olives and other fruit bearing trees and the distribution of their profits from these crops between the state and those citizens who planted the trees" (Forrest and Konijnendijk 2005, 24).

The Asian, specifically Chinese, history is not so different except for its time frame; the Chinese city systematically included nature and trees in the city far before the European city conceived of such notions. *The Rituals of the Zhou Dynasty* (1100–770 BC) verifies that tree planting and maintenance by designated officials along moats of city walls was obligatory. The book as also documents tree planting along riparian corridors in relation to flood protection and soil erosion as well as the strong tradition of street planting in Chinese cities. Initially, capital city streets and imperial highways were planted to provide separated royal passage, shelter against wind, provide shade, protect roads from flooding, and perform specific visual functions. Whenever trees died, they had to be quickly replaced (Yu et al. 2006). "Tree plantings along city streets and country roads were considered as good moral behavior and a blessing to the local people, and state officials were always memorialized for their contribution to the construction of greenways" (Yu et al. 2006, 230). The legacy of street tree planting continued during socialism as national campaigns. In the 1950s, new cities were realized with Soviet assistance, first with boulevards of street trees. The now-matured trees have become part of the rich heritage of many Chinese cities. Even in the barrack-like housing estates integrated in the "production units" (*danwei*) of the Mao times, tree planting was a prominent feature. Their landscape plans might a posteriori indeed be considered more important than the urbanistic plan. Nowadays, the mature trees are the main neighborhood quality of the remaining estates in cities such as Shanghai (Zhao 2007).

21.3 SHIFTING PARADIGMS

In today's era of environmental awareness and management, afforestation in proximity to urban areas and tree planting in cities has become a policy priority. The enthusiasm for tree-planting spans from the countryside, to peri-urban areas to the heart of urban areas—as witnessed by New York City's Department of Parks and Recreation 2007 launch of its MillionTreesNYC initiative or the Mayesbrook "Climate Change Park" in Barking, east London. A decade ago, Flanders also launched a program of new urban forests in the vicinity of its main cities in an attempt to combine ecological repair and development of (soft) recreational areas for urbanites—all this in one of the most densely urbanized areas of Europe. The climate change card has been playing strongly into policy matters worldwide, and pressure is on for nations to increase forest cover and green in urban areas. However, shifting to implementation remains problematic. The contestations between economy and ecology are at the core of the issue, and countries at different levels of development argue accordingly. The Food and Agriculture Organization (FAO) has been issuing "State of the World's Forests" reports beginning in 1995. Since 2005–2010, afforestation has been the trend in all of Europe, and although deforestation threatens many countries in Asia, China, Vietnam, India, the Philippines, and Turkey reported significant increases in forest area (FAO 2012, 17).

Forests and Trees in the City   **431**

European forestry underwent a series of paradigm shifts since the preindustrial era when forests were multifunctional, providing wood, fuel, grazing, fodder, and food. They were integrated with agriculture and often part of a communal regime, when they were not property of nobility. But even when land was royal or other domain, farmers might have rights by tradition (harvesting fruits, etc.). Often, an independent legislation regulated the rights in and on forests (separate from the urban legislation on civil rights and the rural one that governed the countryside). During the industrial era, however, forests were reduced to primarily monofunctional wood production engines. Geometrical definition and monoculture became increasingly dominant. As they solely served the purpose of industrial wood, they were transferred to the realm of the private regime as so many domains that originally belonged to the commons. Since the postindustrial paradigm, forests became multifunctional (recreational, wildlife, amenity, wood) and worked within a regulated regime, balancing between desired public benefits and private ownership, often involving a mixture of incentives and restrictions (Mather 2000).

In Vietnam, forestry must be understood within the broader context of the country's long history of human disturbance, development, politics, and principles of socialist land policies, in which all land, "including forests are the property of the entire people and have to be managed by the State" (from Ministry of Agriculture and Resource Development, quoted in Dang et al. 2012, 31). Until *doi moi* (the era of socioeconomic reforms begun in 1986), forestry in Vietnam was primarily for wood exploitation, and Vietnam's forests became rapidly depleted, suffering massive biodiversity loss. Forestry management shifted with *doi moi*, and the 1993 Land Law facilitated the allocation of both forest and agricultural land to individuals, households, and organizations such as a "forestry socialization" program. In the mid-1990s, forest loss was also associated with the frequent occurrence of natural disasters and species extinction. "Sustainable forestry" became an important component of Vietnam's new forestry discourse (Dang et al. 2012). After a forest cover drop of 25%–31% in 1991–1993 and an increase to 32%–37% in 1999–2001, it was 38% in 2005 due to natural forest regeneration and plantation (Meyfroidt and Lambin 2008, 1334).

21.4   SOUTHWEST FLANDERS

In Flanders, the forest index is only 10.8% (146 ha), and many policy plans, including the regional "Long-term Forestry Plan (2008–2017)" and "Forestry Action Plan (1998–2003)," the "Spatial Structure Plan of Flanders (1997)" sought to expand the forest by 10,000 ha (or a 0.07% increment of the forest index) between 1994 and 2007. This was not met. With the current afforestation rate, reaching the goal requires 127 years. From 1994 to 2000, the forest area declined by 3700 ha (Van Gossum et al. 2008, 515; Van Gossum et al. 2012, 30). The intense and dense occupation of the territory, which is almost completely privately owned, explains a significant share of the difficulties. Ownership rights are divided over almost uncountable amounts of (small) landowners, and the problem is owning agencies, such as the Nature Agency, Land Agency, and local leaders, have all their own instruments, mandates, and priorities that are not necessarily all in line with afforestation. Often, forest expansion and agriculture policies contradict one another. In Flanders, the majority of agricultural land is leased, and the land can only be transferred to forestry by a court order. The compensation to farmers and private landowners to afforest their land is often considered too low. Only 13% of the Flemish farmers take this opportunity. Insecurity concerning "land destination," fear of a decreasing land value, strong beliefs that reconversion to agriculture land will be legally impossible with time, fear of game damage, and the "long" rotation time (15 to 20 years for poplar) explain the hostility of the farmers toward afforestation. On the other hand, short-rotation forestry for energy purposes seems to gain interest (Van Gossum et al. 2008, 519).

Two design research projects in Southwest Flanders, developed for Leiedal,* an inter-municipal organization, address afforestation when conflicts between private interests and the public good are reduced. The region itself is known as the "Texas of Flanders"—a fragmented and diffused territory that is characterized by the simultaneity of differences. The projects focused around the area of Kortrijk (population 75,000), the region's largest city, situated on the Leie River, which is part of the 1.9 million inhabitants Eurométropole (Lille-Kortrijk-Tournai). Medieval Kortrijk received its wealth from flax and wool; it is now known as an entrepreneurial city with a diverse economy. There were two designs for afforestation sites developed for Leiedal—one in a wealthy housing district called Marke and a second in Hoog Kortrijk, the car-based, post-war city extension south of the historic core that developed along functionalistic lines of activity separated and across the E17 highway that is used by the vast majority of users and visitors to Hoog Kortrijk.

Marke is, in many ways, representative of the notorious Belgian dispersed urbanization that is generated by incremental and endogenous development dating already to the middle ages as part of a densely occupied and finely meshed agricultural territory (De Meulder et al. 1999). To a certain degree, it has more in common with the occupation processes in fertile deltas all over the world than with the generic sprawl that came with the advent of the car. Or is it the superimposition of both processes that typify the Belgian territory? Already more then a decade ago, the government designated the remaining open land of Marke for an afforestation program. Concerned farmers protested vehemently, and inhabitants joined the resistance of what was considered an attack on a cultural landscape—a rare piece of open land in the region.

The design proposal by RUA, coproduced with Leiedal, the administration of Kortrijk and stakeholders, including the Nature and Forest Agencies and members of the village council, formulated an alternative form of afforestation on the rich, fertile land that farmers would not, by definition, perceive as a provocation and that inhabitants could appreciate as an improvement to their environment instead of a breach of their peace. A new urban fringe forest was consciously conceived as a new counterpart of Marke (as in the classicist urban designs, in which city and forest are each other's

---

* The Leiedel Landscape Urbanism workshops were commissioned by Stadsbestuur Kortrijk in September 2010 for Marke and January 2012 for Hoog Kortrijk. They were run by B. De Meulder and K. Shannon. The Marke project was made in collaboration with B. Sandra, E. Jacobs, J. Garcia Galindo, N. Plevoets, A. Curtoni, H. Po-Ju, R. Habibi, W. Chen-wei, P. Wen, V. Cox, J. Schreurs, M. Goethals, B. Carron, W. Vandeghinste, B. Lattré, C. Vilquin, and Hoog Kortrijk with V. Cox, B. de Carli, S. Hoornaert, G. Lannoo, I. Llach, K. Lokman, M. Luegening, M. Motti, T. Ono, J. Provoost, P. Russo, C. Van der Zwet, and E. Vanmarke. Follow up on Hoog Kortrijk was made in 2012 by B. De Meulder, K. Shannon, M. Motti, and I. Llach with S. Hoornaert and a team from Leiedal.

Forests and Trees in the City

counterparts). The design for an urban fringe forest simultaneously functions as the new spatial frame in which Marke could densify while simultaneously turning into a more sophisticated quarter of Kortrijk. In hindsight, Marke is not only composed of a mediocre and dispersed built environment, but also, surprisingly, of a quite interesting, intense, and varied deciduous tree structure. Quite systematically, wild chestnut tree (*Castanea sativa*) planted streets mark the territory while woods are dominated by red oaks (*Quercus rubra*), beech (*Fagus sylvatica*), and Norwegian maples (*Acer planancides*); artificial woods tend to be Canadian poplars (*Populus alba*). The tree lines along streets are relatively recent and enliven the monotonous suburban residential areas, while those along regional roads sometimes date from the 18th century. Pockets of remaining woods, tree lines and patches of poplars in the valleys collectively form the skeleton of a structure that merges seamlessly into regular agricultural patterns. These half regular agricultural patterns, which are dominated by the extreme small-scale of the historically produced land division, blend with effortlessly with the softly undulating topography and form the main quality the project builds upon. The pattern of tree lines along roads and streets unite into a structure that to a major extent is simply the conversion of the previous rural hamlet structure. It is the complex existing logic that, becomes the design principle to strengthen. Indeed, to a large extent, the is nothing else than articulating the existing landscape structure. It uses the afforestation program—translated into tree lines and forest pockets—as a means to that end.

Manifold infrastructural ruptures (the E17, A17, A19, and R9 highways) isolate Marke from both the city center and its surroundings and disconnect it from adjacent municipalities. A majestic green counter-figure with the densely planted cloverleaf of the E17 and A17 as its center, encircled with a solid bike path, connects across the ruptures with soft but simultaneously robust connections for bikes and pedestrians and spatially integrates territorial armatures. Key to the connections is the solid bike ring around the forest figure from which panoramic views unfold, themselves views to the contiguous landscapes. Marke would therefore no longer be a peripheral enclave but rather, through its green counter-figure, a central element around which the villages, such as Lauwe, Aalbeke, and the business park Hoog Kortrijk, would be organized (Figure 21.2).

On a larger scale, the urban fringe forest initiates a structure that (re)positions Marke by visually relating and spatially integrating it in territorial armatures. A specific forest type was proposed for all areas between the 35–40 m contour lines, exactly the height of the best-suited clay layer for tiles. Afforestation in these long zones that diagonally cross the Leie-Scheldt interfluvium is, in other words, an operation of ecological healing. In these areas, afforestation occurs in the land less suitable for agriculture because the land is too steep. The forests here work as a major buffer to halt erosion. Evidently, such a ribbon forest would be interrupted to allow views across the valley, panoramic perspectives on particularities save pockets of productive farming land and suburban housing; its realization would be incremental, step by step placing a crown on the hilltops that mark the horizon of the interfluvium (Figure 21.3).

The crown forest is shared by the project for Marke and Hoog Kortrijk, the highway-related part of the city. Dispersed urban functions are relocated here, centralized and rationalized, which give way to an archipelago of monofunctional elements. Today, the area needs densification, a new vitality, and spatial cohesion. Large-scale big boxes mark the territory and, to date, the visitors to the 55,000 m^2 Xpo, regional

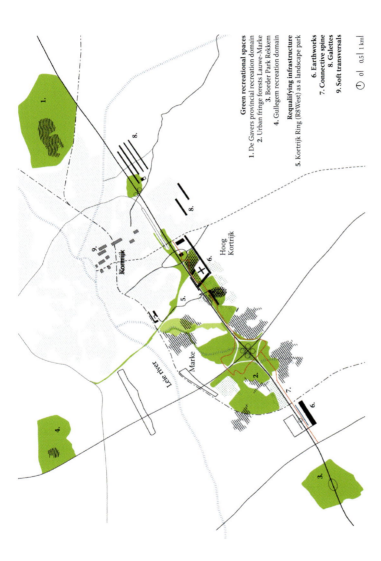

FIGURE 21.2 Southwest Flanders regional landscape and infrastructure strategy. Large recreational spaces serve as compensation spaces in the heavily urbanized region and function as cornerstones of a regional landscape structure in the making, a counter-figure of the urban, inducing quality and structure in the substandard (sub)urban territory. Infrastructure is requalified and the "valley section" is accentuated, emphasizing the spatial quality of the landscape and "repairing" the brutal intervention of the highway. Earthworks anchor a connective spine that creates a new civic space and is linked to an urban fringe forest. This way, the isolation of Hoog Kortrijk south of the highway is broken. East-west oriented galettes structure Hoog Kortrijk with monumental open spaces while soft routes realize transversals connecting to the Leie River. (From RUA, 2010.)

Forests and Trees in the City

FIGURE 21.3 Marke: forest/landcape typologies. The crown forest sets the omnipresent spatial frame that is filled with a wide variety of valley forest types and open spaces that create a diversity of atmospheres that are appropriate for a wide range of uses of varying intensity and scale. The rich palette types (including orchards, flowering trees, etc.) and ground textures create a new public space system for Marke and its surroundings, complement the existing recreational network, and generate new ecologies. (From RUA, 2010.)

hospital, health facilities, business park, retail, regional colleges, and university visited by private car. No one lingers in the area. Other than the few fragmented residential enclaves, there is no everyday life in the suburban area that leaches off the centrality of the historical city while simultaneously being suffocated by its own success. The outdated infrastructure of the 1970s has become insufficient to cope with the ever-increasing attraction to the area. A radical modal shift is required to allow further development of the area. A new tramway system with a hub for the regional hospital and at the university forms the backbone of the proposed mobility on which is anchored a high standard (e-)bicycle path network. This radical shift allows the downgrading of the car infrastructure and, in the long run, the requalification of important parts of the extensive road system into public spaces for the exclusive use by pedestrians, bicycles, and public transport. The central spine of Hoog Kortrijk is one the elements that is envisioned to become a majestic public space framed by monumental tree lines. Amongst others, the oversized Kennedy Avenue that nowadays has a profile of a regional express road (two-lanes each way) is recuperated for that purpose. As a large-scale horizontal and longitudinal element it articulates and structures the undulating landscape, it becomes a collector on which new developments are anchored—not the least programs that enliven and interact with the public space—and remains a connector—be it now for soft modes of traffic such as the bike and the previously mentioned tramway. This reprogramming of car infrastructure goes hand in hand with a simple, be it drastic, shift in spatial character by a systematic planting of multiple tree lines, that simultaneously stress the unity of this majestic public space—an eternally long cathedral of beaches, and adapts itself to the different contexts it passes through: a monumental balcony above the expo halls (that potentially functions as extension of the exhibition space for larger scale manifestations) in the north-east, a regular clearing in the forest in the center, and a plateau carved out of the landscape in the south-west that organizes a vast planted parking space for the regional hospital.

The radical repositioning of Hoog Kortrijk in mobility terms goes hand in hand with a fundamental reformulation of its environment. The ribbon forest is only one of the afforestation strategies intertwined in the plan that makes Hoog Kortrijk part of a regional green structure that links the forest (in the making) of Marke (in the west) and the large nature and recreation area of the Gaevers to the east of Kortrijk city. As systematic as a varied forest figure would be generated by strengthening existing forest structures—including the Kennedy forest (a dump area from the highway construction, which was planted in the 1970s), castle park domain—and developing new ones, including new step-stone forest pockets (such as on a big centrally located parking areas that is taken out of use), planted areas in the university campus, a systematic plantation of open spaces in the college campus, etc. Strategic monumental tree lines (as along the central spine that was liberated from car traffic) simultaneously demarcate public spaces on which can be anchored new activities and link up the mentioned forest figure into a robust spatial framework.

Further development of Hoog Kortrijk basically comes down to densification of the existing and the selective use of clearings within the forest figures (Figure 21.4). Some clearings remain open as internal decompression zones for the hectic activities situated around them. On the larger scale, connections between these clearings connect the inner city with the open countryside while passing through Hoog Kortrijk.

Forests and Trees in the City 437

FIGURE 21.4 Hoog Kortrijk: trees in the city. A rich variety of existing and new forest typologies with a broad palette of tree types and plant species create a wide eco-tone, allowing for diverse atmospheres and to generate new ecologies and recreational settings. (From RUA, 2012.)

FIGURE 21.5 Hoog Kortrijk: "smart" densification. "Smart" densification of the university, college, and student housing would occur along the new public infrastructure and afforested areas. Although urban development in the past was steered through development of road infrastructures, future development would now be guided by the landscape structure—defined by afforested areas, water, and other green structures—defining a robust frame in which densification is possible without new road building. (From RUA, 2010.)

Forests and Trees in the City

The afforestation of Hoog Kortrijk serves different roles. It defines a spatial frame that allows for a requalification—meaning enriching in quality as well as in use—densification, and hybridization of a boring monofunctional suburban environment. It allows the buffering of what needs separation (such as the highway) and uniting by relating and or sharing. The afforestation interweaves an ecological corridor through the area, gives quality to the environment, generates a better microclimate, and plays a role in the water management of Hoog Kortrijk while intertwining with the blue network that systematically revises the current water-management practices. Last, but not least, it challenges conventional planning policies that, until today, solely rely on outdated (and often interpreted as absolute and pure) land-use concepts. The projects for Hoog Kortrijk and Marke challenge the sterile and often unproductive concepts by proposing overlapping layers of forestation and urbanization (Figure 21.5). In a certain sense, Marke and Hoog Kortrijk are both embedded within a forest structure that in the long run, as ecological processes go on, will absorb them. They become complex, multilayered realities. Unlike cities, nature is predominant in their domains. The (reconstructed) "natural landscape" anchors itself in general on the geological conditions, soil composition, climate, hydrology, etc. As such, this is landscape urbanism. It uses the capacity of the (underlying) landscape (structures) to accommodate the urbanization of tomorrow.

21.5 MEKONG DELTA

Meanwhile, in Vietnam's Mekong Delta, "as found" nature has been continually transformed ever since the watery swamp has been drained and inhabited by humankind. Cities in the delta are rapidly urbanizing and modernizing and, at same time, facing severe environmental consequences. Ca Mau is the southernmost city of the delta—one of the lowest lowlands (0.5–1.3 m) in the world and a locale facing extreme effects of climate change, particularly sea-level rise. Cantho is the de facto delta capital, its largest city, and roughly in its geographical center.

Historically, Ca Mau was protected by two massive mangrove forests—a unique freshwater mangrove (U Minh Forest) along the Gulf of Thailand and a saltwater mangrove (Mui Ca Mau) at the southern tip of the country. Originally, these mangroves reached Ca Mau, which was founded only around 1800. After centuries of provincial tranquility, the city entered a dynamic development era while the size of the mangroves simultaneously dropped dramatically. As well, the condition of the mangrove ecosystem deteriorated.

Land domestication in the Ca Mau peninsula implies a process of topographical manipulation, alternatively bringing fresh water from the Mekong Delta and draining the swamplands through a complex canal system. This system was realized piecemeal and interacts with and transforms the natural river and creek system into a hybrid complex that, to some extent, changes nature (salt, fresh, brackish water) with the rhythm of the tides. Internal Viet and French colonization respectively in the 17th and 19th centuries generated the sophisticated water system that turned the territory into a part of Vietnam's notorious "rice basket." The main factor that determines differentiation in the territory is the (always temporary) long gradient of wetness, ranging from (sea and river) water with sedimentation—or, rather, vice versa, sediments with a high degree of wetness—to earth with a low degree of wetness. Each degree of wetness (or solidity) corresponds with natural variations and differentiations in height and suitability for flora and fauna. Forests and salt-tolerant mangroves, in particular, organically, infiltrated areas between water and land and created a process of land-making (Figure 21.6).

RUA (Research Urbanism Architecture) in collaboration with SIUP (Southern Institute of Urban Planning) is currently undertaking design research investigations into Ca Mau's potential future. By 2030, the urban population (presently 270,000) is expected to almost double and reach 450,000. Meticulous documentation of all remaining mangrove and forest relics and of the intertwined and dynamic soil, water, and topography conditions and a forecast of climate change impacts form the starting point of a development scenario. In essence, the landscape urbanism scenario investigates how far the fundamental characteristics of the territory—its "coastal" DNA: land/vegetation/fauna variations in the long gradient of wetness and small height differences—leads to and contains the essential keys and lessons of how to inscribe man-made activities within its majestic natural setting. As it becomes clear that hard engineering alone no longer suffices to fight (with feasible means and at reasonable costs) the forces of nature, the question indeed becomes, rather, how far working *with* (instead of against) the forces of nature can lead.

For Ca Mau, afforestation delivers a frame of coastal protection that works *with* the natural dynamics of the delta (Figure 21.7), heals the salinated landscape in the peri-urban areas south of the city, increases the quality of life in the existing, densifies urban areas, and provides new urban areas in the form of "clearings" in an extended forest. The city's rich tradition of tree-lined boulevards is rejuvenated and simultaneously adapted with the planting of forests from coast to coast. New vegetation would graft onto existing systems—the boulevard network, the freshwater U Minh mangrove forest, the scattered orchard, an urban agriculture mosaic, and the relicts of the coastal mangroves. Interventions would build and canalize existing afforestation programs that fight erosion, rebalance water salinity, and increase food production. Mangrove afforestation along waterways is a necessary "insulator" for the delta landscape and leaves the more inland patches free for sustainable aqua-agriculture (including cleansing oyster and mussel production). "Fallow and unused land is replanted by 'social forestry.'" In particular "extension forestry" (planting alongside canals, roads, and railways) is interesting as it also generates economies and expands the public realm. The forest system resulting from these varied strategies regenerates a forest that protects the land against the turbulent effects of monsoons while simultaneously delivering an endlessly renewable resource.

The design exploits the micro-topographical differences in the territory to choreograph an intertwined system of mangroves and forest, creeks, river branches, canals, and other water elements that can be exploited as shrimp, oyster, clams, scallop, crab, and various fish farms depending on water flows, depths, degrees of salt, etc. This choreography inscribes itself, as much as possible, in the delta dynamics as determined both from nature and from man (including the drastic conversion to shrimp farming). In short, planting becomes planning. A robust forest structure is

Forests and Trees in the City 441

FIGURE 21.6 Ca Mau: a mosaic of water surfaces. Through domestication of the mangrove landscape, a radical distinction is created between lowlands for paddy fields and higher land along the waterways for housing and orchards. Tree planting stabilizes the higher land, but ultimately the landscape remains fluid. The monumental scale of conversion to shrimp and fish ponds transforms the territory instantly and drastically into an endless mosaic of water surfaces that are separated by variations of narrow and wide tree-planted strips. (Courtesy of V. C. Nguyen s.d.)

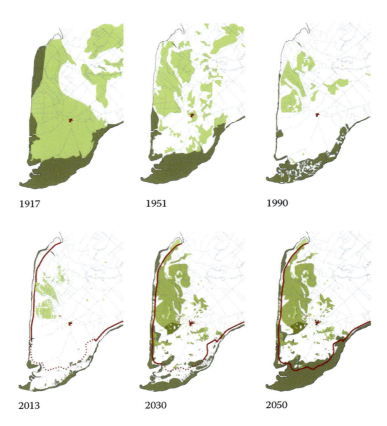

FIGURE 21.7 Ca Mau: demise and afforestation of mangrove forests. In the early 20th century, the Ca Mau Peninsula's alluvial territory consisted of and was generated by saltwater mangroves (Mui Ca Mau) and fresh water mangroves (the *Melaleuca* U Minh Ha). During the last decades, the reduction in forested area is alarming while shrimp farming exploits the opportunities that increased salination offers. Strategies of intensive afforestation are proposed. (From RUA, 2013.)

created that reconceptualizes mangroves and simultaneously generates a frame that embeds the city and its extremely productive countryside (or should one say shore-sides) and that is resilient to climate change (sea-level rise and, with it, saline intrusion). Simultaneously, it generates varied and rich environments for everyday life. It frames a new tropical urbanity (Figure 21.8).

As in Ca Mau, the domestication of the liquid delta landscape in Cantho has everything to do with the making of canals. Also Cantho is in turmoil even more than Ca Mau (unprecedented demographic growth, substantial economic development, large-scale investment programs for infrastructure). The majestic bridges over the upper and lower branches of the Mekong River are only one item on the endless list of new infrastructures that integrate Cantho's economy into contemporary and emerging markets.

Forests and Trees in the City

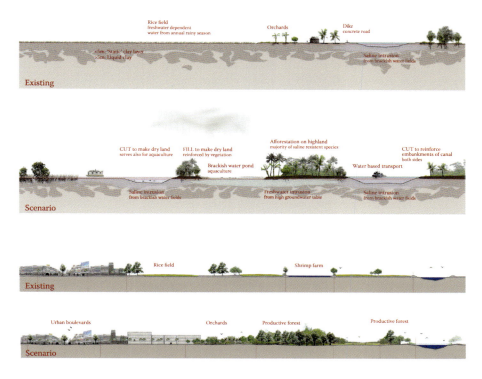

FIGURE 21.8 Ca Mau: profiling the countryside and city. (Re)profiling—always in respect to the cut-and-fill principle—can be a main key in the strategy to adapt to climate change. It can transform the territory by developing new sections. The wetness and soil conditions (re) created by sectional operation could be systematically exploited by replanting forests and mangroves, particularly seen in the before and after countryside sections (top two). In the before and after rearticulation of city sections, a bold, new structure is established, in which water and forest structures, two main topographical landscapes, form a robust, embedded blue/green mesh that protects existing and new urban development. Ca Mau, the blue and green city *par excellence*, is (re)defined by this balanced topographical interplay. (From RUA, PF. Blom, S. Ahi 2013.)

In a 2010 RUA, together with WIT and LATITUDE,* was commissioned a revision of the master plan for Cantho to 2030 when the city of 1.2 million is expected to reach a population of more than two million inhabitants. The domesticated landscape of Cantho arose, as everywhere in the delta, from an enormous swamp. Riverbanks form the highest land in the territory that is created by

---

* The Cantho Master Plan revision team included Belgian team: RUA, WIT Architecten, LATITUDE (K. Shannon, B. De Meulder, G. Geenen, C. Vilquin, P. Dudek, D. Derden, A. De Nijs, and R. Van Durme); Vietnamese team: N. Q. Hùng, L. P. Hiên, T. N. Bình (leaders), T. T. N. Sương, N. L. T. P. Thảo, Đ. T. Quang (planning), T. N. Bình (transport and land form), Đ. T. Thanh Mai (water supply), T. Q. Ninh (waste water and sanitation), P. Q. Khánh (electric supply), N. V. Thắng (planning management), and K. S. P. H. Thảo (infrastructure managment). Client Cantho People's Committee/Cantho Department of Construction. 2010–2013.

sedimentation. Land is nothing more than the alluvial deposits brought by moving water flows. The larger the river, the more sedimentation, the higher the riverbanks. The conflux of rivers induces turbulent flow patterns. Consequently, irregular patterns of sedimentation occur as in Phong Dien, an enormous agricultural area just southwest of urban Cantho that resembles a neuro-image, laced with winding rivers and green. Located between the highlands and the lowlands, this fertile area was naturally taken over by trees and transformed into a mosaic of high-return orchards (Figure 21.9). It forms a unique kind of forest area. O Mon, south of the main island in the Hau River, has a height and similar rich alluvial soil condition on which a "forest" is "natural." Together these two "forests" form the base for the major landscape frame that will, hand in hand with the water, form the structure of Cantho in the new master plan.

Water necessarily is omnipresent in Cantho. From the 16th century onward, canals were established to domesticate the wet lowlands (Shannon 2004, 440–450); the applied cut-and-fill principle mechanically reproduced what river flows created organically: water bodies, banks, and highlands. In general, dyke planting with trees stabilizes dykes, a not unnecessary precaution in such tropical conditions. Occasional flooding, regular monsoons, irregular typhoons, dredging cycles, erosion and silting of the canals and river dynamics perpetuate the never stabilized, perpetually fluid character of the landscape.

Four fundamental land conditions, defined by slight topographical differences, determine the territory of Cantho in the revised master plan: (1) natural and artificial water elements produced by the rhythms of the tides and seasons; (2) lowlands that remain elevated just above or just beneath the water surface amid components of varied scales; (3) "wrinkles or ridges" of plains, squeezed by tectonic forces, that become banks and dykes; and (4) the level between oscillating plains and ridges where orchards flourish. In this "intermediate" landscape, neither low nor high, adaptation recurs in accordance with the ever-fluctuating natural conditions. These four land categories are marginal in relation to the size of the endless low-lying Cantho plain, which remains necessary to maintain in order to give space to the monsoon flows of flood waters that regularly overwhelm the territory. The spreading as wide as possible of the flood waters in the plain can tame water flows of immeasurable force and manipulate and make use of the fertility that the water carries.

The revised master plan for Cantho reorganizes the interplay between these four fundamental landscape conditions, mainly by manipulation of topography, and a cut-and-fill balance is a condition *sine qua non*. In a contemporary interpretation of the indigenous system of organized dispersal, a highland network of roads and platforms for both urban and rural development is counterbalanced at the intermediary level by lower-land waterways, floodplains, and vegetation meshes so that the growth of Cantho is distributed through an archipelago of highland platforms. These are predetermined safety levels of safety (rising to 2.70 m).

A highland 60-km-long "civic spine" links the different centers. At certain occasions, the linear park and integrated water management merge with the civic spine that is conceptualized as a huge new ridge. The spine's giant scale gives concrete shape to the backbone of the city as an enormous platform along the Hau River,

Forests and Trees in the City

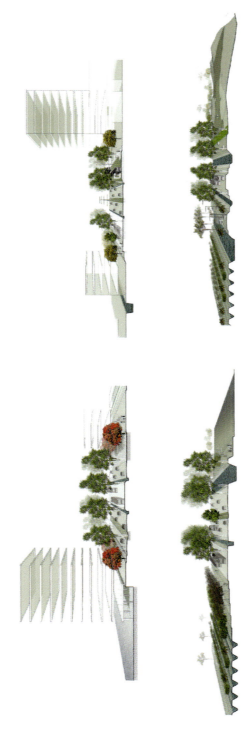

FIGURE 21.9 Cantho: tree planting variation in continuity. A generous profile, as is in use throughout Vietnam for express roads, is the reference for the complete width of the civic spine, that is, composed as a complex assemblage and applies cut-and-fill principles. The profile is systematically planted with trees and incorporates storm water management, parking, and a hierarchy and types of circulation levels. The profile, while keeping some elements strictly fixed, adapts itself to different contexts it passes through (existing city, new centers, agricultural land in between) and thereby creates complementary sets of atmospheres and microclimates generated by the massive tree planting on the civic spine—opening invitations for all types of planned and unplanned, formal and informal public uses. (From RUA, 2010.)

marking rhythms and defining new higher lands for building. The spine interplays with a new layer of multifunctional waterworks that complete and integrate a waterscape for retention, discharge, and treatment but also productive use. It is structurally woven within topography, hydrology, soil conditions, and a new urban morphology.

The master plan defines the new highland (basically the spine and adjacent platforms) and compensates the formation of this topographical wrinkle with the extensive creation of water bodies. It also defines the large area that takes an intermediary height position to safeguard the island in the Hau River and a domain of relatively high land along the riverbank. The master plan saves these fertile resources for the development of a high-tech agro park, which complements the Phong Dien natural orchard area as the main "forest" structure of the city. This resilient high-tech agro park will combine new upscale orchard tissues with various fields for agricultural research and experimentation. The high-tech agro park, Phong Dien, and the city-scale linear park, which is intertwined with the civic spine, form the major green structure of the future city (Figure 21.10). Together with the waterscape, for which the majestic Hau River defines the backbone, this green-blue mesh that extends for 60 km and forms the "counter-figure" of the city. In opposition to classic and earlier mentioned nature counter-figures, this blue/

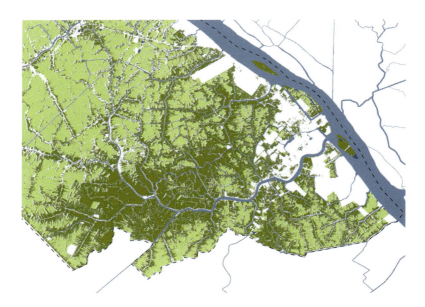

FIGURE 21.10 Cantho: "natural" forests interwoven at confluences. The existing green structure is in direct relation to the topography and hydraulic management: paddy fields on the lowlands while the intermediate and very fertile level is generally planted with orchards. Phong Dien, in the southwest, straddles a tight network of Cantho River tributaries with banks of intermediate height that assemble a most extraordinary collection of orchards. Turbulence in water flows at the confluence of the Hau and Cantho Rivers generates both density as the form of this finely mazed river system and banks that naturally host trees. (From RUA, 2010.)

Forests and Trees in the City

FIGURE 21.11 Cantho: choreographed flooding and green network. Fertile higher land along the Hau River, a result of sedimentation, is as much as possible safeguarded from further urbanization and strategically embedded into the green framework. The Hau River Park is developed as a high-tech agricultural park and closely linked to new technology in agriculture and aquaculture methods. When flooding occurs, settlement and orchards are "safe," and low paddy land temporarily accommodates water. (From RUA, 2010.)

green counter-figure is not so much a mirror or complement but rather a park-waterscape in which the fragile archipelago of urban centers is embedded. As understood by now, the green is positioned on an intermediate, in-between level—neither lowland, nor highland—which can strengthen Cantho's identity and yield a new lucrative economy via orchards and high-tech agro-industry and, more importantly, which can temporarily be taken over by water, the element to which it here all comes in the end (Figure 21.11).

All these proposed landscape urbanism strategies enhance the creation of a climate-resilient and adaptable city in the heart of the delta.

21.6 FORESTS AND TREES IN DELTA CITIES

Southwest Flanders and the Mekong are both delta regions and share the fertility that comes with such a condition, which allows for intensive exploitation of the territory and results in a dense occupation of the territory. They do not accidently share an extreme entrepreneurial attitude, which, in turn, leads to what Dennis Cosgrove once qualified as "promiscuous landscapes" while discussing the Veneto (Cosgrove 2006). They are both landscapes that, without too much consideration, continuously adapted through ad hoc, self-organized, man-made manipulation with only immediate (economic) gain as their primary goal. It goes without saying that such a spirited and utilitarian attitude leads to an almost complete erasure of almost all relicts of the indigenous, "as found," natural environment.

At the beginning of the 21st century, multifarious, organic, bottom-up processes that steer development results in monotonous environments, ecological crises, and lack of resilience as well as a possible economic impasse. Both in Southwest Flanders as in the Mekong Delta, substantial afforestation is used as a strategy to reconstruct a "natural" frame for settlements, which is able to embed further urban development, construct a natural but nevertheless very forceful resilience, contribute to water management, improve the microclimate and environmental quality, potentially diversify the economy and generate—why not—a more beautiful city.

The projects discussed above implement afforestation policy through landscape urbanism strategies. They scan their territory's soil conditions, variations in topography, and wetness to deduce optimal locations and structures for afforestation, which, in turn, adapts themselves simultaneously to different natural conditions without losing sight the their assignment as a frame for future qualitative and sustainable urban development.

REFERENCES

Bridel, J. B. (1798). Manuel pratique du forestier. Ouvrage dans lequel on traite de l'estimation, exploitation, conservation, aménagement, repeuplement, des semis & plantations des forêts, avec les moyens de prévenir la disette des bois de construction & de chauffage, &. Paris: Baudelot & Eberhart.

Cosgrove, D. (2006). Los Angeles and the Italian città diffusa: Landscapes of the cultural space economy. In *Landscapes of a New Cultural Economy of Space*, T. S. Terkenli and A.-M. d'Huatesette (eds.), 69–91. Leiden: Springer.

Cuc, L. T. (1999). Vietnam: Traditional cultural concepts of human relations with the natural environment. *Asian Geographer*, 18:1–2, 67–74.

Da Costa Meyer, E. (2013). Mass-producing nature. In *Public Nature: Scenery, History, and Park Design*, E. Carr, S. Eyring, and R. G. Wilson (eds.), 73–86. Charlottesville and London: University of Virginia Press.

Dang, T. K. P., E. Turnhout, and B. Arts. (2012). Changing forestry discourses in Vietnam in the past 20 years. *Forest Policy and Economics*, 25: 31–41.

De Meulder, B., J. Schreurs, A. Cock, and B. Notteboom. (1999). Patching up the Belgian Landscape/Sleutelen aan het Belgische stadslandschap. Betreffende de politieke economie van een nonchalante wooncultuur. *Oase. Architectural Journal/Tijdschrift voor Architectuur*, 52: 78–112.

De Meulder, B. and K. Shannon. (2010). Traditions of Landscape Urbanism. *Topos* 71: 62–67.

De Meulder, B. and K. Shannon. (2012). Dalat: Archetype of a city embedded in nature. *Conference proceedings on Revision of the Masterplan of Da Lat Conference*, organized by the People's Committee of Da Lat, July 27, 2012, Da Lat, Vietnam.

De Meulder, B. and K. Shannon. (2013). Mangroving Ca Mau, Vietnam. Water and forest as development frames. In *Water Urbanisms East*, B. De Meulder and K. Shannon (eds.), 118–137. Zurich: Park Books, UFO Explorations of Urbanism 3.

FAO (2012). *State of the World's Forests*, Rome: Food and Agriculture Organization of the United Nations.

Forests and Trees in the City

Forrest, M. and C. Konijnendijk. (2005). A history of urban forests and trees in Europe. In *Urban Forests and Trees*, C. Konijnendijk, K. Nilsson, T. Randrup, and J. Schippernjn, (eds.), 23–48. Berlin and Heidelberg: Springer-Verlag.

Mather, A. (2000). Afforestation: Progress, trends and policies. In *NEWFOR—New Forests for Europe: Afforestation at the Turn of the Century, Proceedings of the Scientific Symposium, February 16–17, 2000, Freiburg, Germany*, N. Weber (ed.), 12–20. EFI Proceedings No. 35, Joensuu: European Forest Institute.

Meyfroidt, P. and E. F. Lambin. (2008). Forest transition in Vietnam and its environmental impacts. *Global Change Biology*, 14: 1319–1336.

Shannon, K. (2004). *Rhetorics & Realities. Addressing Landscape Urbanism. Three Cities in Vietnam.* Ph.D. diss., KU Leuven.

Van Gossum, P., Arts, B., and Verheyen, K. (2012). "Smart regulation": Can policy instrument design solve forest policy aims of expansion and sustainability in Flanders and the Netherlands? *Forest Policy and Economics* 16: 23–34.

Van Gossum, P., Ledene, L., Arts, B., De Vreese, and Verheyen, K. (2008). Implementation failure of the forest expansion policy in Flanders (Northern Belgium) and the policy learning potential. *Forest Policy and Economics* 10: 515–522.

Yu, K., Li, D., and Li, N. (2006). The evolution of greenways in China. *Landscape and Urban Planning*, 76: 223–239.

Zhao, C. (2007). *Socio-Spatial Transformation in Mao's China. Settlement Planning and Dwelling Architecture Revisited (1950s–1970s)*, unpublished doctoral dissertation. Leuven: University of Leuven. Department of Architecture.

SUGGESTED READING

http://www.milliontreesnyc.org/html/about/forest.shtml.

Nielsen, A. and R. B. Jensen. (2007). Some visual aspects of planting design and silviculture across contemporary forest management paradigms—Perspectives for urban afforestation. *Urban Forestry & Urban Greening* 6: 143–158.

Oldfield, E., Warren, R. J., Felson, A. J., and Bradford, M. A. (2013). Challenges and future directions in urban afforestation. *Journal of Applied Ecology*, 50:5, 1169–1177.

Shannon, K. and B. De Meulder. (2013). Revising the Cantho Masterplan (Vietnam). Pilotage of a civic spine in a blue-green landscape mesh. In *Water Urbanisms East*, B. De Meulder and K. Shannon (eds.), 138–163. Zurich: Park Books, UFO Explorations of Urbanism 3.

Tyrväinen, L., Pauleit, S. Seeland, K., and de Vries, S. (2005). Benefits and uses of urban forests and trees. In *Urban Forests and Trees*, C. Konijnendijk, K. Nilsson, T. Randrup, and J. Schippernjn, (eds.), 81–114. Berlin and Heidelberg: Springer-Verlag.

Index

Page numbers followed by f and t indicate figures and tables, respectively.

A

Aabach river, 215
ABSP, *see Arten- und Biotopschutzprogramm*
 (ABSP)
Accacia sp., 318–319
Acropora cervicornus, 407
Acropora palmata, 407, 408
Adaptive design tests, 390; *see also* Green
 infrastructure
Aeropuerto de Ciudad Real, 55
Aesthetica, 80
Aesthetic and utilitarian landscape, 140–141
Aesthetic aspect, 95–96
Aesthetic engagement, 80
Aesthetic functionalism, 3
 mechanistic functionalism, 3
 organicistic functionalism, 3
Aesthetic mediation, 80
Affordances, 96
Afforestation, 430, 432
 defined, 302
 of Hoog Kortrijk, 439
Africa, great green wall in, 316–318, 317f
Agence Ter project, 52
Agility, 62–65
 defined, 62
Agroforestry systems, 336
"Aid-embargo," 324
Airport crossing, design proposition for, 36f
Alberti, Leon Battista, 230
Allen, Stan, 72, 195
The All London Green Grid, 20
Allopoiesis, 242
Alphand, Jean-Charles Adolphe, 429
American Civil War, 6
American Great Plains, 307
Amorphous stream landscape, 186
Analogy
 meaning of, 227
 origin of, 227
 types of, 225
Anarchitectural incisions, 200
Andean mountains, 356
Antropotechnique, 96
Architecture, 47
 and urbanism, 37

Architecture of the Well-Tempered Environment, 233
Architecture of Vitruvius, 230
Aridity, 308
Arten- und Biotopschutzprogramm (ABSP), 295
Artificial ecologies, 72
Artistic intuition, 11
Atlantic Ocean, 316
Aubreville, André, 305
"Auto-construction," 328, 333
The Autopoeisis of Architecture, 242
Autopoiesis, 242
 defined, 117
Avicennia germinans, 421

B

Bach, J. S., 233
Baden-Württemberg, network planning on,
 293–295
Baer, William C., 48
Banham, Reyner, 233
Banking policies, 48
Baraccopoli, 183
Barreiro's historic settlements, 179
Barros, V. R., 392
The Basic Unit of Life, 226
Baumgarten, Alexander, 80
Bavaria, network planning on, 295–297, 296f
Beautification, 92
Beauty, 3–4
Beethoven's Fifth Symphony, 233
Beijing, dust storms in, 311
Bélanger, Pierre, 29, 72, 141
Bénard cells, 121
Benevolo, Leonardo, 37
Benjamin, Walter, 221
Berger, Alan, 59
Berlin
 plans from, 257–258, 257f
 transatlantic lens on, 251–252, 252f
Best management practices (BMP), 399
"Biggest land grab of Latin-America," 324
Biodiversity
 in cities, 210
 conservation in abandoned gravel quarries,
 218f
 in monocultural landscapes, 216f

451

Biodiversity Strategy, 252
Biological analogy, 225
Biological diversity, climate change effects on, 286
Biological systems, complexity of, 120
Bioremediation, 40
Bio-retention cells plan, for La Parguera, 413f, 414
Biosphere, endosomatic and exosomatic energies in, 75f
Biosystems, 115
Biotopes, 287
"Bird habitat," 40
Blue City Project, 272, 273f
BMP, *see* Best management practices (BMP)
Bohai Economic Rim, 311
Bohr, Niels, 228
Bolm, Holger, 429
Boltzmann constant, 122
Bonfadio, Jacopo, 144
Bookkeeping model, 94
Borges, Jorge Luis, 39
Braae, Ellen, 140
Brazilian National Institute for Space Research, 392
Brew, Kathy, 203
Brillouin, Leon, 123
Broadacre City, 234
Brunelleschi, Filippo, 230
Bucida buceras, 419
Buffer zones, 288
Building with nature program, 275, 276, 278, 281, 282
Bulwark strategy, 276
Bundesländer, 298
    legally protected biotopes of, 291
    national biotope network, 289
Burkhardt, R., 291

C

Calatrava, Santiago, 55
Calderon, García, 359
Ca Mau, 441f
    afforestation of mangrove forests in, 442f
    land domestication in, 439
    profiling the countryside and city, 443f
    protection of, 439
Caminada, Gion A., 214
Camp Corail, artificial flow channel in, 332
Canaan
    development framework for, 333–334
    emergence of, 327
    environmental risks in, 330
    inclusive city, 335–336
    site characteristics of, 330–333, 331f
Cantho, 444
    choreographed flooding and green network, 447f

land conditions in, 444
master plan for, 444
natural forests, 446f
tree planting variation, 445f
Cantus firmus, 231
Capital flows, 30
Cartesian grid, 228
The Causes and Progression of Desertification, 305
Cavallini, Julio César Moscoso, 363
CCC, *see* Civilian Conservation Corps (CCC)
CEN-SAD, *see* Community of the Sahel-Saharan States (CEN-SAD)
Center for Watershed Protection (CWP), 410
Chery, Jephte, 343
Chillon river, 356
    linear park, 381, 382f, 383f
China, 3 North Shelterbelt Project in, 311–315, 312f, 313f
Chinese Poplar, 315–316
Cicero, M. T., 227, 230
City Climate Development Plan, 252
City of Arts and Sciences in Valencia, 55
City of commerce, 226
"City of thousands pine trees," 427
City planning, 47
"The City That Never Was," 47
"City without the city" model, 6
Civil engineering, 47
Civilian Conservation Corps (CCC), 308
Clement, Frederic, 114
Clément, Gilles, 78
Climate change and ecological networks, 286
"Climate Change Park," 430
Climate-proof design, 92
Climats, Forêts et Désertification de L'Afrique Tropicale, 305
Climax community, 115–116
The Coastal Shelterbelt Development Program, 311
Coexistence and habitats of humans, fauna, and flora, 217f
Colonial silviculture, 305
Communication theory, concept of information, 120–121
Community-based activity, in onaville, 345–346; *see also* Onaville
Community gardening, 336
Community of the Sahel-Saharan States (CEN-SAD), 318
Comunidad de Madrid, 55
Confined Ontic Open System (COOS), 97–98
    visualization of, 99f
Connecting element, 288
Connection element
    defined, 295
Connectivity, defined, 254; *see also* Green infrastructure (GI)
Connolly, Peter, 151

Index

Conocarpus erectus, 419, 421
Constructed wetland, 368
Contrapuntal analogy, 232
Contrapuntal composition, 231–233
Contrapuntal landscapes, 233–237, 234f,
        236f–241f
Contrapuntal thinking, characteristics of, 232
Contrary energy, defined, 80
Contrary motion, 231
Convention of modernism, 209, 211
"Conveyance or distribution network," 72
COOS, *see* Confined Ontic Open System
        (COOS)
"Corail-Cesseless," 325
Corboz, André, 150
Core areas, defined, 294
"CORINE land cover," 291
Corner, James, 39, 52, 143, 194, 198, 214
Cosgrove, Dennis, 447
Cosmopolitan Commons: Sharing Resources and
        Risks across Borders, 43
Counterrevolution, 225
Coup de main, 123
Creative fitting, 13
Cultural landscapes, 5, 13, 15f
        defined, 209
Culture, defined, 125
Cultures, designing, 214–215
        test design gravel quarry, 217–219, 218f
        test design river delta, 215–217, 216f, 217f
        test design urban district, 219–220, 220f
CWP, *see* Center for Watershed Protection
        (CWP)
Cyperus polystachyos, 419

D

Daoistic belief, 95
De Beldomandi, Prosdocimo, 230
De Block, G., 156
Deficiency areas, defined, 295
De Geldersee Poort Project, 271
Degradation, 307
De Jong, T. M., 165
Delauney triangulation, 228
Delta Act, 275
Delta cities, forests and trees in, 447–448
Delta Program, 275
De Natura Deorum, 227
De Oude Scheepswerf Project, 158
De Rodriguez, Rosa Maria Miglio Toledo, 363
Descombes, Georges, 138, 145
Desertification
        definition of, 304
        timeline of, 314f
Design and ecology, 92–93
"Designing landscapes as evolutionary systems," 39

"Design testing and treatment space demand
        estimation charts," 368
"Design testing and water demand calculation
        tool," 367
Design with Nature, 197
Deutsche Technik, 18
Development areas, 288
        defined, 294–295
Deviations, defined, 232
De Vriend, H., 276
De Zaat Project, 158
Direct motion, 231
Disaster, backgrounds of humanitarian, 324–325
Dispositional function, defined, 118
Dissipative structures, 121
        defined, 97, 117
Diurnal pulse, of power flows, 74f
Dobbelsteen, Andy van den, 87
Domestic wastewater treatment parks and
        corridors, in Lima, 371–374, 375f,
        376f; *see also* Lima
Drought, 307
"Dust Bowl," 316
Dust storms, in Beijing, 311
Dynamic infrastructure, 37

E

ECA, *see* Economic Commission for Africa (ECA)
Ecological apocalypse, 34
Ecological arguments, 52
Ecological crisis, 20, 40, 211
Ecological design, 78
Ecological intelligence, 72
Ecological network
        areas with special importance for, 295
        components of, 288
        definitions, 287–289
        establishment of, 287
        Federal Nature Conservation Act and, 285
        multifunctional habitat networks, 288
        regional planning of, 288–289
        species-oriented habitat networks, 288
        types of, 288
        valuation parameters to measure the sites
                for, 292t
        in woodlands, 291
Ecological network planning
        history
                background, 285–286
                planting hedgerows, 286–287
        importance of, 289
        national biotope network, 289–293
        on regional and local levels
                Baden-Württemberg and Karlsruhe,
                        293–295
                Bavaria, 295–297

"Ecological performance," 30
Ecological planning and design, 12
Ecological projects, 52
Ecological restoration, 179
Ecological succession, 115
Ecological systems, remaking of, 40
Ecological urbanism, 29, 37
Ecology-driven design logics, 58
Economic Commission for Africa (ECA), 318
Economic crisis, 47
Economic decline, 252
Economic regimes, neoliberal, 29
The Economics of Ecosystems and Biodiversity
        (TEEB), 250
Economy and green infrastructure, 389; *see also*
        Green infrastructure
Ecoshape Consortium, 276
Ecosystem, 40
    boundaries, 116
    development, 116–117
        principles of, 124
    Eugene Odum's organismic understanding of,
        114–116
    functional units, 118–119
    function of, 115
    homeostasis, 117–118
    Howard Odum's physicalistic understanding
        of, 119–121
        limits of, 121–124
    limits of
        biological systems, 121–122
        concept of information, 122–124
    as ontological units, 116
    organismic understanding of, 114–116
    physicalistic understanding of, 119–121
    and productivity, 120
    self-organization, 117
    services, 92, 119
"Eco-System-Services," 281
Ecosystem services (ES)
    awareness of, 259
    meaning of, 250–251, 250t
    range of, 255–257, 256f
"Ecosystem services assessment toolbox," 401
Edge city, 154
Ehrensberger, Viviane, 214
Eidetic Operations and New Landscapes essay, 199
Eisenman, Peter, 55
Elbe River, 101
The Ems, full hybrid project, 101–103, 104f–110f
    challenges in, 102, 102f
    and tidal polder, 108f
Ems estuary, adaptation of, 104f
Energese, defined, 119
Energy
    defined, 119
    designing to catch and store, 78f

infrastructure, 72–76
landscapes, aesthetic qualities of, 80
Energy, in landscape infrastructure
    design, 76–77
    gardening, 78–80
Entropy, 59, 121, 123
    defined, 200
    thermodynamic concept of, 120
Environment, defined, 274
Environment, Power and Society, 73
Environmental art, 81
Environmental concerns, 29
Environmental decline, 252
Environmental planning, 254
"Environmental Risks, Gardening, Soil- and
        Watershed Protection," 341
Environmental risks, in Canaan, 330
Equilibrium, 117
ES, *see* Ecosystem services (ES)
Eucalipto grandis, 359
European Commission, 248, 401
European Environment Agency, 291
European Natura, 285
Evolutionary thermodynamics, 97
Exergy, defined, 121
Extension forestry, 440
Extrapolation, 92

F

Faidherbia albida, 318
"Famous First Bubbles," 48
FAO, *see* Food and Agriculture Organization (FAO)
The Farmland Shelterbelt Network of the Plains,
        311
Federal Ministry of Education and Research, 361
Federal Nature Conservation Act of 2002,
        285–287
Federal Plan for Wildlife Paths, 297
Federal Transport Infrastructure Plan, 289
Feng shui forests, 429
Fertile Crescent, 227
Fertility, 59–62
Ficus benjamina, 359
Field Operations, 195
Field Operations project, 52
First law of thermodynamics, 119, 121
Fiselier, J., 275
Flanders
    light-rail projects in, 154
    relationship between infrastructure and
        urbanization in, 154
Flanders, southwest, 431–439, 434f, 435f
    forest index of, 431
    short-rotation forestry, 432
Flevoland Polder, 267
Flood control, 393

Index

Florida Land Rush, 48
  advertisement and photograph, 48, 50f
Flows of energy, in environmental systems, 73
Food and Agriculture Organization (FAO), 318, 430
Forebay, in La Parguera, 421–422, 421f–423f
"Forestry Action Plan (1998–2003)," 431
Forestry management, 431
"Forestry socialization" program, 431
Forests and trees, in delta cities, 447–448
Foucault, Michel, 37
Fournier, Colin, 30
The Fractal Geometry of Music, 231
Frederick Law Olmsted, 5
Free movement, 8
Fresh Kills landfill, 194, 194f
  intervention, 200–202, 201f
  residency, 202–204, 203f
  variances in, 195–200, 195f–197f, 199f
Functional design, 92
Functionalistic aesthetics, 3
Functional units, 118–119
Fundamentals of Ecology, 114
Fux, Johann, 230

G

Galicia City of Culture, 55
Garber, Peter M., 48
Garden
  beauty and management of, 4–5
  concept of, 138
  defined, 145
  designing with nature, 13
  importance of, 151
  as machine, 143, 144f
  as a third nature, 144–145
    garden of Lancy, 146–147, 146f, 147f
    garden of River Aire, 147–148, 148f–150f
Gardening, 78–80; *see also* Landscape
    infrastructure
Gardens in movement, 79, 79f
Geist, Helmut, 305
Geographic information system (GIS), 225, 291,
    363
Geometric Networks Landscape, 186
German Academic Exchange Service, 330
German Advisory Council, 286
German Federal Agency for Nature
    Conservation, 289
Germany
  connectivity axes and core areas of, 290f
  ecological network in, 285
  establishment of ecological network in, 287
  green functionalism in, 14–19, 15f
  problem of network planning in, 297
Gethmann, Daniel, 215
GGW, *see* Great green wall (GGW)

Ghost airports, 55
GI, *see* Green infrastructure (GI)
Gibson, James, 96
GIS, *see* Geographic information system (GIS)
Glaeser, Ed, 66
The Global Campaign on Urban Governance, 335
Global urban spatial products image matrix, 49f
Golden age of polyphony, 230
Goudplevier Project, 271
Gradus Ad Parnassum, 230
Grasslands, role of, 308
Gray infrastructures, 20, 249
*Greater Perfections—The Practice of Garden
    Theory*, 144
Great green wall (GGW), 301, 316–318, 317f
Green areas, water demand for the irrigation of, 361
Green functionalism, 4
  aim of organicistic, 5
  in Germany, 14–19
    cultural landscape, 15, 15f
  in United States, 5–14, 10f
Green infrastructure (GI), 5, 30, 171–172, 301, 362
  acceptance, challenges for, 387–388
    appearance, 389
    economy, 389
    performance, 388–389
    working method, 389–390
  adaptive design
    and designed experiments, 390, 391f, 392f
    and planning research agenda for,
      400–401
  characteristics of, 20–21
  climate change and rainfall in São Paulo,
    392–393
  core elements of, 254–255
  coverage of, 257–259, 257f, 258f
  defined, 387
  and ecosystem services, 247–248
    application of principles, 255–259
    methodological approach, 253–255, 254t
    planning practice and scientific
      discourses, 259–261
  field experiment in São Paulo, 396–398, 397f,
    398t, 399t
  introduction of, 388
  versus landscape, 19–20
  meaning of, 248–249
  planning, 23
  planning principles, 249t
  urban flood management in São Paulo,
    393–396, 394f, 395f
Green infrastructure, in La Parguera, 413f, 416,
    416t–417t
  green street bio-retention cell, 418f, 419
  mangrove restoration and forebay, 421–422,
    421f–423f
  sand filter, 419–420, 419f–420f

456   Index

Green investments, 92
"Green lungs," 429
Green space structures and district structure, 220f
Green storm-water infrastructure project, 261
Green street bio-retention cell, in La Parguera, 418f, 419
"Green wall" projects, 308
Green-washing, 41
Greifensee lake, 215, 217
Grensmaas Project, 271, 272f
Grimm, A. M., 392
Gross, Matthias, 214
Gross Max project, 52
Groundlab, for Shenzhen, 30
Guerra, Roberto, 203
Guin, Raphaela, 345

H

"Habitat Haïti 2020," 326
HABITAT-NET, 291
The Hague, 276
Haiti
    aftermath of quake, 325
    disaster, backgrounds of, 324–325
    earthquake and after effects in, 323–324
    16/6 Program in, 337, 338f
    research project, background of, 330
    residents and academics in, 339
    rise of new city, 325–327, 327f
    showcases from, 337–339, 337f
    urban crisis in, 327–328, 329f
Hampe, Michael, 210, 211, 212, 214
Hannemann, Johann-Christian, 339
Happy Isles, 237
Hartberg, Martin, 325
Hau River, 444, 446
Hauser, Susanne, 215
Healthy open space, 5–6
Heimhuber, Valentin, 331
Heraclitus, 241
Hibiscus pernambucensis, 421
"History of collective ideas," 269
Holistic worldview, 4
Homeostasis, 115, 117–118
Hoog Kortrijk, 433, 436
    afforestation of, 439
    densification in, 438f
    development of, 436
    trees in, 437f
    water management of, 439
Hough, M. H., 93
Hubig, C., 125
Human ecology, 34
Human experience and judgment, affect of environment on, 95
"Humankind's greatest invention," 66

Human territory and natural surfaces, 37–39
Humboldt current, 356
Hunt, John Dixon, 138, 144
Hurricane Sandy, 276
Hydro-urban typologies, in Lima, 369–371, 370f
    domestic wastewater treatment parks and corridors, 371–374, 375f, 376f
    irrigation channel water treatment parks and corridors, 374–379, 377f, 378f
    polishing parks and corridors, 371, 372f–374f

I

IDP, *see* Internally Displaced Persons (IDP)
IHRC, *see* Interim Haiti Recovery Commission (IHRC)
Illegal settlements, 183
IMP, *see* Metropolitan Planning Institute (IMP)
"Inclusive City," defined, 335
Industrial cities, 72
Industrial development, 177
Industrial revolution, 37, 429
Ineichen, Stefan, 210
"Infrastructural urbanism" article, 72
Infrastructure
    construction and consequences of, 10
    defined, 72, 171
    as landscape, 406
    terms of nature as, 173t, 174t
Infrastructure policy, 38
Infrastructure systems, 29
Instrumental knowledge, 126
Integrated Economical Zone, 326
INTERACTION—Journal of the Tureck Bach Research Foundation, 232
Intergovernmental Panel on Climate Change, 392
Interim Haiti Recovery Commission (IHRC), 325
Intermediating structure
    defined, 175
    terms of nature as, 174t
Internally Displaced Persons (IDP), 325
"International community," 326, 347–348
Intersection net, 77
An Introduction to the Performance of Bach, 232
InVEST tool, 261
Irrigation channel water treatment parks and corridors, in Lima, 374–379, 377f, 378f; *see also* Lima
The Isolated State, 226, 228
i-Tree tool, 261
Iyicil, Gokce, 339

J

Jackson, J. B., 150
Jackson, John Brinckerhoff, 173, 175
Jay Gould, Stephen, 231

Index

457

K

Karlsruhe, network planning on, 293–295, 294f
Kennedy Avenue, 436
Kennedy forest, 436
Kerwin, Sean, 345
Kirchhoff, Thomas, 34
Klein-Brabant region
    analysis of landscape structure of, 158–159, 158f
    Scheldt and Rupel rivers in the test case of, 156–159, 157f, 159f
Koolhaas, Rem, 38

L

Laboratory Madrid, 55–59
    agility, 62–65
    entropy, 59
    fertility, 59–62
    utility, 65
LAF, *see* Landscape Architecture Foundation (LAF)
Laguncularia racemosa, 421
Lake Sahara, 317
Lally, Sean, 86
Lancy, garden of, 146–147, 146f, 147f
Land domestication, in Ca Mau peninsula, 439
Landespflege, 17, 19
Landesplanung
    aim of, 17
    establishment of, 16
    objective of, 17
Land Law 1993, 431
Land Law of 1998, 56
Landscape, 173
    beyond infrastructure and superstructure, 190
    defined, 209, 269
    design, 151
    design and ecology of, 92–93
    as design medium, 194
    development and human sense, relationship between, 95
    development of, 268
    functional design, 92
    as infrastructure, 141–143
    as intermediating level, 173–175
    making to making nature, 269–278, 270f, 272f, 273f, 277f–279f
    objectives and concepts, 176
    planners, 173
    qualities, 175, 176t
    strategies
        integrating landscapes, 183–185, 184f, 185f
        PEATLANDscape, 179–183, 181f–183f
        two landscapes, 186–190, 186f–189f
        urban ensemble, 176–179, 177f–181
    theory, 209

Landscape, aesthetic qualities of
    closing energy-nature loop, 81–86, 84f, 85f
    energy exchanges and connections within, 80–81
Landscape and Species Protection Program, 258
Landscape architects
    as ecological planners, 13
    role of, 12
Landscape architecture, 47, 53–54, 319, 388
Landscape Architecture Foundation (LAF), 388
    categories of, 388
Landscape as energy infrastructure
    design, 76–77
    energy and potential energies, 73–76
    energy flows in, 72–73
    gardening, 78–80
    horizontal nature of, 72
"Landscape as Urbanism" essay, 52
Landscape-based design concepts
    characteristics of, 21–23, 22f
    objective of, 21
Landscape-based urbanization patterns, 158
Landscape cyborg
    defined, 406
    functions of, 406
Landscape designers, role of, 5
Landscape ecological infrastructure, 34
Landscape ecology, 34
    and infrastructure, 29–30
Landscape filters plan, for La Parguera, 413f, 415
Landscape garden, 4–6
Landscape infrastructure, 52
    importance of, 73
Landscape infrastructure systems, 30
Landscape machines, 30, 93–95
    definition of, 94
    metabolism of, 96–100
        cultural meaning, 96
    stages of, 94
Landscape matrix, surrounding, 288
Landscape Performance Series, 388, 389
Landscapes, contrapuntal, 233–237
Landscapes of Holistics and Consistency, 176
Landscape urbanism, 14, 29, 34, 52, 92, 214
    emergence of, 53
The Landscape Urbanism Reader, 30, 52
Landscape Urbanist, 52
Languages, parts of, 225
La Parguera, 407–408, 408f, 424f
    coastal resources in, 408
    green infrastructure in, 413f, 416, 416t–417t
        green street bio-retention cell, 418f, 419
        mangrove restoration and forebay, 421–422, 421f–423f
        sand filter, 419–420, 419f–420f

green infrastructure strategies for, 413f
  landscape context, 409–410, 409f
  methodology and analysis, 410–413, 410f–412f
  pollutants in, 408
  polycyclic aromatic hydrocarbon plume in, 409f
  proposed strategies, 413
    filtering practices/sand filter, 415
    mangrove/wetland restoration and buffer systems, 413f, 414
    permeable pavement/road stabilization, 415
    rain gardens, 415
    rainwater harvesting, 415
    vegetated swales/bio-retention cells, 414
    wet ponds, 414
  setting, 407–408
  tourism industry in, 408
La Parguera Plan, 424
LAREG, role of, 176
Latour, Bruno, 40, 209, 210, 211, 213
Least-social-cost/maximum-social-benefit, 197
Lefèbvre, Henri, 173, 175, 185
Leibniz, Gottfried Wilhelm, 126
Leie River, 432
Leie-Scheldt interfluvium, 433
LEIS, *see* Lima Ecological Infrastructure Strategy (LEIS)
Le Jardin en Mouvement, 78
Le Manifeste du Tiers Paysage, 80
Lenné, Peter Josef, 429
Lewis, Phil, 235
LID, *see* Low-impact development (LID)
Lifescape, 195
  design for Fresh Kills Park, 197
  master plan, 196f
Lifescape: Fresh Kills Draft Master Plan, 204
Light-rail connection, landscape-embedded, 160–165
  industrial patch regeneration, 162–164, 163f
  Polder edge, 160–161, 161f
  Polder island, 160, 160f
  postproductive landscape, 164–165, 164f
  waterfront, 161–162, 162f
Light-rail technology, 154
Light-rail trajectories, 168
Ligue de Protection des Oiseaux Project, 88f
Lima
  design method for integrated solutions, 366–367
  hydro-urban typologies, 369–371, 370f
    domestic wastewater treatment parks and corridors, 371–374, 375f, 376f
    irrigation channel water treatment parks and corridors, 374–379, 377f, 378f
    polishing parks and corridors, 371, 372f–374f

  treating polluted water, 367–368
  urban landscape development in, 356
  urban water cycle and open space, 356–359, 358f
    decorative approach of open space, 359, 360f
    local ecology, open space system, and hydrological infrastructure, 361
    open space with high demand on water quality, 361
    open space with high water demand, 359–361
  water saving design, 367
  water-sensitive urban design approach of, 365–366, 365f
Lima Ecological Infrastructure Strategy (LEIS), 361–362
  manual, 363
  methodology of, 362–363
  principles, 362–363
  theoretical framework and adaptation, 362
  tool, 363, 364f
Lima Regional Concerted Development Plan 2012–2025 (PRDC), 363
Limited self-organization, 39
Limits of Control, 58
Lisbon, urban development of, 176–179, 177f–181f
Living landscapes, 72
Living organisms, thermodynamic properties of, 98
LiWa research project, 366
Lomas, 356, 357f
"Long-term Forestry Plan (2008–2017)," 431
Los Angeles Land Boom and Bust, 48
Lotka, Alfred, 119, 120
Lower Chillon River Watershed, 362
  strategic projects in, 379
    Chillon River linear park, 381, 382f, 383f
    water-sensitive future, 379–381, 380f, 381f
Low-impact development (LID), 406
Luckner, A., 125
Lurin river, 356
Lynn Margulis, 226

M

MacArthur, R. H., 286
The Machine in the Garden, 137
Madrid metropolitan region, speculative development in, 57f
Magic disappearing act, 199
Magnus Larsson project, 237
"Making of Landscape," 268
Mandelbrot, Benoit, 231

Index

Mangle of Practice, 215
Mangrove afforestation, 440
Mangrove restoration, in La Parguera, 413f, 414, 421–422, 421f–423f
Manifesto for Maintenance Art, 202
Man-made landscapes, 269
"Map of Desertification," 304f
Marot, Sébastien, 150, 225
Marx, Leo, 137–139
Marx's economic theory, 172
MASP, *see* Metropolitan Area of São Paulo (MASP)
Matta-Clark, Gordon, 200
"Maximum power principle," 120
McGrath, Brian, 30
McHarg, I. L., 93, 195
McHargian planning, 54
MEA, *see* Millennium Ecosystem Assessment (MEA)
Mekong Delta, 439–447
Mekong River, 442
Metabolism, of landscape machines, 96–100
Method in Social Science: A Realist Approach, 228
Metropolitan Area of São Paulo (MASP), 392–393
Metropolitan Park Authority, 362, 379
Metropolitan Planning Institute (IMP), 362
Metz, Tracy, 274
Meydag, Herman, 201
Meyer, Elizabeth, 389, 406
Miasma, 6
Michel, Yvonne, 214
Middle landscape, 139
Millennium Ecosystem Assessment (MEA), 250
MillionTreesNYC, 430
Mississippi Bubble, 48
Modernists' conception, of nature, 209–210
Modern Pastoralism and the machine in the garden, 138–139
  aesthetic and utilitarian landscape, 140–141
Mollison, Bill, 77
Monitoring, 92
Moral treatment movement, 7
Moringa leaves, usage of, 343
Morphing Timelines: Energy Project, 204
Mossop, Elisabeth, 30
Motion, types of, 231
Multifunctional habitat networks, 288; *see also* Ecological network
Multifunctionality, defined, 254; *see also* Green infrastructure (GI)
Muscular metropolitan reforestation strategy, 63f
Musical analogy, 225
  origins of, 230–231, 231f
MVVA project, 52

N

NABATEC's plans, 326
Narbonnaise Regional Natural Park, 87
Nash, John, 429
National and Local Water Authority, 363
National biotope network, 289–293
National Ecological Network, 270
The National Program against Desertification, 311
National Socialist landscape, 19
Natural-artificial park, 217
Natural cultural patterns, 229f
"Natural environment," 427
The Natural Forest Conservation Program, 311
Naturalism
  defined, 114, 125
  of the Odum brothers, 124–125
Naturalistic fallacy, defined, 125
"Natural landscape," 439
Nature
  as infrastructure, 171, 172
  modernists' conception and its contradictions, 209–210
  pattern in understanding of, 278–281
  realistic to modernists' conception of, 210–213
  reserves, 210
  terms as infrastructure, superstructure and intermediating structure, 174t
  terms as infrastructure and superstructure, 173t
Nature and Biodiversity Conservation Union Germany, 297
Nature and Forest Agencies, 432
Nature development project, 268
Nature of an area as a beautiful landscape, 212
Nature Policy Plan, 270–271
Natuurmonumenten, 274, 275t
Netherlands
  Delta Program in, 275
  flood protection and water supply, 275
  nature protection organization member in, 275t
  new nature in, 267
  "Plan Ooievaar" project, 270
  "Room for the River" program in, 276
Network core, 288
Network planning, on Baden-Württemberg and Karlsruhe, 293–295
New city, rise of, 325–327, 327f
New Deal Agency, 308
New nature, 267–268
  defined, 268
  making landscape to making nature, 269–278, 270f, 272f, 273f, 277f–279f
  mentality of making, 268–269
New port facilities, 178
New Urbanism, 38, 41

Index

New York City's Department of Parks and Recreation, 430
New York's Staten Island, 194
NGO, *see* Nongovernmental organizations (NGO)
NOAA, *see* U.S. National Oceanic and Atmospheric Administration (NOAA)
Noise reduction, 257, 259
Nolf, C., 156
Nongovernmental organizations (NGO), 324
Nonlinear organization, 58
North Sea, renewable energy production in, 42f–44f
North Sea Master Plan, 41
3 North Shelterbelt Project, 301, 311–315, 312f, 313f

O

Oakeshott, Michael, 242
Oblique motion, 231
Occidental culture, 268
Odum, Eugene, 114, 115, 124
Odum, Howard T., 73, 76, 77, 98, 114, 119, 120, 124, 125
"Odumian" ecosystem theory, 114
Odumian naturalism, critique of, 125–126
Oil rigs
    decommissioned, 105f
    disassembled, 107f
    usage of, 103
Olmsted, Frederic Law, 406
Onaville
    academic action, 346–347
    community-based activity, 345–346
    flood hazard zoning of, 332f
    general assembly and first community workshops in, 340–345, 342f, 345f
    handmade public square in, 348f
Onticness, concept of, 98
Oosterling, Henk, 39
Oostvaardersplassen, 267
    rediscovery of nature in, 270
Open country, 291
Openness, concept of, 98
Open space, 172
Open Space 2100, 258
ORDERin'F project, *see* Organizing Rhizomic Development along a Regional pilot network in Flanders (ORDERin'F project)
Organicism, 114
Organismic holism, 115
Organisms
    and ecosystem, 115
    and spatial boundaries, 116
Organizing Rhizomic Development along a Regional pilot network in Flanders (ORDERin'F project), 155

Ouroboric Urbanism, 61f
Overgrazing, 311

P

PAGGW, *see* Pan African Agency of the Great Green Wall (PAGGW)
PAH, *see* Polycyclic aromatic hydrocarbons (PAH)
Palimpsest, 150–151
Pan African Agency of the Great Green Wall (PAGGW), 318
PAN GmbH Project, 289
The Panic of 1873, 48
Paradigm shifts, 430–431
Parallel urbanism, 237f
Parametric infrastructure design, 31f–33f
Parametric urbanism, 30
Park
    defined, 7
    functionalistic principles of, 8
Parks and Recreation Strategic Action Plan, 259
Pastoralism, 139
Patchworks or Nonlinear organization, 58
Peat land cultures, 179
PEATLANDscape, 179–183, 181f–183f
    concept, 179, 182f, 183f
    strategies, 183
Peck, Raoul, 325
Penetration and Transparency Morphed Project, 203
Pennisetum spp., 419
Penrose, Roger, 231
Pereyra, Maribel Zapater, 363
Permaculture: A Designer's Manual, 77
Permeable pavement plan, for La Parguera, 415
Phase transition, 121
Phillips, John, 115
Phong Dien, 444, 446
Physicalism, 114
Pickett, S. T. A., 407
Picon, Antoine, 37
Picturesque enclaves, 8
Planetary urbanization, 30
Planning, new practice of, 213–214
Plan Ooievaar Project, 270
Polasky, J., 156
Polemics, harnessing, 226–227
"Polishing of treated wastewater," 371
Political-economic projects, 52
Pollak, L., 198, 199
Pollination, 259
Polluted water, treating, 367–368; *see also* Lima
Polycyclic aromatic hydrocarbons (PAH), 408
Polytechniciens, 37
Ponts et Chaussées, 37
Pope, A., 206
Poplar trees, advantages of, 315

Index

Populus simonii, 315–316
Postwar engineering, 29
Potential energy, 73
Praille-Acacias-Vernets, 148
PRDC, *see* Lima Regional Concerted Development Plan 2012–2025 (PRDC)
Preservation, 92
Prigogine, Ilya, 97
Principle of periodicity, 228
Priority areas, defined, 295
16/6 Projects, in Haiti, 337–338, 338f
Prominski, Martin, 39, 213
Public transport, 154
    development of, 168
Puget Sound region, 253
Puget Sound Salmon Recovery Plan, 258
Pythagoras, 241

Q

Quantification, 92
Quasi-Orwellian operational models, 30

R

Rahm, Philippe, 82, 83
Rain gardens plan, for La Parguera, 413f, 415
Rainwater harvesting plan, for La Parguera, 413f, 415
Ravine Lan Couline, 331
Re, M., 392
Recurrent design criteria, emergence of, 165–168, 166f, 167f
Re-design of infrastructure, 29
Red Sea, 316
Reflexive knowledge, defined, 127
Regional public transport networks, 154
    catalyst for urban development, 155
Reich, M., 298
Reichslandschaftsanwalt, 286
Relational design, 215
Re-Raw Recovery Project, 202
"Research by design," 165
Research project, background of, 330
Research Urbanism Architecture (RUA), 440
Resettlement programs, 179
Resilience/Adaptability, 58
Resistance, 117
Resource, defined, 77
Restoration, 311
Rhizophora mangle, 421
"Rice basket," 439
Rimac river, 356
Riparian vegetation, 177
"Rising Currents" exhibition, 406–407
The Rituals of the Zhou Dynasty, 430
River Aire, garden of, 147–148, 148f–150f

Roach, Stephen S., 54
Road stabilization plan, for La Parguera, 415
Roberts, Wallace McHarg, 195
"Room for the River" program, 276, 279f
Roosevelt, Franklin D., 307
Rowe, Peter G., 139
RUA, *see* Research Urbanism Architecture (RUA)
Ruckstuhl, Max, 210
Rupel river, in the test case of Klein-Brabant, 156–159, 157f, 159f
Rural-urban hybridity, 29
Rurbanization, 154

S

Salmon Recovery Plan, 259
Sand filter usage, in La Parguera, 413f, 415, 419–420, 419f–420f
Sandström, U. G., 260
Sansevieria spp., 419
Santo André green infrastructure study, 393–396, 394f, 395f
São Paulo
    green infrastructure
        climate change and rainfall in, 392–393
        and urban flood management in, 393–396, 394f, 395f
    interdisciplinary green infrastructure field experiment in, 396–398, 397f, 398t, 399t
    water quality analysis, 399–400
Sayer, Andrew, 228
Scenery, 9, 10
Scheldt river, in the test case of Klein-Brabant, 156–159, 157f, 159f
Schinus molle, 359
Schumacher, Patrick, 242
Search areas, defined, 295
Seattle
    green structure, 258–259
    plans from, 257–258, 257f
    transatlantic lens on, 252–253, 253f
Secchi, B., 156
Second law of thermodynamics, 97, 98, 119–121, 200
Second nature, 227–230, 229f
Sectorial planning, 254
Seifert, Alwin, 286
Self-assembly, defined, 117
Self-organization, 58
Self-organizing systems, 117
Settlement landscapes, 210
SFA, *see* State Forestry Administration (SFA)
Shannon, Claude, 122
Shannon's information, 122–123
Shelterbelt planting, 307–311, 308, 309f, 310f
    components of, 308

462   Index

Siedlungsverband Ruhrkohlenbezirk (SVR), 16
Sijmons, Dirk, 92
Simon's poplar, 315–316
SIUP, *see* Southern Institute of Urban Planning
    (SIUP)
Sloterdijk, Peter, 39, 96
Slum clearings, 11
"Slumification," 326
Smithson, Robert, 81, 193, 200
Social acceptance, 81
Societal modernization, 34
Society, 209
Socio-political arguments, 52
SODIS, *see* Solar Disinfection (SODIS)
Soil erosion, 10, 303, 311
    in Onaville, 341
Solar Disinfection (SODIS), 345
Solar system and atomic structure, analogy
    between, 228
Sorkin, Michael, 226
Southern Institute of Urban Planning (SIUP),
    440
South Sea Bubble, 48
Space and energy management, 82
Spain
    highway and rail track development, 55
    investment in urbanization, 56
    spatial products, 49f
    urbanized territory expansion, 55
Spatial organization, 177
Spatial Structure Plan of Flanders, 431
Spatial turn, defined, 173
Species-oriented habitat networks, 288; *see also*
    Ecological network
Stahlia monosperma, 421
State Forestry Administration (SFA), 315
Staten Island, 197
"State of the World's Forests," 430
Stationstraat, 156
Steadman, Philip, 225
Storm-water management, 261, 405–406
Strang, Gary, 72, 142, 406
Street tree planting, 430
Stremke, Sven, 87
Structural compass, defined, 187
Subjective Landscape, 4f
Suburbanization, 176, 252
Suburban villages, 8
Succession, parameters of, 115
Sudden state-change, 58
SUDS, *see* Sustainable Urban Drainage Systems
    (SUDS)
Sullivan, Louis, 233, 241
Supergardens, 190
Superstructure, 172
    terms of nature as, 173t, 174t
Surface systems, 143

Sustainable Design, Ecology, Architecture and
    Planning, 76
Sustainable development, 92
"Sustainable forestry," 431
Sustainable Sites Initiative, 388–389
Sustainable Urban Drainage Systems (SUDS), 259
SVR, *see* Siedlungsverband Ruhrkohlenbezirk
    (SVR)
SWA project, 52
Swyngedouw, Erik, 41
System modeling, 92

T

Taihang Mountain Afforestation Program, 311
Taliesin West, 234–235, 234f
Tansley, Arthur, 115
Target areas, 295, 297
Technoecosystems, defined, 125
TECHO organization, 330
TEEB, *see* The Economics of Ecosystems and
    Biodiversity (TEEB)
Tejo estuary, traces along, 176–179, 177f–181f
Tennessee Valley Authority (TVA), 10
"Terra Fluxus" essay, 52
Territorial organization, 34
Terza natura, 144
Terzidis, Kostas, 227
Texas Highway Department (THD), 10
"Texas of Flanders," 432
THD, *see* Texas Highway Department (THD)
Theory of Island Biogeography, 286
Thespesia populnea, 421
Third Reich, 286
Third voice, 23–25
Thompson, Ian, 34, 95
Tiber river, 184
Tidal polder, dynamics of, 109f–110f
Tiezzi, Enzo, 97, 98
"Tippy Tap" techniques, 345, 346f
Topotek project, 52
Topsoil degradation, 331
Topsoil erosion, 341
Touch Sanitation Project, 202
Traffic-calming practices, 393
Transandean tunnel, 356
Transatlantic lens, 251
    on Berlin, 251–252, 252f
    on Seattle, 252–253, 253f
Transdisciplinarity, defined, 255; *see also* Green
    infrastructure (GI)
"Transition zone," 316
Transit-oriented development, defined, 154
Transport technology, 156
Tropical areas, green infrastructure in, 413f, 416,
    416t–417t
    green street bio-retention cell, 418f, 419

Index

mangrove restoration and forebay, 421–422,
421f–423f
sand filter, 419–420, 419f–420f
Tulsa riverfront plan, 238f
Tunnel bridge, 146–147
Tureck, Rosalyn, 231
Turrenscape project, 52
TVA, *see* Tennessee Valley Authority (TVA)

U

UCLBP, *see* Unit for Housing Construction and
Public Buildings (UCLBP)
Ukeles, Mierle Laderman, 202
UNCCD, *see* United Nations Convention to
Combat Desertification (UNCCD)
UN earth summit, 317
Uneven development, 227
United Nations Convention to Combat
Desertification (UNCCD), 304, 317
United Nations Peacekeeping Mission, 325
United States
green functionalism in, 5–14, 10f
Unit for Housing Construction and Public
Buildings (UCLBP), 327
Unity of art and life, 3
Unity of opposites, 241
"Urban Acupuncture," 347
Urban agriculture, 336
Urban crisis, 327–328, 329f
Urban design disciplines, 47, 52
Urban design experiments, defined, 390
Urban development
regional public transport networks as catalyst
for, 155
Urban expansions, 48
Urban farming, 211
Urban forests and trees
functions of, 429
history of, 429–430
Urban form, structuring elements of, 30
Urban fringe forest, 433
Urban gardening, 211
Urban green
role of, 257
for tourism, 257
Urban growth, liberal capitalist model of, 6–7
Urban infrastructure
continuum, 386f, 387
defined, 72
Urbanization
Chinese, 54
contemporary, 52
and economic production, 51
and growth, 48
horizontal, 29
hyper-dense, 54

rates, 52
speculative, 48, 50–51
Urban landscape, 212
of associations, 212
Urban landscape development, in Lima, 356;
see also Lima
Urban Landscape Strategy Berlin, 252
Urban "organism," 34
Urban or landscape planners, 173
Urban park, design principles for, 7
Urban planning, liberal, 11
Urban resilience, 387
Urban sanitation, 336
Urban self-sufficiency, 226
Urban settlement and ecology-driven design
logics, 58
Urban sprawl and nature, 209
Urban Strategies for Onaville Project, 330, 339
Urban water cycle and open space, in Lima,
356–359, 358f
decorative approach of open space, 359, 360f
local ecology, open space system, and
hydrological infrastructure, 361
open space with high demand on water
quality, 361
open space with high water demand, 359–361
U.S. Endangered Species Act, 407
U.S. Environmental Protection Agency (USEPA),
405–406
U.S. Forest Services, 307
U.S. National Oceanic and Atmospheric
Administration (NOAA), 408
USEPA, *see* U.S. Environmental Protection
Agency (USEPA)

V

Value-neutral technology, 126
Value of bigness, 58
Van der Voordt, D. J. M., 165
Van Koningsveld, M., 276
Van Valkenburgh, Michael, 407
Vegetated swales plan, for La Parguera, 413f, 414
"Venice of the East," 427
Vernacular-mobile landscape, 173
Vietnam, forestry in, 431
Viganò, P., 156
Ville territoire, 154
The Voice of Liberal Learning, 242
Voiret, 146
Volgermeerpolder, 272, 274f
von Haaren, C., 298
von Neumann, John, 123
von Thünen, Johann Heinrich, 226
von Thünen's model, 226f
Voronoi/Delaunay dual network, 228, 228f
Voronoi diagrams, 228

Index

W

Waldheim, Charles, 52
Walking, 96
Wall, Alex, 151
War of the worlds, 211
Washington State Growth Management Act, 253
Waste decomposition, 336
Wastewater, polishing parks and corridors for insufficiently treated, 371–374, 375f, 376f
Wastewater infrastructure, 361
Wastewater treatment, 257, 259
Wastewater treatment plants (WWTP), 359
Waterfronts, 53
Water landscapes, 406
Water pollution, sources of, 400
Water quality analysis, in São Paulo, 399–400; *see also* São Paulo
Water sensitive urban design (WSUD), 362
Water supply systems and human mortality, 35f
WEATHERS project, 86
Weaver, Warren, 122
The Well-Tempered Clavier, 233
West 8's projects, 52, 235, 237
Wetland restoration plan, for La Parguera, 413f, 414

Wet ponds plan, for La Parguera, 413f, 414
Williams, Daniel, 76
Wilson, E. O., 286
Woodlands, ecological network in, 291
Works Progress Administration (WPA), 10
WPA, *see* Works Progress Administration (WPA)
Wright, Frank Lloyd, 233
WSUD, *see* Water sensitive urban design (WSUD)
WWTP, *see* Wastewater treatment plants (WWTP)

X

Xinyang, massive unnamed satellite development of, 51f

Y

"The Yellow Dragon" storm, 311

Z

Zandmotor Project, 276, 278f
Zurich University of Applied Science, 215